KU-692-855

Methods in Enzymology

Volume 120

CUMULATIVE SUBJECT INDEX

Volumes 81–94, 96–101

METHODS IN ENZYMOLOGY

EDITORS-IN-CHIEF

Sidney P. Colowick Nathan O. Kaplan

Methods in Enzymology

Volume 120

Cumulative Subject Index

Volumes 81–94, 96–101

1986

ACADEMIC PRESS, INC.

Harcourt Brace Jovanovich, Publishers

Orlando San Diego New York Austin
London Montreal Sydney Tokyo Toronto

ACADEMIC PRESS, INC.
Orlando, Florida 32887

United Kingdom Edition published by
ACADEMIC PRESS INC. (LONDON) LTD.
24–28 Oval Road, London NW1 7DX

LIBRARY OF CONGRESS CATALOG CARD NUMBER: 54-9110

ISBN 0-12-182020-3

PRINTED IN THE UNITED STATES OF AMERICA

86 87 88 89 9 8 7 6 5 4 3 2 1

Table of Contents

Preface

The idea for a cumulative index was recognized by the founding editors who prepared one for Volumes I through VI of *Methods in Enzymology* by weeding and interpolating from the entries that had been indexed in the individual volumes. As the series developed in both number and complexity, different individuals with different backgrounds served as volume indexers. Subsequently, the series was fortunate in having Dr. Martha G. Dennis and Dr. Edward A. Dennis accept the challenge of computerizing the data available from these individual indexes, and this effort resulted in Volumes 33, 75, and 95, which cover Volumes 1 through 80.

Although each of the three books produced with the aid of computerization provided an appropriate cumulative index, major problems were encountered. One was time, both expensive computer time and lag time before such efforts resulted in publication. The most important difficulty was that the compilers were hampered by the lack of uniformity in the indexing of the individual volumes, resulting in the need for much hand editing to achieve a reasonable collation. The products were very decent, if uneven, indexes that also contributed to the methodology of computerized indexing, albeit with much delay and great expense.

Could one produce a relatively simplified, hopefully adequate index at a reasonable price and produce it in a timely manner? A different approach was used for this volume which, although it did not do well in terms of speed of production, served as an experiment; publishers also experiment and this is one of them. The remainder of the question must be answered by the users. This is meant literally because the publisher welcomes the feedback that will be necessary to improve the cumulative indexes which are so important to a series that presents methods.

This cumulative index has been produced by the staff of Academic Press under the guidance of Mrs. Evelyn Sasmor, and the product has received the comments of Dr. Frank Eisenberg, Jr. The indexers have gone back to the text rather than to the individual volume indexes and, using a uniform set of guidelines, have culled the major topics, leading to five to ten entries for each article. Thus, if searching by name for one of the dozen substrates of an enzyme, the isolation of which is being presented, it probably will not be found in the index, although the assay substrate may be there. Nor will the specific inhibitors of the enzyme be itemized, although the topic of the enzyme's inhibition will form an entry. The index, then, is not complete, but should lead to the broad subject

headings. Since there is a tendency to identify specific topics and methods with particular individuals, a contributor index is included. Finally, the complete table of contents of each of the volumes indexed is included.

If this type of index is viewed favorably, it should be possible to produce such cumulative indexes, covering a somewhat larger number of volumes, in relatively rapid fashion.

Contents of Volumes 81–94, 96–101

VOLUME 81

Biomembranes (Part H: Visual Pigments and Purple Membranes, I)

Section I. Overview: Morphology of Visual Photoreceptors

1. Morphology of Vertebrate Photoreceptors — Eichi Yamada — 3
2. Morphology of Invertebrate Photoreceptors — Richard M. Eakin — 17

Section II. Isolation of Organelles and Membranes Containing Visual Pigments

3. Isolation and Mounting of Whole Vertebrate Retinas for Optical Studies — Christian Baumann — 29
4. Isolation of Crayfish Rhabdoms — Denis Larrivee and Timothy H. Goldsmith — 34
5. Dissection of Retinas for Biochemical Studies — Richard N. Lolley — 37
6. Culture of Mammalian Retinal Pigment Epithelium and Neural Retina — Ross B. Edwards — 39
7. Isolation and Purification of Squid Rhabdoms — Yuji Kito, Takaharu Seki, and Frances M. Hagins — 43
8. Preparation of Retinal Rod Outer Segments — David S. Papermaster — 48
9. Preparation and Characterization of Sealed Bovine Rod Cell Outer Segments — William F. Zimmerman and Walter Godchaux III — 52
10. Preparation of Osmotically Intact Rod Outer Segment Disks by Ficoll Flotation — H. Gilbert Smith, Jr. and Burton J. Litman — 57
11. Isolation of Intact Disks by Concanavalin A Columns — Alois J. Adams, Minoru Tanaka, and Hitoshi Shichi — 61
12. Use of Lectins to Investigate Photoreceptor Membranes — C. D. B. Bridges and Shao-Ling Fong — 65
13. Isolation and Characterization of the Interphotoreceptor Matrix — Elaine R. Berman — 77
14. Microdissection of Retina and Measurement of Cyclic Nucleotides — G. W. De Vries, J. A. Ferrendelli, and O. H. Lowry — 85

15. Column Method for the Isolation of Pigment Epithelium — A. N. Siakotos, G. Aguirre, L. Aders, and R. J. Schnell — 90

16. Methods for Isolating and Fractionating Pigment Epithelial Cells — Elaine R. Berman and Lynette Feeney-Burns — 95

17. Isolation and Characterization of Osmotically Sealed Bovine Rod Outer Segments — Paul P. M. Schnetkamp and Frans J. M. Daemen — 110

18. Preparation of Isolated Osmotically Intact Bovine Rod Outer Segment Disk Membranes — Peter P. Nemes and Edward A. Dratz — 116

19. Purification of Rat Retinal Rod Outer Segment Membranes — Christopher C. Farnsworth and Edward A. Dratz — 124

Section III. Characterization of Visual Pigments

20. Detergents for Extraction of Visual Pigments: Types, Solubilization, and Stability — Shao-Ling Fong, Andrew T. C. Tsin, C. D. B. Bridges, and Gregory I. Liou — 133

21. Purification of Rhodopsin on Agarose — Paul J. O'Brien — 141

22. Purification of Rhodopsin on Hydroxyapatite Columns, Detergent Exchange, and Recombination with Phospholipids — Keelung Hong, Peter J. Knudsen, and Wayne L. Hubbell — 144

23. Purification of Rhodopsin by Concanavalin A Affinity Chromatography — Burton J. Litman — 150

24. Preparation of Chicken Iodopsin — Hiroyuki Matsumoto and Tôru Yoshizawa — 154

25. Chromatographic Separation of Rod and Cone Pigments from Chicken Retinas — Lei Yen Fager and Roger S. Fager — 160

26. Purification of Squid and Octopus Rhodopsin — Yuji Kito, Takayuki Naito, and Kazuko Nashima — 167

27. Some Non-Rhodopsin-Like Properties of a Gecko Visual Pigment — Frederick Crescitelli — 171

28. Insect Visual Pigments — Joachim Schwemer and Helmut Langer — 182

29. Cephalopod Retinochrome — Tomiyuki Hara and Reiko Hara — 190

30. Purification of Bovine Rhodopsin over Concanavalin A Sepharose — Willem J. De Grip — 197

Section IV. Rhodopsin Protein Chemistry

31. Retinyl Peptide Isolation and Characterization — PAUL A. HARGRAVE, DERIC BOWNDS, JANET K. WANG, AND J. HUGH MCDOWELL — 211

32. Structural Analysis of the Carbohydrate Moiety of Bovine Rhodopsin — MICHIKO N. FUKUDA, DAVID S. PAPERMASTER, AND PAUL A. HARGRAVE — 214

33. Sulfhydryl Chemistry of Rhodopsin — WILLEM J. DE GRIP AND FRANS J. M. DAEMEN — 223

34. Proteolysis of Rhodopsin — JORDAN S. POBER — 236

35. Preparation of Antibodies to Rhodopsin and the Large Protein of Rod Outer Segments — DAVID S. PAPERMASTER — 240

36. Identification of the N-Terminal Residue of Rhodopsin — SUSUMU TSUNASAWA, KOZO NARITA, AND HITOSHI SHICHI — 247

37. Preparation of Peptides from Bovine Rhodopsin — PAUL A. HARGRAVE, J. HUGH MCDOWELL, DONNA R. CURTIS, SHAO-LING FONG, AND ELIZABETH JUSZCZAK — 251

38. Thermal Stability of Rhodopsin and Opsin in Some Novel Detergents — WILLEM J. DE GRIP — 256

Section V. Rhodopsin Chemical Modification

39. Iodination of Rhodopsin and Transmembrane Topology — BERNARD KWOK-KEUNG FUNG AND WAYNE L. HUBBELL — 269

40. Chemical Modification of Rhodopsin with Imidoesters: Synthesis of Reagents, Membrane Permeability of Reagents, and Modification Methods — PETER P. NEMES, GEORGE P. MILJANICH, AND EDWARD A. DRATZ — 275

41. Tryptophan Reactivity — ALAN COOPER — 285

42. Cyanoborohydride Reduction of Rhodopsin — ROGER S. FAGER — 288

43. Borane Dimethylamine Reduction of the Retinal–Opsin Linkage in Rhodopsin — MICHAEL O. HALL AND DEAN BOK — 290

Section VI. Rod Outer Segment Membrane Lipids

44. Determination of Molecular Species of Rod Outer Segment Phospholipids — REX D. WIEGAND AND ROBERT E. ANDERSON — 297

45. Methods for Determination of Gangliosides in Retinas and Rod Outer Segments — H. DREYFUS, N. VIRMAUX-COLIN, S. HARTH, AND P. MANDEL 304

46. Methods for Determining Rod Outer Segment Disk Phospholipid Transmembrane Topology — BURTON J. LITMAN 309

47. Lipids of Invertebrate Photoreceptors — TOYOAKI AKINO 315

48. Phospholipases as Tools for Studying Structure and Function of Photoreceptor Membranes — ERIK H. S. DRENTHE AND FRANS J. M. DAEMEN 320

Section VII. Spectral Responses of Visual Pigments to Light

49. Low-Temperature Spectrophotometry of Intermediates of Rhodopsin — TÔRU YOSHIZAWA AND YOSHINORI SHICHIDA 333

50. Picosecond Spectroscopy of Visual Pigments — M. L. APPLEBURY AND P. M. RENTZEPIS 354

51. Bleaching Intermediate Kinetics of Rhodopsin: Picosecond Kinetics for Squid Rhodopsin — T. KOBAYASHI AND S. NAGAKURA 368

52. Bleaching Intermediate Kinetics of Rhodopsin, Metarhodopsin I, and Metarhodopsin II — THEODORE P. WILLIAMS AND BARBARA N. BAKER 374

53. Lipid Dependence of Rhodopsin Kinetics — DAVID F. O'BRIEN 378

54. Photosensitivity and Quantum Efficiency of Photoisomerization in Rhodopsin and Retinal — ALLEN KROPF 384

55. Spectral Changes in the Photolysis of Invertebrate Rhodopsin by Rapid Scan Spectrophotometry — MOTOYUKI TSUDA 392

Section VIII. Electrical Response

56. Electrical Signaling in Vertebrate Photoreceptors — D. A. BAYLOR AND B. NUNN 403

57. Relation of Bleaching to Sensitivity of Vertebrate Photoreceptors — WILLIAM L. PAK 423

58. Electrical Responses to Light: Fast Photovoltages of Rhodopsin-Containing Membrane Systems and Their Correlation with the Spectral Intermediates — H.-W. TRISSL 431

Section IX. Regeneration of Pigments from Bleached Rhodopsin

59. Kinetics and Mechanism of Rhodopsin Regeneration with 11-*cis*-Retinal — MICHAEL A. CUSANOVICH 443

60. Technique for Introducing Retinol Analogs into the Isolated Retina — S. YOSHIKAMI AND G. N. NÖLL — 447

61. Generation of Rhodopsin and "Artificial" Visual Pigments in Electrophysiologically Active Photoreceptors — DAVID R. PEPPERBERG — 452

62. Photoregeneration — HITOSHI SHICHI AND ROBERT L. SOMERS — 459

63. Measurement of the Vitamin A Cycle — C. D. B. BRIDGES AND RICHARD A. ALVAREZ — 463

Section X. Rod Outer Segment Enzymes and Enzymatic Responses to Light

64. Assay of Phosphorylation of Rhodopsin *in Vitro* and *in Vivo* — HERMANN KÜHN AND URSULA WILDEN — 489

65. Purification and Properties of Rod Outer Segment Kinases — REHWA H. LEE, DEBORA B. FARBER, AND RICHARD N. LOLLEY — 496

66. Preparation of Bovine Rod Outer Segment Membranes Capable of Regenerating Visual Pigment with Added 11-*cis*-Retinol — ROBERT A. SACK — 506

67. Assays and Characterization of Mg^{2+}-ATPase in the Rod Outer Segment of Vertebrate Photoreceptors — RAINER UHL, TOM BORYS, AND E. W. ABRAHAMSON — 509

68. Characterization of the Light-Activated Mg^{2+}-ATPase in Rod Outer Segments — SCOTT M. THACHER — 514

69. Characterization and Localization of Adenylate and Guanylate Cyclases — G. W. DE VRIES AND J. A. FERRENDELLI — 517

70. Localization and Assay of Guanylate Cyclase — DARRELL FLEISCHMAN — 522

71. Purification and Characterization of a Light-Activated Rod Outer Segment Phosphodiesterase — AKIO YAMAZAKI, NAOMASA MIKI, AND MARK W. BITENSKY — 526

72. Real Time Assay of Rod Disk Membrane cGMP Phosphodiesterase and Its Controller Enzymes — PAUL A. LIEBMAN AND ANN TAVORMINA EVANCZUK — 532

73. Isolation and Assay of a Phosphodiesterase Inhibitor from Retinal Rod Outer Segments — JAMES B. HURLEY — 542

74. Measurement of Retinal cGMP Changes with Quick-Freeze Techniques — PAUL KILBRIDE — 547

75. Measurement of Cyclic Nucleotides in Retina — DEBORA B. FARBER AND RICHARD N. LOLLEY — 551

76. Light-Regulated Binding of Proteins to Photoreceptor Membranes and Its Use for the Purification of Several Rod Cell Proteins — HERMANN KÜHN — 556

77. Ambiquitous GTPase of Bovine Rod Cell Outer Segments — WALTER GODCHAUX III AND WILLIAM F. ZIMMERMAN — 564

78. Detection and Properties of Rapid Calcium Release from Binding Sites in Isolated Rod Outer Segments upon Photoexcitation of Rhodopsin — U. BENJAMIN KAUPP AND WOLFGANG JUNGE — 569

79. Light Release of ^{45}Ca Trapped in Sonicated Bovine Disk Vesicles — ROGER S. FAGER, BURTON J. LITMAN, H. GILBERT SMITH, JR. — 577

Section XI. Physical Studies on Retinal Photoreceptors

80. Isolation, Rapid Freezing, and Freeze-Fracture Methods for Frog Retinal Photoreceptors — JOSEPH M. CORLESS AND M. JOSEPH COSTELLO — 585

81. X-Ray and Neutron Diffraction of Retinal Rod Outer Segments — MARC CHABRE AND D. L. WORCESTER — 593

82. Linear Dichroism Studies in the Visible, UV, and IR on Oriented Rod Suspensions — MARC CHABRE, J. BRETON, M. MICHEL-VILLAZ, AND H. SAIBIL — 605

83. Size and Shape of Isolated Rod Disks by Light Scatterings — HYUK YU — 616

84. Ultraviolet Circular Dichroism of Rhodopsin in Disk Membranes and Detergent Solution — BURTON J. LITMAN — 629

85. Low-Temperature Circular Dichroism of Intermediates of Rhodopsin — TÔRU YOSHIZAWA AND YOSHINORI SHICHIDA — 634

86. Polarized Microspectrophotometry for Pigment Orientation and Concentration — FERENC I. HÁROSI — 642

87. A Field-Portable Microspectrophotometer — ELLIS R. LOEW — 647

88. Birefringence and Birefringence Gradients in Rod Outer Segments — MICHAEL W. KAPLAN — 655

89. Lateral Diffusion of Visual Pigment in Rod Disk Membranes — PAUL A. LIEBMAN, HENRY L. WEINER, AND ROBERT E. DRZYMALA — 660

90. Fluorescence Energy Transfer Measurements of Distances in Rhodopsin and the Purple Membrane Protein — LUBERT STRYER, DAVID D. THOMAS, AND WILLIAM F. CARLSEN — 668

91. Fluorescent Probe: Diphenylhexatriene — BURTON J. LITMAN AND Y. BARENHOLZ — 678

92. Analysis of Rod Outer Segment Disk Membrane Phospholipid Organization Using Parinaric Acid Fluorescent Probes — LARRY A. SKLAR AND EDWARD A. DRATZ — 685

93. Fluorine-Labeled Retinals and Rhodopsins — ROBERT S. H. LIU AND HIROYUKI MATSUMOTO — 694

94. ^{13}C NMR Spectroscopy of the Chromophore of Rhodopsin — JOHN W. SHRIVER, GHEORGHE D. MATEESCU, AND E. W. ABRAHAMSON — 698

95. Lipid–Protein Interactions: Saturation Transfer Electron Paramagnetic Resonance of Spin-Labeled Rhodopsin — PHILIPPE F. DEVAUX — 703

96. Proton, Carbon-13, and Phosphorus-31 NMR Methods for the Investigation of Rhodopsin–Lipid Interactions in Retinal Rod Outer Segment Membranes — MICHAEL F. BROWN, ALAN J. DEESE, AND EDWARD A. DRATZ — 709

97. Microspectrophotometric Investigation of Insect Visual Pigments — HELMUT LANGER, PETER SCHLECHT, AND JOACHIM SCHWEMER — 729

98. Use of Hydrogenated and Deuterated Detergents in the Study of Rhodopsin by Neutron Small-Angle Scattering — H. BEVERLEY OSBORNE AND MARC CHABRE — 742

99. Noninvasive Optical Techniques for Probing Insect Photoreceptors — GARY D. BERNARD — 752

Section XII. Biogenesis

100. Renewal of Photoreceptor Cells — DEAN BOK — 763

101. Rhodopsin Biosynthesis in Isolated Retinas — SCOTT F. BASINGER AND ROSEMARY T. HOFFMAN — 772

102. Glycosylation of Rhodopsin — PAUL J. O'BRIEN — 783

103. Lipid Synthesis in Retinas — HAYDÉE E. PASCUAL DE BAZÁN AND NICOLÁS G. BAZÁN — 788

104. Altered Biosynthesis in Retinal Degeneration — H. WILLIAM READING — 794

105. Renewal of Lipids in Rod Outer Segments — ROBERT E. ANDERSON — 800

106. Composition and Fatty Acid Pairing in Phospholipids of Rod Outer Segments — GEORGE P. MILJANICH AND EDWARD A. DRATZ — 806

Section XIII. Other Retinal Proteins

107. Isolation of Cellular Retinoid-Binding Proteins from Bovine Retina with Bound Endogenous Ligands — JOHN C. SAARI — 819

108. Cephalopod Retinochrome — TOMIYUKI HARA AND REIKO HARA — 827

109. Vitamin A Storing Cells — EICHI YAMADA — 834

110. Isolation and Characterization of Serum Retinol-Binding Protein — YASUTOSHI MUTO, YOSHIHIRO SHIDOJI, AND YOSHIKAZU KANDA — 840

VOLUME 82

STRUCTURAL AND CONTRACTILE PROTEINS (PART A: EXTRACELLULAR MATRIX)

Section I. Collagen

A. The Multiple Types and Forms of Collagen

1. Collagen: An Overview — EDWARD J. MILLER AND STEFFEN GAY — 3

2. Preparation and Characterization of the Different Types of Collagen — EDWARD J. MILLER AND R. KENT RHODES — 33

3. Preparation and Characterization of Invertebrate Collagens — LOUANN W. MURRAY, J. HERBERT WAITE, MARVIN L. TANZER, AND PETER V. HAUSCHKA — 65

4. Preparation and Characterization of Procollagens and Procollagen–Collagen Intermediates — HELENE SAGE AND PAUL BORNSTEIN — 96

5. Characterization of Fibrous Forms of Collagen — BARBARA BRODSKY AND ERIC F. EIKENBERRY — 127

6. Solid State NMR Studies of Molecular Motion in Collagen Fibrils — D. A. TORCHIA — 174

7. Characterization of Soluble Collagens by Physical Techniques — ARTHUR VEIS — 186

B. Biosynthesis of Procollagen

8. Preparation and Translation of Interstitial (Calvaria) Procollagen mRNA — JANET M. MONSON — 218

9. Cell-Free Synthesis of Noninterstitial (CHL Cell) Procollagen Chains — MICHAEL A. HARALSON — 225

10. Posttranslational Enzymes in the Biosynthesis of Collagen: Intracellular Enzymes — KARI I. KIVIRIKKO AND RAILI MYLLYLA — 245

11. Posttranslational Enzymes in the Biosynthesis of Collagen: Extracellular Enzymes — DARWIN J. PROCKOP AND LEENA TUDERMAN — 305

C. Preparation of Other Proteins Containing Collagen-like Sequences

12. C1q — K. B. M. REID — 319

13. Acetylcholinesterase — TERRONE L. ROSENBERRY, PHILIP BARNETT, AND CAROL MAYS — 325

D. Characterization of Special Features in the Chemistry of Collagens

14. Hydroxylysine Glycosides — WILLIAM T. BUTLER — 339

15. Asparagine-Linked Glycosides — CHARLES C. CLARK — 346

16. Covalent Cross-Links in Collagen — NICHOLAS D. LIGHT AND ALLEN J. BAILEY — 360

17. Determination of 3- and 4-Hydroxyproline — RICHARD A. BERG — 372

18. Urinary Metabolites of Collagen — JERE P. SEGREST — 398

19. Estimation of the Size of Collagenous Proteins by Electrophoresis and Gel Chromatography — RALPH J. BUTKOWSKI, MILTON E. NOELKEN, AND BILLY G. HUDSON — 410

E. Proteolysis of Collagen

20. Vertebrate Collagenases — EDWARD D. HARRIS, JR. AND CAROL A. VATER — 423

21. Bacterial Collagenase — BEVERLY PETERKOFSKY — 453

F. Immunology of Collagen

22. Antibodies to Collagens and Procollagens — RUPERT TIMPL — 472

G. Interaction of Cells with Collagen

23. Cell–Collagen Interactions: Overview — FREDERICK GRINNELL — 499

24. Fibroblast Adhesion to Collagen Substrates — HYNDA K. KLEINMAN — 503

25. Platelet–Collagen Adhesion — SAMUEL A. SANTORO AND LEON W. CUNNINGHAM — 509

26. Hepatocyte–Collagen Adhesion — BJÖRN ÖBRINK — 513

27. Biosynthetic Matrices from Cells in Culture — CARL G. HELLERQVIST — 530

28. Ultrastructural Studies of Cell–Collagen Interactions — FREDERICK GRINNELL AND MARYLYN HOY BENNETT — 535

29. Collagen as a Substrate for Cell Growth and Differentiation — STEPHEN C. STROM AND GEORGE MICHALOPOULOS — 544

Section II. Elastin

A. Elastin Structure and Biosynthesis

30. Elastin Structure and Biosynthesis: An Overview — JUDITH ANN FOSTER — 559

31. Elastin Isolation and Cross-Linking — MERCEDES A. PAZ, DAVID A. KEITH, AND PAUL M. GALLOP — 571

32. Proteolysis of Insoluble Elastin — P. J. STONE, C. FRANZBLAU, AND H. M. KAGAN — 588

33. Primary Structure of Insoluble Elastin — RASHID A. ANWAR — 606

34. Biosynthesis of Insoluble Elastin in Cell and Organ Cultures — CARL FRANZBLAU AND BARBARA FARIS — 615

35. Lysyl Oxidase: Preparation and Role in Elastin Biosynthesis — HERBERT M. KAGAN AND KATHLEEN A. SULLIVAN — 637

36. Isolation of Soluble Elastin from Copper-Deficient Chick Aorta — ROBERT B. RUCKER — 650

37. Production and Isolation of Soluble Elastin from Copper-Deficient Swine — L. B. SANDBERG AND T. B. WOLT — 657

38. Isolation of Soluble Elastin-Lathyrism — CELESTE B. RICH AND JUDITH ANN FOSTER — 665

39. Characterization of Soluble Peptides of Elastin by Physical Techniques — DAN W. URRY — 673

40. Biosynthesis of Soluble Elastin in Organ and Cell Culture — JOEL ROSENBLOOM — 716

41. Cell-Free Translation of Elastin mRNAs — JUDITH ANN FOSTER, CELESTE B. RICH, STEVEN B. KARR, AND ALAN PRZYBYLA 731

B. Immunology of Elastin

42. Antibodies to Insoluble and Solubilized Elastin — ROBERT P. MECHAM AND GARY LANGE 744

43. Antibodies to Desmosine — BARRY C. STARCHER 759

44. Preparation of Antiserum to Tropoelastin — JUDITH ANN FOSTER, CELESTE B. RICH, SU CHEN WU, AND MARIE-THERESE MARZULLO 762

Section III. Proteoglycans

45. Proteoglycans: Isolation and Characterization — VINCENT C. HASCALL AND JAMES H. KIMURA 769

Section IV. Other Connective Tissue Proteins

46. Fibronectin: Purification, Immunochemical Properties, and Biological Activities — ERKKI RUOSLAHTI, EDWARD G. HAYMAN, MICHAEL PIERSCHBACHER, AND EVA ENGVALL 803

47. Laminin — RUPERT TIMPL, HEILWIG ROHDE, LEILA RISTELI, URSULA OTT, PAMELA GEHRON ROBEY, AND GEORGE R. MARTIN 831

48. Structural Glycoproteins — LADISLAS ROBERT AND MADELEINE MOCZAR 839

VOLUME 83

COMPLEX CARBOHYDRATES (PART D)

Section I. Analytical Methods

1. Structural Analysis of Complex Carbohydrates Using High-Performance Liquid Chromatography, Gas Chromatography, and Mass Spectrometry — MICHAEL MCNEIL, ALAN G. DARVILL, PER ÅMAN, LARS-ERIK FRANZÉN, AND PETER ALBERSHEIM 3

2. Gas Chromatography and Mass Spectrometry of Hexosamine-Containing Oligosaccharide Alditols as Their Permethylated, *N*-Trifluoroacetyl Derivatives	Bo Nilsson and David Zopf	46
3. ^{13}C NMR Analysis of Complex Carbohydrates	R. Barker, H. A. Nunez, P. Rosevear, and A. S. Serianni	58
4. Analysis of Glycosphingolipids by High-Resolution Proton Nuclear Magnetic Resonance Spectroscopy	Janusz Dabrowski, Peter Hanfland, and Heinz Egge	69
5. Polysaccharide Structure by X-Ray Fiber Diffraction	William T. Winter	87
6. Analysis of Oligosaccharides by Gel Filtration	Katsuko Yamashita, Tsuguo Mizuochi, and Akira Kobata	105
7. Resolution of Carbohydrates by Lectin Affinity Chromatography	Diane A. Blake and Irwin J. Goldstein	127
8. Resolution of Acetylated Oligosaccharides by Reverse-Phase High-Pressure Liquid Chromatography	Gerald B. Wells, Voula Kontoyiannidou, Salvatore J. Turco, and Robert L. Lester	132
9. Resolution of Dolichylpyrophosphoryl Oligosaccharides by High-Pressure Liquid Chromatography	Gerald B. Wells, Salvatore J. Turco, Barbara A. Hanson, and Robert L. Lester	137
10. Gangliosides: Structure, Isolation, and Analysis	Robert W. Ledeen and Robert K. Yu	139
11. Radioisotopic Assay for Galactocerebrosides and Sulfatides	Harmeet Singh and Julian N. Kanfer	191
12. The Ultrastructural Visualization of Cell Surface Glycoconjugates	Edward A. Bayer, Ehud Skutelsky, and Meir Wilchek	195
13. Immunological Characterization of Cartilage Proteoglycans	John R. Baker, Bruce Caterson, and James E. Christner	216
14. Detection of Glycolipid Ligands by Direct Binding of Carbohydrate-Binding Proteins to Thin-Layer Chromatograms	John L. Magnani, Manfred Brockhaus, David F. Smith, and Victor Ginsburg	235

15. Isolation of Specific Sugar Sequences Using Carbohydrate-Binding Proteins — DAVID F. SMITH — 241

16. Organic Esters and Ethers of Polysaccharides — GARY R. GRAY — 248

Section II. Preparations

17. Hydrazinolysis of Asparagine-Linked Sugar Chains to Produce Free Oligosaccharides — SEIICHI TAKASAKI, TSUGUO MIZUOCHI, AND AKIRA KOBATA — 263

18. Preparation and Fractionation of Glycopeptides — JUKKA FINNE AND TOM KRUSIUS — 269

19. Preparation of Neoglycoproteins Using 2-Imino-2-methoxyethyl 1-Thioglycosides — CHRISTOPHER P. STOWELL AND YUAN CHUAN LEE — 278

20. Preparation of Neoglycoproteins Using ω-Aldehydoalkyl 1-Thioglycosides — REIKO T. LEE AND YUAN CHUAN LEE — 289

21. Preparation of Polyacrylamide Gels Containing Copolymerized ω-Acrylamidoalkyl Glycosides — PAUL H. WEIGEL, RONALD L. SCHNAAR, SAUL ROSEMAN, AND YUAN CHUAN LEE — 294

22. Preparation of Polyacrylamide Gels Containing Thioglycoside — REIKO T. LEE AND YUAN CHUAN LEE — 299

23. Preparation of Polyacrylamide Gels Containing Active Esters — RONALD L. SCHNAAR, PAUL H. WEIGEL, SAUL ROSEMAN, AND YUAN CHUAN LEE — 306

24. Preparation of High Molecular Weight Glycopeptides and Oligosaccharides (Erythroglycan) from Human Erythrocytes — JOHAN JÄRNEFELT, JEFFREY S. RUSH, JAWEED ASHRAF, JUHA VIITALA, AND ROGER A. LAINE — 311

25. Isolation of the Major Envelope Glycoprotein of Avian Myeloblastosis Virus — DANTE J. MARCIANI — 320

26. Cloning Genes for Bacterial Glycosyltransferases — ELVA S. CREEGER AND LAWRENCE I. ROTHFIELD — 326

27. Isolation and Synthesis of Sugar Nucleotides — OTHMAR GABRIEL — 332

Section III. Carbohydrate-Binding Proteins

28. The Toxic Lectin Modeccin — SJUR OLSNES, THOMAS HAYLETT, AND KIRSTEN SANDVIG — 357

29. Tomato (*Lycopersicon esculentum*) Lectin — Martin S. Nachbar and J. D. Oppenheim 363

30. *Datura stramonium* Lectin — Jane F. Crowley and Irwin J. Goldstein 368

31. *Aleuria aurantia* Hemagglutinin — Naohisa Kochibe and Ken Furukawa 373

32. *Pseudomonas aeruginosa* Lectins — Nechama Gilboa-Garber 378

33. *Escherichia coli* Surface Lectins — Yuval Eshdat and Nathan Sharon 386

34. Membrane Receptor for Phosphomannosyl Residues — G. Gary Sahagian, Jack J. Distler, and George W. Jourdian 392

Section IV. Biosynthesis

35. *N*-Asparagine-Linked Oligosaccharides: Biosynthesis of the Lipid-Linked Oligosaccharides — Marsha R. Rosner, S. Catherine Hubbard, Raymond J. Ivatt, and Phillips W. Robbins 399

36. *N*-Asparagine-Linked Oligosaccharides: Transfer of Oligosaccharides to Peptides and Proteins *in Vitro* — James J. Elting and William J. Lennarz 408

37. *N*-Asparagine-Linked Oligosaccharides: Processing — Ira Tabas and Stuart Kornfeld 416

38. *N*-Asparagine-Linked Oligosaccharides: Glucosidase-1 from Hen Oviduct — James J. Elting and William J. Lennarz 429

39. Inhibition of the Dolichol Pathway of Protein Glycosylation — Ralph T. Schwarz and Roelf Datema 432

40. Isolation of Somatic Cell Glycoprotein Mutants — Raymond M. Baker, Carlos B. Hirschberg, William A. O'Brien, Tamara E. Awerbuch, and Douglas Watson 444

41. Purification of Mammalian Glycosyltransferases — J. Evan Sadler, Thomas A. Beyer, Catherine L. Oppenheimer, James C. Paulson, Jean-Paul Prieels, James I. Rearick, and Robert L. Hill 458

42. UDP-*N*-Acetyl-D-glucosamine 2′-Epimerase from *Escherichia coli* — TAKASHI KAWAMURA, NOBUTOSHI ISHIMOTO, AND EIJI ITO 515

43. UDP-*N*-Acetyl-D-mannosamine Dehydrogenase from *Escherichia coli* — TAKASHI KAWAMURA, NOBUTOSHI ISHIMOTO, AND EIJI ITO 519

44. GDP-D-Mannose : GDP-L-Galactose Epimerase from *Chlorella pyrenoidosa* — GEORGE A. BARBER AND PATRICIA A. HEBDA 522

45. 3-Deoxy-D-*manno*-octulosonate-8-phosphate (KDO-8-P) Synthase — PAUL H. RAY 525

46. 3-Deoxy-D-*manno*-octulosonate-8-phosphate (KDO-8-P) Phosphatase — PAUL H. RAY AND CHARLES D. BENEDICT 530

47. CTP : CMP-3-Deoxy-D-*manno*-octulosonate Cytidylyltransferase (CMP-KDO Synthetase) — PAUL H. RAY AND CHARLES D. BENEDICT 535

48. Synthesis of Capsular Polymers Containing Polysialic Acid in *Escherichia coli* 07-K1 — FREDERIC A. TROY, INDER K. VIJAY, MICHAEL A. MCCLOSKEY, AND THOMAS E. ROHR 540

49. Isolation of *Escherichia coli* Structural Genes Coding for the Glycogen Biosynthetic Enzymes — THOMAS W. OKITA, RAYMOND L. RODRIGUEZ, AND JACK PREISS 549

Section V. Degradation

50. Enzymic Diagnosis of the Genetic Mucopolysaccharide Storage Disorders — HANS KRESSE, KURT VON FIGURA, UDO KLEIN, JOSEF GLÖSSL, EDUARD PASCHKE, AND REGINA POHLMANN 559

51. Iduronate Sulfatase from Human Plasma — ÅKE WASTESON AND ELIZABETH F. NEUFELD 573

52. α-L-Iduronidase from Human Kidney — LEONARD H. ROME 578

53. Uptake and Binding of α-L-Iduronidase — LEONARD H. ROME 582

54. Protein Activators for the Hydrolysis of G_{M1} and G_{M2} Gangliosides — SU-CHEN LI AND YU-TEH LI 588

55. Protein Activator (Coglucosidase) for the Hydrolysis of β-Glucosides — NORMAN S. RADIN AND SUSAN L. BERENT 596

56. Endo-β-*N*-acetylglucosaminidase L from *Streptomyces plicatus* — ROBERT B. TRIMBLE, ANTHONY L. TARENTINO, GEORGINA EVANS AUMICK, AND FRANK MALEY 603

57. Endo-β-galactosidase from *Escherichia freundii* — Yu-Teh Li, Hiroki Nakagawa, Manabu Kitamikado, and Su-Chen Li — 610

58. Endo-β-galactosidase from *Flavobacterium keratolyticus* — Manabu Kitamikado, Makoto Ito, and Yu-Teh Li — 619

59. α-L-Fucosidases from Almond Emulsin — Akira Kobata — 625

60. β-D-Xylosidase from *Bacillus pumilus* — Hilda Kersters-Hilderson, Marc Claeyssens, Elisa Van Doorslaer, Eric Saman, and Clement K. De Bruyne — 631

61. β-D-Xylosidase from *Penicillium wortmanni* — François Deleyn, Marc Claeyssens, and Clement K. De Bruyne — 639

VOLUME 84

Immunochemical Techniques (Part D: Selected Immunoassays)

Section I. Oncofetal Proteins

1. Radioimmunoassay of α-Fetoprotein with Polyclonal and Monoclonal Antibodies — Erkki Ruoslahti, Marjatta Uotila, and Eva Engvall — 3

2. Enzyme Immunoassay of Human α_1-Fetoprotein — Luc Bélanger and René Masseyeff — 19

3. Radioimmunoassay of Carcinoembryonic Antigen — Mark J. Krantz, Suzanne Laferté, and Naomi Ariel — 32

Section II. Proteins and Peptides of the Blood Clotting System

4. Immunoradiometric Assays for Factor VIII Antigens: Coagulant Protein (Antihemophilic Factor) and Factor VIII-Related Protein (Von Willebrand Factor) — Leon W. Hoyer and Norma C. Trabold — 51

5. Conformation-Specific Antibodies: Approach to the Study of the Vitamin K-Dependent Blood Coagulation Proteins — Bruce Furie, Rita A. Blanchard, David J. Robison, Mindy M. Tai, and Barbara C. Furie — 60

6. Radioimmunoassay of Platelet Factor 4 — KAREN L. KAPLAN AND JOHN OWEN — 83

7. Radioimmunoassay of β-Thromboglobulin — KAREN L. KAPLAN AND JOHN OWEN — 93

8. Immunoassay of Human Fibrinopeptides — VINCENT P. BUTLER, JR., HYMIE L. NOSSEL, AND ROBERT E. CANFIELD — 102

Section III. Metal and Heme Binding Proteins

9. Radioimmunoassay of Metallothioneins — JUSTINE S. GARVEY, RONALD J. VANDER MALLIE, AND CHIN C. CHANG — 121

10. Radioimmunoassay of Calmodulin — JAMES G. CHAFOULEAS, JOHN R. DEDMAN, AND ANTHONY R. MEANS — 138

11. Immunoassay of Ferritin in Plasma — JUNE W. HALLIDAY — 148

12. Radioimmunoassay of Myoglobin — MARVIN J. STONE, JAMES T. WILLERSON, AND MICHAEL R. WATERMAN — 172

Section IV. Nucleic Acids and Their Antibodies

13. Quantitation of Submicrogram Amounts of DNA by Rocket Electrophoresis — CHARLES R. STEINMAN — 181

14. Detection and Semiquantitation of DNA by Counterimmunoelectrophoresis (CIE) — CHARLES R. STEINMAN — 187

15. Enzyme-Linked Immunosorbent Assay for Antibodies to Native and Denatured DNA — JOAN L. KLOTZ — 194

16. Passive Hemagglutination and Hemolysis Tests for Anti-DNA Antibody — TAKESHI SASAKI — 201

17. Radioimmunoassay of Antibody to DNA·RNA Hybrids of Naturally Occurring Sequence — HIROSHI NAKAZATO — 210

Section V. Toxins

18. The Enzyme-Linked Immunosorbent Assay (ELISA) for the Detection and Determination of *Clostridium botulinum* Toxins A, B, and E — S. NOTERMANS, A. M. HAGENAARS, AND S. KOZAKI — 223

19. Determination of *Escherichia coli* Enterotoxin and Cholera Toxin by Radioimmunoassay — MIROSLAV CESKA — 238

20. Radioimmunoassay of Staphylococcal Enterotoxin C — N. DICKIE AND M. AKHTAR — 254

21. Solid-Phase Radioimmunoassay for Bacterial Lipopolysaccharide — DIANE M. JACOBS AND JAN A. GUTOWSKI — 264

22. Radioimmunoassays of Thyroxine (T_4), 3,5,3′-Triiodothyronine (T_3), 3,3′,5′-Triiodothyronine (Reverse T_3, rT_3), and 3,3′-Diiodothyronine (T_2) — WILMAR M. WIERSINGA AND INDER J. CHOPRA — 272

23. Chemiluminescence Immunoassay for Serum Thyroxine — HARTMUT R. SCHROEDER — 303

Section VI. Endogenous Compounds of Low Molecular Weight

24. Radioimmunoassays of Serum Bile Acids — P. E. ROSS — 321

25. Radioimmunoassays of Ecdysteroids — M. A. DELAAGE, M. H. HIRN, AND M. L. DE REGGI — 350

26. Radioimmunoassays of Motilin — JOHN C. BROWN AND JILL R. DRYBURGH — 359

27. Radioassay of Vitamin B_{12} and Other Corrinoids — J. J. PRATT AND M. G. WOLDRING — 369

Section VII. Drugs

A. Antineoplastic Agents

28. Radioimmunoassay of Methotrexate, Leucovorin, and 5-Methyltetrahydrofolate — JOHN J. LANGONE — 409

29. Preparation of ^{125}I-Labeled Methotrexate and Its Use in Magnetizable Particle Solid-Phase Radioimmunoassays — R. S. KAMEL AND J. LANDON — 422

30. Competitive Protein Binding Assay of Methotrexate — CHARLES ERLICHMAN, ROSS C. DONEHOWER, AND CHARLES E. MYERS — 447

31. Ligand-Binding Radioassay of N^5-Methyltetrahydrofolate and Its Application to N^5-Formyltetrahydrofolate — MARIA DA COSTA AND SHELDON P. ROTHENBERG — 458

32. Radioimmunoassay of Bleomycin — ALAN BROUGHTON — 463

33. Radioimmunoassay of 1-β-D-Arabinofuranosylcytosine — TADASHI OKABAYASHI AND JOHN G. MOFFATT — 470

34. Radioimmunoassay of Neocarzinostatin, a Small Cytotoxic Protein Used in Cancer Chemotherapy — VIC RASO — 483

B. Drugs Active on the Nervous System and Neurotransmitters

35. Radioimmunoassay of Benzodiazepines — ROSS DIXON — 490

36. Radioimmunoassay of N-Substituted Phenylpiperidine Carboxylic Acid Esters and Dealkylated Metabolites — DAVID S. FREEMAN AND HILDA B. GJIKA — 516

37. Radioimmunoassay for Fentanyl — ROBERT P. SCHLEIMER — 526

38. Radioimmunoassay of Haloperidol — ROBERT T. RUBIN, BARBARA B. TOWER, SALLY E. HAYS, AND RUSSELL E. POLAND — 532

39. Radioimmunoassay of Pimozide — M. MICHIELS, R. HENDRIKS, AND J. HEYKANTS — 542

40. Radioimmunoassay for Morphine — SYDNEY SPECTOR — 551

41. Radioimmunoassay for Acetylcholine — SYDNEY SPECTOR — 555

C. Other Drugs

42. Immunoassay of Digoxin and Other Cardiac Glycosides — VINCENT P. BUTLER, JR. AND DORIS TSE-ENG — 558

43. Radioimmunoassay of *n*-Butylbiguanide — F. ÖTTING — 577

44. Radioimmunoassay of Metyrapone and Reduced Metyrapone — A. WAYNE MEIKLE — 585

45. Radioimmunoassay of Pyrazolone Derivatives — TAKEHIKO TAKATORI — 596

Section VIII. Environmental Agents

46. Radioimmunoassay for 2-Acetylaminofluorene–DNA Adducts — MIRIAM C. POIRIER AND ROBERT J. CONNOR — 607

47. Radioimmunoassay for Chlorinated Dibenzo-*p*-dioxins — PHILLIP W. ALBRO, MICHAEL I. LUSTER, KUN CHAE, GEORGE CLARK, AND JAMES D. MCKINNEY — 619

48. Radioimmunoassay of Nicotine, Cotinine, and γ-(3-Pyridyl)-γ-oxo-*N*-methylbutyramide — JOHN J. LANGONE AND HELEN VAN VUNAKIS — 628

49. Radioimmunoassay for N′-Nitrosonornicotine and N′-Acylnornicotine Analogs — ANDRE CASTONGUAY AND HELEN VAN VUNAKIS 641

50. Radioimmunoassay of Nicotinamide Nucleotide Analogs of Nicotine and Cotinine — WEI-CHIANG SHEN 650

Section IX. Summary

51. Previously Published Articles from Methods in Enzymology Related to Immunochemical Techniques — HELEN VAN VUNAKIS AND JOHN J. LANGONE 657

VOLUME 85

STRUCTURAL AND CONTRACTILE PROTEINS (PART B: THE CONTRACTILE APPARATUS AND THE CYTOSKELETON)

1. Introduction — JOHN GERGELY 1

Section I. Methods for Striated Muscle Chemistry

2. Preparation of Myofibrils — PETER J. KNIGHT AND JOHN A. TRINICK 9

3. Preparation of Native Thin Filaments — PETER J. KNIGHT AND STEPHEN J. LOVELL 12

4. Hybridization and Reconstitution of the Thin Filament — JOHN M. MURRAY 15

5. Preparation of Native Thick Filaments — JOHN A. TRINICK 17

6. Hybridization and Reconstitution of Thick Filament Structure — JANE F. KORETZ 20

7. Preparation of Myosin and Its Subfragments from Rabbit Skeletal Muscle — SARKIS S. MARGOSSIAN AND SUSAN LOWEY 55

8. Preparation and Fractionation of Myosin Light Chains and Exchange of the Essential Light Chains — PAUL D. WAGNER 72

9. Preparation of Light Chains from Scallop Myosin — ANDREW G. SZENT-GYÖRGYI AND REGINA NIEBIESKI 81

10. Sulfhydryl Modification and Labeling of Myosin — EMIL REISLER 84

11. Chemical Modification of Myosin by Active-Site Trapping of Metal-Nucleotides with Thiol Crosslinking Reagents — JAMES A. WELLS AND RALPH G. YOUNT 93

12. Myosin Active-Site Trapping with Vanadate Ion	CHARLES C. GOODNO	116
13. Assays for Myosin	THOMAS D. POLLARD	123
14. Preparation of C-Protein, H-Protein, X-Protein, and Phosphofructokinase	ROGER STARR AND GERALD OFFER	130
15. Preparation and Properties of M-Line Proteins	HANS M. EPPENBERGER AND EMANUEL E. STREHLER	139
16. Preparation and Assay of Paramyosin	RHEA J. C. LEVINE, MYRA J. ELFVIN, AND VITALY SAWYNA	149
17. Preparation Problems Unique to *Mercenaria* Paramyosin	SONJA KRAUSE AND NANCY LETKO MUNSON	160
18. Purification of Muscle Actin	JOEL D. PARDEE AND JAMES A. SPUDICH	164
19. Methods to Measure Actin Polymerization	JOHN A. COOPER AND THOMAS D. POLLARD	182
20. Methods to Characterize Actin Filament Networks	THOMAS D. POLLARD AND JOHN A. COOPER	211
21. Preparation and Identification of α- and β-Tropomyosins	L. B. SMILLIE	234
22. Preparation of Troponin and Its Subunits	JAMES D. POTTER	241
23. Purification of Titin and Nebulin	KUAN WANG	264

Section II. Methods for Smooth Muscle Chemistry

24. Preparation of Isolated Thick and Thin Filaments	PETER COOKE	277
25. Preparation of Individual Smooth Muscle Cells from the Stomach of *Bufo marinus*	F. S. FAY, R. HOFFMANN, S. LECLAIR, AND P. MERRIAM	284
26. Preparation of Smooth Muscle Myosin	DIXIE W. FREDERIKSEN AND DIANNE D. REES	292
27. Purification of Smooth Muscle Myosin Light Chain Kinase	ROBERT S. ADELSTEIN AND CLAUDE B. KLEE	298
28. Purification of Smooth Muscle Phosphatases	MARY D. PATO AND ROBERT S. ADELSTEIN	308

29. Preparation of Smooth Muscle α-Actinin — SUSAN W. CRAIG, CHRISTINE L. LANCASHIRE, AND JOHN A. COOPER 316

30. Purification and Properties of Avian and Mammalian Filamins — PETER J. A. DAVIES, YUTAKA SHIZUTA, AND IRA PASTAN 322

Section III. Methods for Study of the Motility Apparatus in Nonmuscle Cells

31. Preparation of Nonmuscle Myosins — THOMAS D. POLLARD 331

32. Myosins from *Acanthamoeba castellanii* — EDWARD D. KORN, JIMMY H. COLLINS, AND HIROSHI MARUTA 357

33. Phosphorylation of Nonmuscle Myosin and Stabilization of Thick Filament Structure — JOHN KENDRICK-JONES, K. A. TAYLOR, AND J. M. SCHOLEY 364

34. Preparation of Cytoplasmic Actin — JERRY P. WEIR AND DIXIE W. FREDERIKSEN 371

35. Cytoplasmic Tropomyosins — ISAAC COHEN 373

36. Preparation of Tubulin from Brain — ROBLEY C. WILLIAMS, JR. AND JAMES C. LEE 376

37. Isolation of Tubulin from Nonneural Sources — KEVIN W. FARRELL 385

38. Physical Properties of Purified Calf Brain Tubulin — GEORGE C. NA AND SERGE N. TIMASHEFF 393

39. Purification and Assay of Microtubule-Associated Proteins (MAPs) — ROGER D. SLOBODA AND JOEL L. ROSENBAUM 409

40. Characterization of Tubulin and Microtubule-Associated Protein Interactions with Guanine Nucleotides and Their Nonhydrolyzable Analogs — DANIEL L. PURICH, BRIAN J. TERRY, ROBERT K. MACNEAL, AND TIMOTHY L. KARR 416

41. Techniques for the Study of Microtubule Assembly *in Vitro* — FELICIA GASKIN 433

42. Microtubule Disassembly: A Quantitative Kinetic Approach for Defining Endwise Linear Depolymerization — DANIEL L. PURICH, TIMOTHY L. KARR, AND DAVID KRISTOFFERSON 439

43. Preparation and Purification of Dynein — Christopher W. Bell, Clarence L. Fraser, Winfield S. Sale, Wen-Jing Y. Tang, and I. R. Gibbons — 450

44. Preparation of Spectrin — W. B. Gratzer — 475

45. Macrophage Actin-Binding Protein — John H. Hartwig and Thomas P. Stossel — 480

46. Preparation and Assay of the Intermediate Filament Proteins Desmin and Vimentin — E. Lazarides and B. L. Granger — 488

Section IV. Special Techniques for the Study of the Contractile Protein Complex and the Cytoskeleton

47. Optical Diffraction and Filtering of Electron Micrographs of Contractile Proteins — Pauline M. Bennett — 511

48. Fluorescent Localization of Contractile Proteins in Tissue Culture Cells — Kuan Wang, James R. Feramisco, and John F. Ash — 514

49. Preparation of Contractile Proteins for Photon Correlation Spectroscopic and Classical Light Scattering Studies — Charles Montague and Francis D. Carlson — 562

50. Hydrodynamic Approaches to the Study of High Molecular Weight Proteins and Large Macromolecular Assemblies — Walter F. Stafford III — 570

51. Fluorescence as a Probe of Contractile Systems — Roger Cooke — 574

52. Electron Paramagnetic Resonance of Contractile Systems — John C. Seidel — 594

53. Phosphorus-31 Nuclear Magnetic Resonance of Contractile Systems — Michael Bárány and Thomas Glonek — 624

54. Optical Activity Measurements for Elucidating Structure–Function Relationships in Muscle Protein Systems — William D. McCubbin and Cyril M. Kay — 677

55. Special Instrumentation and Techniques for Kinetic Studies of Contractile Systems — Howard D. White — 698

56. Interaction of Actin and Myosin in the Presence and Absence of ATP — Lois E. Greene and Evan Eisenberg — 709

VOLUME 86

PROSTAGLANDINS AND ARACHIDONATE METABOLITES

Section I. Enzymes and Receptors: Purification and Assay

1. Preparation of Selectively Labeled Phosphatidylinositol and Assay of Phosphatidylinositol-Specific Phospholipase C — SUSAN ERIKA RITTENHOUSE — 3

2. Characterization and Assay of Diacylglycerol Lipase from Human Platelets — PHILIP W. MAJERUS AND STEPHEN M. PRESCOTT — 11

3. Purified Human Placental Arylsulfatases: Their Actions on Slow-Reacting Substance of Anaphylaxis and Synthetic Leukotrienes — AKIRA KUMAGAI AND HISAO TOMIOKA — 17

4. The 5-Lipoxygenase and Leukotriene Forming Enzymes — BARBARA A. JAKSCHIK, TIMOTHY HARPER, AND ROBERT C. MURPHY — 30

5. γ-Glutamyl Transpeptidase, a Leukotriene Metabolizing Enzyme — KERSTIN BERNSTRÖM, LARS ÖRNING, AND SVEN HAMMARSTRÖM — 38

6. Arachidonic Acid-15-Lipoxygenase from Rabbit Peritoneal Polymorphonuclear Leukocytes — SHUH NARUMIYA AND JOHN A. SALMON — 45

7. Arachidonic Acid-12-Lipoxygenase from Bovine Platelets — D. H. NUGTEREN — 49

8. Purification and Assay of PGH Synthase from Bovine Seminal Vesicles — SHOZO YAMAMOTO — 55

9. Purification of PGH Synthase from Sheep Vesicular Glands — F. J. G. VAN DER OUDERAA AND M. BUYTENHEK — 60

10. Preparation of PGH Synthase Apoenzyme from the Holoenzyme — BRADLEY G. TITUS AND WILLIAM E. M. LANDS — 69

11. Purification of PGH–PGD Isomerase from Rat Brain — TAKAO SHIMIZU, SHOZO YAMAMOTO, AND OSAMU HAYAISHI — 73

12. Isolation of PGH–PGD Isomerase from Rat Spleen — E. CHRIST-HAZELHOF AND D. H. NUGTEREN — 77

13. Purification of PGH–PGE Isomerase from Sheep Vesicular Glands — P. MOONEN, M. BUYTENHEK, AND D. H. NUGTEREN — 84

14. Preparation and Assay of Prostacyclin Synthase — JOHN A. SALMON AND RODERICK J. FLOWER — 91

15. Assay of Prostacyclin Synthase Using [5,6-^{3}H]PGH_2 — HSIN-HSIUNG TAI AND CHARLES J. SIH — 99

16. Partial Purification and Assay of Thromboxane A Synthase from Bovine Platelets — TANIHIRO YOSHIMOTO AND SHOZO YAMAMOTO — 106

17. Assay of Thromboxane A Synthase Inhibitors — HSIN-HSIUNG TAI — 110

18. Purification and Assay of 9-Hydroxyprostaglandin Dehydrogenase from Rat Kidney — HSIN-HSIUNG TAI AND BARBARA YUAN — 113

19. Purification of PGD_2 11-Ketoreductase from Rabbit Liver — PATRICK Y-K. WONG — 117

20. Isolation and Properties of an NAD^+-Dependent 15-Hydroxyprostaglandin Dehydrogenase from Human Placenta — JOSEPH JARABAK — 126

21. Assay of NAD^+-Dependent 15-Hydroxyprostaglandin Dehydrogenase Using (15*S*)-[15-^{3}H]PGE_2 — HSIN-HSIUNG TAI — 131

22. Microassay Procedure for NAD^+-Dependent 15-Hydroxyprostaglandin Dehydrogenase — CLINTON N. CORDER — 135

23. Purification of $NADP^+$/NADPH-Dependent 15-Hydroxyprostaglandin Dehydrogenase and Prostaglandin 9-Ketoreductase from Porcine Kidney — HSIN-HSIUNG TAI AND DAVID GUEY-BIN CHANG — 142

24. Isolation of $NADP^+$-Dependent PGD_2-Specific 15-Hydroxyprostaglandin Dehydrogenase from Swine Brain — TAKAO SHIMIZU, KIKUKO WATANABE, HIDEKADO TOKUMOTO, AND OSAMU HAYAISHI — 147

25. Isolation and Properties of an $NADP^+$-Dependent PGI_2-Specific 15-Hydroxyprostaglandin Dehydrogenase from Rabbit Kidney — JEFFREY M. KORFF AND JOSEPH JARABAK — 152

26. Purification and Assay of 15-Ketoprostaglandin Δ^{13}-Reductase from Bovine Lung — HARALD S. HANSEN — 156

27. Isolation and Properties of a 15-Ketoprostaglandin Δ^{13}-Reductase from Human Placenta — JOSEPH JARABAK — 163

28. Measurement of Prostaglandin ω-Hydroxylase Activity — WILLIAM S. POWELL — 168

29. Receptors for PGI_2 and PGD_2 on Human Platelets — ADELAIDE M. SIEGL — 179

30. Distribution of PGE and $PGF_{2\alpha}$ Receptor Proteins in the Intracellular Organelles of Bovine Corpora Lutea — CH. V. RAO AND S. B. MITRA — 192

31. A Receptor for Prostaglandin $F_{2\alpha}$ from Corpora Lutea — SVEN HAMMARSTRÖM — 202

Section II. Immunochemical Assays of Enzymes and Metabolites

32. Characteristics of Rabbit Anti-PGH Synthase Antibodies and Use in Immunocytochemistry — WILLIAM L. SMITH AND THOMAS E. ROLLINS — 213

33. Radioimmunoassay for PGH Synthase — GERALD J. ROTH — 222

34. Monoclonal Antibodies against PGH Synthase: An Immunoradiometric Assay for Quantitating the Enzyme — DAVID L. DEWITT, JEFFREY S. DAY, JOHN A. GAUGER, AND WILLIAM L. SMITH — 229

35. Monoclonal Antibodies against PGI_2 Synthase: An Immunoradiometric Assay for Quantitating the Enzyme — DAVID L. DEWITT AND WILLIAM L. SMITH — 240

36. Radioimmunoassay and Immunochromatography of 12-L-Hydroxyeicosatetraenoic Acid — RICHARD A. MORGAN AND LAWRENCE LEVINE — 246

37. Radioimmunoassay of the 6-Sulfido-Peptide-Leukotrienes and Serologic Specificity of the Anti-Leukotriene D_4 Plasma — LAWRENCE LEVINE, ROBERT A. LEWIS, K. FRANK AUSTEN, AND E. J. COREY — 252

38. Problems of PGE Antisera Specificity — FERNAND DRAY, SUZANNE MAMAS, AND BISTRA BIZZINI — 258

39. Enzyme Immunoassay of $PGF_{2\alpha}$ — YOKO HAYASHI AND SHOZO YAMAMOTO — 269

40. A Radioimmunoassay for 6-Keto-$PGF_{1\alpha}$ — JACQUES MACLOUF — 273

41. A Radioimmunoassay for Thromboxane B_2 — F. A. FITZPATRICK — 286

42. Iodinated Derivatives as Tracers for Eicosanoid Radioimmunoassays — FERNAND DRAY — 297

43. Radioimmunologic Determination of 15-Keto-13,14-dihydro-PGE_2: A Method for Its Stable Degradation Product, 11-Deoxy-15-keto-13,14-dihydro-11β,16ξ-cyclo-PGE_2 — ELISABETH GRANSTRÖM, F. A. FITZPATRICK, AND HANS KINDAHL — 306

44. Radioimmunoassay for the Major Plasma Metabolite of $PGF_{2\alpha}$, 15-Keto-13,14-dihydro-PGF_2 — ELISABETH GRANSTRÖM AND HANS KINDAHL — 320

45. Radioimmunoassay of 5α, 7α-Dihydroxy-11-ketotetranorprostane-1,16-dioic Acid, a Major Prostaglandin F Metabolite in Blood and Urine — ELISABETH GRANSTRÖM AND HANS KINDAHL — 339

Section III. Substrates, Reagents, and Standards

46. Synthesis of Radiolabeled Fatty Acids — HOWARD SPRECHER AND SHANKAR K. SANKARAPPA 357

47. Preparation of Deuterated Arachidonic Acid — DOUGLASS F. TABER, MARK A. PHILLIPS, AND WALTER C. HUBBARD 366

48. Preparation and Analysis of Radiolabeled Phosphatidylcholine and Phosphatidylethanolamine Containing ^{3}H- and ^{14}C-Labeled Polyunsaturated Fatty Acids — HARUMI OKUYAMA AND MAKOTO INOUE 370

49. Preparation of PGG_2 and PGH_2 — GUSTAV GRAFF 376

50. Preparation of 15-L-Hydroperoxy-5,8,11,13-eicosatetraenoic Acid (15-HPETE) — GUSTAV GRAFF 386

51. Preparation of [acetyl-^{3}H]Aspirin and Use in Quantitating PGH Synthase — GERALD J. ROTH 392

52. Synthesis of Stable Thromboxane A_2 Analogs: Pinane Thromboxane A_2 (PTA_2) and Carbocyclic Thromboxane A_2 (CTA_2) — K. C. NICOLAOU AND RONALD L. MAGOLDA 400

53. Purification and Characterization of Leukotrienes from Mastocytoma Cells — ROBERT C. MURPHY AND W. RODNEY MATHEWS 409

54. Production and Purification of Slow-Reacting Substance (SRS) from RBL-1 Cells — CHARLES W. PARKER, SANDRA F. FALKENHEIN, AND MARY M. HUBER 416

55. Preparation, Purification, and Structure Elucidation of Slow-Reacting Substance of Anaphylaxis from Guinea Pig Lung — PRISCILLA J. PIPER, J. R. TIPPINS, H. R. MORRIS, AND G. W. TAYLOR 426

56. Physical Chemistry, Stability, and Handling of Prostaglandins E_2, $F_{2\alpha}$, D_2, and I_2: A Critical Summary — RANDALL G. STEHLE 436

57. Synthesis of Prostacyclin Sodium Salt from $PGF_{2\alpha}$ — ROY A. JOHNSON 459

Section IV. General Separation Procedures

58. Rapid Extraction of Arachidonic Acid Metabolites from Biological Samples Using Octadecylsilyl Silica — WILLIAM S. POWELL 467

59. Extraction and Thin-Layer Chromatography of Arachidonic Acid Metabolites — JOHN A. SALMON AND RODERICK J. FLOWER 477

60. Two-Dimensional Thin-Layer Chromatography of Prostaglandins and Related Compounds — ELISABETH GRANSTRÖM — 493

61. Separation of Arachidonic Acid Metabolites by High-Pressure Liquid Chromatography — THOMAS ELING, BETH TAINER, ARIFF ALLY, AND ROBERT WARNOCK — 511

62. High-Pressure Liquid Chromatography of Underivatized Fatty Acids, Hydroxy Acids, and Prostanoids Having Different Chain Lengths and Double-Bond Positions — MIKE VAN ROLLINS, MARTA I. AVELDAÑO, HOWARD W. SPRECHER, AND LLOYD A. HORROCKS — 518

63. Argentation–High-Pressure Liquid Chromatography of Prostaglandins and Monohydroxyeicosenoic Acids — WILLIAM S. POWELL — 530

Section V. Gas Chromatography–Mass Spectrometry of Prostaglandin Derivatives

64. Preparation of ^{18}O Derivatives of Eicosanoids for GC–MS Quantitative Analysis — ROBERT C. MURPHY AND KEITH L. CLAY — 547

65. Preparation of Deuterium-Labeled Urinary Catabolites of $PGF_{2\alpha}$ as Standards for GC–MS — C. R. PACE-ASCIAK AND N. S. EDWARDS — 552

66. Quantification of the PGD_2 Urinary Metabolite 9α-Hydroxy-11,15-dioxo-2,3,18,19-tetranorprost-5-ene-1,20-dioic Acid by Stable Isotope Dilution Mass Spectrometric Assay — L. JACKSON ROBERTS II — 559

67. Quantitation of 15-Keto-13,14-dihydro-PGE_2 in Plasma by GC–MS — WALTER C. HUBBARD — 571

68. Quantitation of the Major Urinary Metabolite of $PGF_{2\alpha}$ in the Human by GC–MS — ALAN R. BRASH — 579

69. Quantitation of Two Dinor Metabolites of Prostacyclin by GC–MS — PIERRE FALARDEAU AND ALAN R. BRASH — 585

70. Quantitative Assay of Urinary 2,3-Dinor Thromboxane B_2 by GC–MS — RICHARD L. MAAS, DOUGLASS F. TABER, AND L. JACKSON ROBERTS II — 592

71. Preparation of 2,3,4,5-Tetranor-Thromboxane B_2, the Major Urinary Catabolite of Thromboxane B_2 in the Rat — C. R. PACE-ASCIAK AND N. S. EDWARDS — 604

72. Measurement of 5-Hydroxyeicosatetraenoic Acid (5-HETE) in Biological Fluids by GC–MS — MARTIN L. OGLETREE, KENNETH SCHLESINGER, MARY NETTLEMAN, AND WALTER C. HUBBARD — 607

73. Open Tubular Glass Capillary Gas Chromatography for Separating Eicosanoids — JACQUES MACLOUF AND MICHEL RIGAUD — 612

Section VI. Biological Methods

74. Use of Microwave Techniques to Inactivate Brain Enzymes Rapidly — CLAUDIO GALLI AND GIORGIO RACAGNI — 635

75. Platelet Aggregation and the Influence of Prostaglandins — JON M. GERRARD — 642

76. Pharmacologic Characterization of Slow-Reacting Substances — CHARLES W. PARKER, MARY M. HUBER, AND SANDRA F. FALKENHEIN — 655

VOLUME 87

ENZYME KINETICS AND MECHANISM (PART C: INTERMEDIATES, STEREOCHEMISTRY, AND RATE STUDIES)

Section I. Enzyme Intermediates

1. Criteria for Evaluating the Catalytic Competence of Enzyme–Substrate Covalent Compounds — DANIEL L. PURICH — 3

2. Galactose-1-Phosphate Uridylyltransferase: Detection, Isolation, and Characterization of the Uridylyl Enzyme — PERRY A. FREY, LEE-JUN WONG, KWAN-FU SHEU, AND SUE-LEIN YANG — 20

3. The Trimethylacetyl–Transglutaminase Complex — J. E. FOLK — 36

4. Intermediates in the Phosphoglycerate Mutase and Bisphosphoglycerate Synthase Reactions — ZELDA B. ROSE — 42

5. Covalent Chemistry of Pyruvate, Orthophosphate Dikinase — NEIL H. GOSS AND HARLAND G. WOOD — 51

6. Preparation and Properties of Covalent Compounds and Intermediates of Serine Proteases — R. J. COLL, P. D. COMPTON, AND A. L. FINK — 66

7. Covalent Reaction of Substrates and Antimetabolites with Formylglycinamide Ribonucleotide Amidotransferase — JOHN M. BUCHANAN — 76

8. Enzymology of Enol Intermediates — IRWIN A. ROSE — 84

Section II. Stereochemistry

9. Stereospecificities of the Pyridine Nucleotide-Linked Enzymes — KWAN-SA YOU — 101

10. Preparation, Analysis, and Biochemical Applications of Chiral Methyl Groups	HEINZ G. FLOSS	126
11. Preparation of Chromium(III) and Cobalt(III) Nucleotides as Chirality Probes and Inhibitors	W. WALLACE CLELAND	159
12. NMR and EPR Studies of Chromium and Cobalt Nucleotides and Their Interactions with Enzymes	JOSEPH J. VILLAFRANCA	180
13. Stereochemistry of Enzymic Phosphoryl and Nucleotidyl Transfer	FRITZ ECKSTEIN, PAUL J. ROMANIUK, AND BERNARD A. CONNOLLY	197
14. Stereochemistry of Selected Phosphotransferases and Nucleotidyltransferase	PERRY A. FREY, JOHN P. RICHARD, HSU-TSO HO, RICHARD S. BRODY, R. DOUGLAS SAMMONS, AND KWAN-FU SHEU	213
15. Use of $^{31}P(^{18}O)$, $^{31}P(^{17}O)$, and ^{17}O NMR Methods to Study Enzyme Mechanisms Involving Phosphorus	MING-DAW TSAI	235
16. Chiral [^{16}O, ^{17}O, ^{18}O]Phosphoric Monoesters as Stereochemical Probes of Phosphotransferases	STEPHEN L. BUCHWALD, DAVID E. HANSEN, ANNEMARIE HASSETT, AND JEREMY R. KNOWLES	279
17. The Stereochemical Course of Nucleoside Triphosphatase Reactions	MARTIN R. WEBB	301

Section III. Initial Rate and Inhibitor Methods

18. Kinetic Examination of Enzyme Mechanisms Involving Branched Reaction Pathways—A Detailed Consideration of Multifunctional Glucose-6-phosphatase	ROBERT C. NORDLIE	319
19. Initial Velocity Analysis for Terreactant Mechanisms	RONALD E. VIOLA AND W. WALLACE CLELAND	353
20. An Analysis of Haldane Relationships	W. WALLACE CLELAND	366
21. Regression Analysis, Experimental Error, and Statistical Criteria in the Design and Analysis of Experiments for Discrimination between Rival Kinetic Models	BENGT MANNERVIK	370
22. The Use of pH Studies to Determine Chemical Mechanisms of Enzyme-Catalyzed Reactions	W. WALLACE CLELAND	390
23. Buffers of Constant Ionic Strength for Studying pH-Dependent Processes	KEITH J. ELLIS AND JOHN F. MORRISON	405

24. Two-Protonic-State Electrophiles as Probes of Enzyme Mechanism — KEITH BROCKLEHURST 427

25. Application of Affinity Labeling for Studying Structure and Function of Enzymes — BRYCE V. PLAPP 469

26. The Yonetani–Theorell Graphical Method for Examining Overlapping Subsites of Enzyme Active Centers — TAKASHI YONETANI 500

27. Determination of Binding Stoichiometry by the Continuous Variation Method: The Job Plot — CHARLES Y. HUANG 509

28. The Mechanism of Action of Yeast Inorganic Pyrophosphatase — BARRY S. COOPERMAN 526

Section IV. Isotopes as Mechanistic Probes

29. Solvent Isotope Effects on Enzyme Systems — K. BARBARA SCHOWEN AND RICHARD L. SCHOWEN 551

30. Deuterium and Tritium Kinetic Isotope Effects on Initial Rates — DEXTER B. NORTHROP 607

31. The Use of Isotope Effects to Determine Transition-State Structure for Enzymic Reactions — W. WALLACE CLELAND 625

32. Determination of Equilibrium Isotope Effects by the Equilibrium Perturbation Method — W. WALLACE CLELAND 641

33. Isotope Exchange Studies of Regulatory Enzymes: "Fingerprint" Kinetic Patterns for Modifier Action — FREDERICK C. WEDLER AND WILLIAM H. SHALONGO 647

34. Deduction of Kinetic Mechanisms from Primary Hydrogen Isotope Effects: Dopamine β-Monooxygenase—A Case History — SUSAN M. MILLER AND JUDITH P. KLINMAN 711

35. Carbonic Anhydrase: Oxygen-18 Exchange Catalyzed by an Enzyme with Rate-Contributing Proton-Transfer Steps — DAVID N. SILVERMAN 732

VOLUME 88

BIOMEMBRANES (PART I: VISUAL PIGMENTS AND PURPLE MEMBRANES, II)

Section I. Bacteriorhodopsin

A. Purple Membrane Preparations and Protein Structure

1. The Isolation of Deuterated Bacteriorhodopsin from Fully Deuterated *Halobacterium halobium* — HENRY L. CRESPI 3

2. Preparation and Properties of Monomeric Bacteriorhodopsin — NORBERT A. DENCHER AND MAARTEN P. HEYN — 5

3. Reconstitution of the Retinal Proteins Bacteriorhodopsin and Halorhodopsin — DIETER OESTERHELT — 10

B. Reconstituted Systems

4. Bacteriorhodopsin in Phospholipid Vesicles — P. W. M. VAN DIJCK AND K. VAN DAM — 17

5. Phospholipid Substitution of the Purple Membrane — EVERT P. BAKKER AND S. ROY CAPLAN — 26

6. Reconstitution of Monomeric Bacteriorhodopsin into Phospholipid Vesicles — MAARTEN P. HEYN AND NORBERT A. DENCHER — 31

7. A Single Turnover Study of Photoelectric Current-Generating Proteins — V. P. SKULACHEV — 35

8. The Assembly of Bacteriorhodopsin-Containing Planar Membranes by the Sequential Transfer of Air–Water Interface Films — JUAN I. KORENBROT — 45

C. Molecular Structure of Purple Membranes

9. Primary Structure of Bacteriorhodopsin: Sequencing Methods for Membrane Proteins — GERHARD E. GERBER AND H. GOBIND KHORANA — 56

10. Delipidation, Renaturation, and Reconstitution of Bacteriorhodopsin — HAGEN BAYLEY, BO HÖJEBERG, KUO-SEN HUANG, H. GOBIND KHORANA, MEI-JUNE LIAO, CHRISTINA LIND, AND ERWIN LONDON — 74

11. The Identification of Helical Segments in the Polypeptide Chain of Bacteriorhodopsin — D. M. ENGELMAN, A. GOLDMAN, AND T. A. STEITZ — 81

12. Determination of Modified Positions in the Polypeptide Chain of Bacteriorhodopsin — HORST-DIETER LEMKE, JURGEN BERGMEYER, AND DIETER OESTERHELT — 89

13. Lipids of Purple Membrane from Extreme Halophiles and of Methanogenic Bacteria — M. KATES, S. C. KUSHWAHA, AND G. D. SPROTT — 98

14. Preparation of New Two- and Three-Dimensional Crystal Forms of Bacteriorhodopsin — HARTMUT MICHEL AND DIETER OESTERHELT — 111

15. Freeze-Substitution and Freeze-Etching Method for Studying the Ultrastructure of Photoreceptive Membrane — JIRO USUKURA AND EICHI YAMADA — 118

16. Analysis of Bacteriorhodopsin Structure by X-Ray Diffraction — A. E. BLAUROCK — 124

D. Chemistry, Spectroscopy, and Photochemistry

17. Rapid Kinetic Optical Absorption Spectroscopy of Bacteriorhodopsin Photocycles — RICHARD H. LOZIER — 133

18. The Photoreaction Cycle of Bacteriorhodopsin: Low-Temperature Spectrophotometry — FUMIO TOKUNAGA AND TATSUO IWASA — 163

19. Photochemistry and Isomer Determination of 13-*cis*- and *trans*-Bacteriorhodopsin — NORBERT A. DENCHER, CHARLES N. RAFFERTY, AND WALTER SPERLING — 167

20. Spin Labeling of Bacteriorhodopsin — ROSALIE K. CROUCH — 175

21. Heavy Atom Labeling of Retinal in Bacteriorhodopsin — MICHAEL G. MOTTO AND KOJI NAKANISHI — 178

22. Analysis of Photocycle and Orientation in Thin Layers — RAFI KORENSTEIN AND BENNO HESS — 180

23. Cooperativity of Photocycle in Purple Membrane — RAFI KORENSTEIN AND BENNO HESS — 193

24. High-Performance Liquid Chromatography Method for Isolation of Membrane Proteins from Halobacterial Membrane — TETSUYA KONISHI — 202

25. Heterobifunctional Cross-Linking of Bacteriorhodopsin by Hydrophobic Azidoarylisothiocyanates — HANS SIGRIST AND PETER ZAHLER — 207

26. Measurement of Light-Regulated Phosphoproteins of *Halobacterium halobium* — ELENA NEGRI SPUDICH AND JOHN LEE SPUDICH — 213

27. Emission Spectroscopy of Rhodopsin and Bacteriorhodopsin — AARON LEWIS AND G. J. PERREAULT — 217

E. Specialized Physical Techniques

28. Preparation of Planar Membrane Monolayers for Spectroscopy and Electron Microscopy — KNUTE A. FISHER — 230

29. Metal Decoration of the Purple Membrane — DOROTHEA-CH. NEUGEBAUER — 235

30. Neutron Scattering of Bacteriorhodopsin — GLEN I. KING AND BENNO P. SCHOENBORN — 241

31. Transient Dichroism of Bacteriorhodopsin — RICHARD J. CHERRY — 248

32. Spectroscopic Methods for Protonation State Determination — DIETRICH KUSCHMITZ AND BENNO HESS — 254

33. Monitoring of Protein Conformation Changes during Photocycle — BRIAN BECHER — 265

34. X-Ray Diffraction Studies of Purple Membranes Reconstituted from Brown Membrane — TOSHIAKI HAMANAKA, KENJI HIRAKI, AND TOSHIO MITSUI — 268

35. Light-Induced Volume Changes — WILLIAM W. PARSON — 272

36. Application of Nanosecond X-Ray Diffraction Techniques to Bacteriorhodopsin — ROBERT D. FRANKEL AND JAMES M. FORSYTH — 276

37. X-Ray Diffraction Studies of Light Interactions with Bacteriorhodopsin — J. STAMATOFF, R. H. LOZIER, AND S. GRUNER — 282

38. Orientation of Purple Membranes by Electric Field — LAJOS KESZTHELYI — 287

39. Effect of Viscosity on the Photocycle of Bacteriorhodopsin — LAURA EISENSTEIN — 297

40. Time-Resolved Protein Fluorescence Measurements of Intermediates in the Photocycle of Bacteriorhodopsin — WILLIAM D. HOPEWELL AND JOSEPH M. FUKUMOTO — 306

41. Recent Advances in the Study of Bacteriorhodopsin Dynamic Structure Using High-Field Solid-State Nuclear Magnetic Resonance Spectroscopy — ERIC OLDFIELD, ROBERT A. KINSEY, AND AUGUSTIN KINTANAR — 310

42. Preparation of Oriented Multilamellar Arrays of Natural and Artificial Biological Membranes — NOEL A. CLARK AND KENNETH J. ROTHSCHILD — 326

F. Ion Transport and Physiology

43. Light-Induced pH Gradients Measured with Spin-Labeled Amine and Carboxylic Acid Probes: Application to *Halobacterium halobium* Cell Envelope Vesicles — R. J. MEHLHORN AND I. PROBST — 334

44. Secondary Ion Movements in *Halobacterium halobium* — GOTTFRIED WAGNER — 344

45. Photophosphorylation and Reconstitution of Photophosphorylation in Halobacterial Cells — DIETER OESTERHELT — 349

46. Simultaneous Measurements of Proton Movement and Membrane Potential Changes in the Wild-Type and Mutant *Halobacterium halobium* Vesicles — NAOKI KAMO, TONI RACANELLI, AND LESTER PACKER — 356

G. Biogenesis, Genetics, and Microorganisms

47. The Classification of Halobacteria — TERJE TORSVIK AND IAN DUNDAS — 360

48. The *Halobacterium* Group: Microbiological Methods — HANS JURGEN WEBER, SYAM SARMA, AND TERRANCE LEIGHTON — 369

49. Plasmids in Halobacteria: Restriction Maps — GOTTFRIED WEIDINGER, FELICITAS PFEIFER, AND WERNER GOEBEL — 374

50. The Isolation of *Halobacterium* Mutant Strains with Defects in Pigment Synthesis — HANS JURGEN WEBER AND ROBERTO A. BOGOMOLNI — 379

51. The Brown Membrane of *Halobacterium halobium:* The Biosynthetic Precursor of the Purple Membrane — MANFRED SUMPER — 391

52. Lipids of *Thermoplasma* — THOMAS A. LANGWORTHY — 396

53. Isolation of White Membrane of Crystalline Bacterio-opsin from *Halobacterium halobium* R_1mW Lacking Carotenoid — YASUO MUKOHATA AND YASUO SUGIYAMA — 407

54. Identification and Detection of Electron Transfer Components in Methanogens — JACK R. LANCASTER, JR. — 412

55. Anaerobic Growth of Halobacteria — DIETER OESTERHELT — 417

H. Light-Dependent Behavioral Responses of the Intact Organism

56. Photobehavior of *Halobacterium halobium* — NORBERT A. DENCHER AND EILO HILDEBRAND — 420

I. Other Retinal Proteins

57. Demonstration of Primary Sodium Transport Activity in *Halobacterium halobium* Envelope Vesicles — EDWARD V. LINDLEY, PHILLIP N. HOWLES, AND RUSSELL E. MACDONALD — 426

58. Assay of Pigment P_{588} and Its Discrimination from Bacteriorhodopsin by Flash Spectroscopy Techniques — ROBERTO A. BOGOMOLNI AND HANS JURGEN WEBER — 434

59. Spectrophotometric Determination of Halorhodopsin in *Halobacterium halobium* Membranes — JANOS K. LANYI — 439

Section II. General Methods for Retinal Proteins

A. Bacteriorodopsin and Rhodopsin Molecular Structure

60. Comparison of Bacteriorhodopsin and Rhodopsin Molecular Structure — B. A. WALLACE — 447

61. Protein–Chromophore Interactions as Spectroscopic and Photochemical Determinants — BARRY HONIG AND THOMAS G. EBREY — 462

62. Molecular Aspects of the Photocycles of Rhodopsin and Bacteriorhodopsin: A Comparative Overview — MICHAEL OTTOLENGHI — 470

B. Model Chromophores

63. Use of High-Performance Liquid Chromatography to Separate and Identify Retinals — RICHARD A. BLATCHLY AND KOJI NAKANISHI — 491

64. Synthetic Analogs of Retinal, Bacteriorhodopsin, and Bovine Rhodopsin — VALERIA BALOGH-NAIR AND KOJI NAKANISHI — 496

65. Synthesis and Photochemistry of Stereoisomers of Retinal — ROBERT S. H. LIU AND ALFRED E. ASATO — 506

66. Synthetic Pigments of Rhodopsin and Bacteriorhodopsin — THOMAS G. EBREY — 516

67. Energy Levels of the Visual Chromophores by Two-Photon Spectroscopy — ROBERT R. BIRGE — 522

68. Photoisomerization of Kinetics of Retinal Analogs — DAVID S. KLIGER — 533

69. Isolation and Purification of Retinals from Purple Membranes for Mass Spectral Analysis — STANLEY SELTZER AND MOW LIN — 542

70. Identification of cis/trans Isomers of Retinal Analogs by High-Performance Proton NMR Method — PAUL TOWNER AND WOLFGANG GÄRTNER — 546

71. Methods for Extraction of Pigment Chromophore — MOTOYUKI TSUDA — 552

C. Physical and Chemical Methods

72. Resonance Raman Spectroscopy of Rhodopsin and Bacteriorhodopsin: An Overview — AARON LEWIS — 561

73. Time-Resolved Chromophore Resonance Raman and Protein Fluorescence of the Intermediates of the Proton Pump Photocycle of Bacteriorhodopsin — M. A. EL-SAYED — 617

74. Resonance Raman Techniques for Photolabile Samples: Pump–Probe and Flow — ROBERT CALLENDER — 625

75. Resonance Raman Spectroscopy of Rhodopsin and Bacteriorhodopsin Isotopic Analogs — RICHARD MATHIES — 633

76. Kinetic Resonance Raman Spectroscopy of Purple Membrane Using a Rotating Sample Cell — PRAMOD V. ARGADE AND KENNETH J. ROTHSCHILD — 643

77. Spinning Sample Raman Spectroscopy at 77°K: Bacteriorhodopsin's Primary Photoproduct — MARK BRAIMAN AND RICHARD MATHIES — 648

78. Kinetic Resonance Raman Spectroscopy with Microsampling Rotating Cells — AARON LEWIS — 659

79. Calorimetric Measurements of Light-Induced Processes — ALAN COOPER — 667

80. Tritium–Hydrogen Exchange Kinetics — NANCY W. DOWNER AND JOAN J. ENGLANDER — 673

81. Infrared Measurement of Peptide Hydrogen Exchange in Rhodopsin — H. BEVERLEY OSBORNE AND ELIANE NEBEDRYK-VIALA — 676

82. A Comparison of Hydrogen Exchange Methods Applied to Rhodopsin — NANCY W. DOWNER AND H. BEVERLEY OSBORNE — 681

83. Spin-Label Probes of Light-Induced Electrical Potentials in Rhodopsin and Bacteriorhodopsin — DAVID S. CAFISO, WAYNE L. HUBBELL, AND ALEXANDRE QUINTANILHA — 682

84. Infrared Absorption of Photoreceptor and Purple Membranes — KENNETH J. ROTHSCHILD, ROSEMARY SANCHES, AND NOEL A. CLARK — 696

85. Effect of Pressure on Visual Pigment and Purple Membrane — MOTOYUKI TSUDA — 714

86. Some Approaches to Determining the Primary Structure of Membrane Proteins — N. G. ABDULAEV AND YU. A. OVCHINNIKOV — 723

87. Kinetic Properties of Rhodopsin and Bacteriorhodopsin Measured by Kinetic Infrared Spectroscopy (KIS) — WERNER MÄNTELE, FRITZ SEIBERT, AND W. KREUTZ — 729

88. Photodichroism and the Rotational Motions of Rhodopsin and Bacteriorhodopsin — PATRICK L. AHL AND RICHARD A. CONE — 741

89. Measurements of Volumes and Electrochemical Gradients with Spin Probes in Membrane Vesicles — ROLF J. MEHLHORN, PEDRO CANDAU, AND LESTER PACKER — 751

90. Diffusible Spin Labels Used to Study Lipid–Protein Interactions with Rhodopsin and Bacteriorhodopsin — DEREK MARSH AND ANTHONY WATTS — 762

91. A Computer Technique for Structural Studies of Bacteriorhodopsin — HAYWARD ZWERLING, ROLF MEHLHORN, LESTER PACKER, AND ROBERT MACELROY 772

VOLUME 89

CARBOHYDRATE METABOLISM (PART D)

Section I. Analytical Methods

1. Resolution of DL-*myo*-Inositol 1-Phosphate and Other Sugar Enantiomers by Gas Chromatography — ALAN L. LEAVITT AND WILLIAM R. SHERMAN 3

2. Determination of Inositol Phosphates by Gas Chromatography — ALAN L. LEAVITT AND WILLIAM R. SHERMAN 9

3. Enzymic Method for Determination of Fructose 6-Phosphate, Xylulose 5-Phosphate, and Sedoheptulose 7-Phosphate — G. A. KOCHETOV 19

4. Enzymic Microdetermination of D-Glucose, D-Fructose, D-Gluconate, 2-Keto-D-gluconate, Aldehyde, and Alcohol with Membrane-Bound Dehydrogenases — MINORU AMEYAMA 20

5. Automated Enzymic Determination of L(+)- and D(−)-Lactic Acid — FLOYD M. BYERS 29

6. Determination of D-Lactate in Plasma — RICHARD B. BRANDT 35

Section II. Enzyme Assay Procedures

7. Determination of Transketolase Activity via Ferricyanide Reduction — G. A. KOCHETOV 43

8. Spectrophotometric Method for Glucose-6-Phosphate Phosphatase — PETER GIEROW AND BENGT JERGIL 44

9. Activation and Assay of Ribulose-1,5-bisphosphate Carboxylase/Oxygenase — JOHN W. PIERCE, STEPHEN D. MCCURRY, R. MICHAEL MULLIGAN, AND N. E. TOLBERT 47

Section III. Preparation of Substrates and Effectors

10. Synthesis of (2*R*)-Glycerol-*o*β-D-Galactopyranoside by β-Galactosidase — WINFRIED BOOS 59

11. Chemical Synthesis of Monosaccharides Enriched with Carbon Isotopes — A. S. Serianni, H. A. Nunez, M. L. Hayes, and R. Barker — 64

12. Chemical Synthesis of Aldose Phosphates Enriched with Carbon Isotopes — A. S. Serianni, J. Pierce, and R. Barker — 73

13. Chemical Synthesis of Aldoses Enriched with Isotopes of Hydrogen and Oxygen — A. S. Serianni, E. L. Clark, and R. Barker — 79

14. Enzymic Synthesis of ^{13}C-Enriched Aldoses, Ketoses, and Their Phosphate Esters — A. S. Serianni, E. Cadman, J. Pierce, M. L. Hayes, and R. Barker — 83

15. D-Galactose 6-Phosphate and D-Tagatose 6-Phosphate — Richard L. Anderson, William C. Wenger, and Donald L. Bissett — 93

16. Enzymic Synthesis of 2-Keto-3-Deoxygluconate 6-Phosphate Using 6-Phosphogluconate Dehydratase — E. L. O'Connell and H. Paul Meloche — 98

17. Chemical Synthesis of Fructose 2,6-Bisphosphate — S. J. Pilkis, M. R. El-Maghrabi, D. A. Cumming, J. Pilkis, and T. H. Claus — 101

18. Synthesis of Ribulose 1,5-Bisphosphate: Routes from Glucose 6-Phosphate (via 6-Phosphogluconate) and from Adenosine Monophosphate (via Ribose 5-Phosphate) — Chi-Huey Wong, Alfred Pollak, Stephen D. McCurry, Julia M. Sue, Jeremy R. Knowles, and George M. Whitesides — 108

19. Isomers of α-D-Apiofuranosyl 1-Phosphate and α-D-Apiose 1,2-Cyclic Phosphate — Joseph Mendicino, Ragy Hanna, and E. V. Chandrasekaran — 121

Section IV. Oxidation–Reduction Enzymes

20. Cellobiose Oxidase from *Sporotrichum pulverulentum* — Arthur R. Ayers and Karl-Erik Eriksson — 129

21. Sorbitol Dehydrogenase from Rat Liver — Nancy Leissing and E. T. McGuinness — 135

22. D-Sorbitol Dehydrogenase from *Gluconobacter suboxydans,* Membrane-Bound — EMIKO SHINAGAWA AND MINORU AMEYAMA 141

23. Hexose Oxidase from *Chondrus crispus* — MIYOSHI IKAWA 145

24. D-Glucose Dehydrogenase from *Pseudomonas fluorescens,* Membrane-Bound — KAZUNOBU MATSUSHITA AND MINORU AMEYAMA 149

25. D-Fructose Dehydrogenase from *Gluconobacter industrius,* Membrane-Bound — MINORU AMEYAMA AND OSAO ADACHI 154

26. D-Glucose Dehydrogenase from *Gluconobacter suboxydans* — OSAO ADACHI AND MINORU AMEYAMA 159

27. Galactose Oxidase from *Dactylium dendroides* — PAUL S. TRESSEL AND DANIEL J. KOSMAN 163

28. Galactose Oxidase from Commercial Samples — MARK W. C. HATTON AND ERWIN REGOECZI 172

29. D-Galactose Dehydrogenase from *Pseudomonas fluorescens* — E. MAIER AND G. KURZ 176

30. Aldose Reductase from Human Tissues — BENDICHT WERMUTH AND JEAN-PIERRE VON WARTBURG 181

31. D-Gluconate Dehydrogenase from Bacteria, 2-Keto-D-gluconate-Yielding, Membrane-Bound — KAZUNOBU MATSUSHITA, EMIKO SHINAGAWA, AND MINORU AMEYAMA 187

32. 2-Keto-D-gluconate Dehydrogenase from *Gluconobacter melanogenus,* Membrane-Bound — EMIKO SHINAGAWA AND MINORU AMEYAMA 194

33. 5-Keto-D-gluconate Reductase from *Gluconobacter suboxydans* — MINORU AMEYAMA AND OSAO ADACHI 198

34. 2-Keto-D-gluconate Reductase from Acetic Acid Bacteria — MINORU AMEYAMA AND OSAO ADACHI 203

35. D-Mannonate and D-Altronate-NAD Dehydrogenases from *Escherichia coli* — RAYMOND PORTALIER AND FRANÇOIS STOEBER 210

36. 2-Keto-3-deoxy-L-fuconate Dehydrogenase from Pork Liver — HARRY SCHACHTER, JULIA Y. CHAN, AND NGOZI A. NWOKORO 219

37. D-Xylose Dehydrogenase — A. STEPHEN DAHMS AND JOHN RUSSO 226

38. D-Apiitol Dehydrogenase from Bacteria — E. V. CHANDRASEKARAN, MARK DAVILA, AND JOSEPH MENDICINO 228

39. D-Erythrulose Reductase from Beef Liver — KIHACHIRO UEHARA AND SABURO HOSOMI 232

40. Glycerol Dehydrogenase from Rabbit Muscle — T. G. Flynn and J. A. Cromlish — 237

41. Glycerol Oxidase from *Aspergillus japonicus* — T. Uwajima and O. Terada — 243

42. L-Sorbose-1-phosphate Reductase — Richard L. Anderson and Ronald A. Simkins — 248

43. Glucose-6-phosphate Dehydrogenase from Mouse — Chi-Yu Lee — 252

44. Glucose-6-phosphate Dehydrogenase, Vegetative and Spore *Bacillus subtilis* — Susumu Ujita and Kinuko Kimura — 258

45. D-Glucose-6-phosphate Dehydrogenases from *Pseudomonas fluorescens* — P. Maurer, D. Lessmann, and G. Kurz — 261

46. Glucose-6-phosphate Dehydrogenase from *Methylomonas* M15 — Rolf A. Steinbach, Horst Schütte, and Hermann Sahm — 271

47. D-Galactitol-6-phosphate Dehydrogenase — Richard L. Anderson and John P. Markwell — 275

48. 6-Phospho-D-gluconate Dehydrogenase from *Pseudomonas fluorescens* — C. Stournaras, F. Butz, and G. Kurz — 278

49. 6-Phosphogluconate Dehydrogenase from *Bacillus stearothermophilus* — Francesco M. Veronese, Enrico Boccù, and Angelo Fontana — 282

50. 6-Phospho-D-gluconate Dehydrogenase from *Gluconobacter suboxydans* — Osao Adachi and Minoru Ameyama — 291

51. *sn*-Glycerol-3-phosphate Dehydrogenase (Soluble) from *Drosophila melanogaster* — David W. Niesel, Glenn C. Bewley, Chi-Yu Lee, and Frank B. Armstrong — 296

52. Glyceraldehyde-3-phosphate Dehydrogenase from Human Tissues — Fritz Heinz and Barbara Freimüller — 301

53. Glyceraldehyde-3-phosphate Dehydrogenase from Rabbit Muscle — R. M. Scheek and E. C. Slater — 305

54. Glyceraldehyde-3-phosphate Dehydrogenase from Pig Liver — Shawky M. Dagher and William C. Deal, Jr. — 310

55. Glyceraldehyde-3-phosphate Dehydrogenase (Glycolytic Form) from Spinach Leaves — M. L. Speranza and G. Ferri — 316

56. Glyceraldehyde-3-phosphate Dehydrogenase from Pea Seeds — Ronald G. Duggleby and David T. Dennis — 319

57. Glyceraldehyde-3-phosphate Dehydrogenase from Yeast — LARRY D. BYERS — 326

58. Glyceraldehyde-3-phosphate Dehydrogenase from *Thermus thermophilus* — TAIRO OSHIMA, SHINOBU C. FUJITA, AND KAZUTOMO IMAHORI — 335

59. Hydroxypyruvate Reductase (D-Glycerate Dehydrogenase) from *Pseudomonas* — LEONARD D. KOHN AND JOHN M. UTTING — 341

60. L-Lactate Dehydrogenase Isozymes from Sweet Potato Roots — KAZUKO ÔBA AND IKUZO URITANI — 345

61. Lactate Dehydrogenase Isozymes from Mouse — CHI-YU LEE, JAMES H. YUAN, AND ERWIN GOLDBERG — 351

62. L(+)-Lactate Dehydrogenase from *Homarus americanus* — R. D. EICHNER — 359

63. L-Lactate Dehydrogenase, FDP-Activated, from *Streptococcus cremoris* — ALAN J. HILLIER AND G. RICHARD JAGO — 362

64. Lactate-Oxaloacetate Transhydrogenase from *Veillonella alcalescens* — S. H. GEORGE ALLEN — 367

65. Pyruvate Dehydrogenase Complex from Bovine Kidney and Heart — FLORA H. PETTIT AND LESTER J. REED — 376

66. Pyruvate Dehydrogenase Complex from *Neurospora* — ROY W. HARDING, DINA F. CAROLINE, AND ROBERT P. WAGNER — 386

67. Pyruvate Dehydrogenase Complex from *Escherichia coli* — JAAP VISSER AND MARIJKE STRATING — 391

68. Pyruvate Dehydrogenase Complex from *Bacillus* — JAAP VISSER, HARRY KESTER, KANDIAH JEYASEELAN, AND RANDY TOPP — 399

69. Pyruvate Dehydrogenase Complex from Broccoli and Cauliflower — DOUGLAS D. RANDALL — 408

70. Pyruvate Dehydrogenase Complex from Pigeon Breast Muscle — SHUICHI FURUTA AND TAKASHI HASHIMOTO — 414

71. Pyruvate Dehydrogenase Complex from *Hansenula* — T. HARADA AND T. HIRABAYASHI — 420

72. Alcohol Oxidase from *Candida boidinii* — HERMANN SAHM, HORST SCHÜTTE, AND MARIA-REGINA KULA — 424

73. Alcohol Dehydrogenase from Horse Liver, Steroid-Active SS Isozyme — REGINA PIETRUSZKO 428

74. Alcohol Dehydrogenase from Horse Liver by Affinity Chromatography — LARS ANDERSSON AND KLAUS MOSBACH 435

75. Alcohol Dehydrogenase from *Drosophila melanogaster* — CHI-YU LEE 445

76. Alcohol Dehydrogenase from Acetic Acid Bacteria, Membrane-Bound — MINORU AMEYAMA AND OSAO ADACHI 450

77. Covalent Enzyme–Coenzyme Complexes of Liver Alcohol Dehydrogenase and NAD — MATS-OLLE MÅNSSON, PER-OLOF LARSSON, AND KLAUS MOSBACH 457

78. Aldehyde Dehydrogenase from Baker's Yeast — NANAYA TAMAKI AND TAKAO HAMA 469

79. Isozymes of Aldehyde Dehydrogenase from Horse Liver — JOHN H. ECKFELDT AND TAKASHI YONETANI 474

80. Aldehyde Dehydrogenase from *Proteus vulgaris* — SHOJI SASAKI AND YASUTAKE SUGAWARA 480

81. Aldehyde Dehydrogenases from *Pseudomonas aeruginosa* — J. P. VANDECASTEELE AND L. GUERRILLOT 484

82. Aldehyde Dehydrogenase from Acetic Acid Bacteria, Membrane-Bound — MINORU AMEYAMA AND OSAO ADACHI 491

83. Aldehyde Dehydrogenase from Bovine Liver — ANDREA WESTERHAUSEN, WOLFGANG LEICHT, AND FRITZ HEINZ 497

84. Aldehyde Reductase (L-Hexonate:NADP Dehydrogenase) from Pig Kidney — T. G. FLYNN, J. A. CROMLISH, AND W. S. DAVIDSON 501

85. Aldehyde Reductase from Human Tissues — JEAN-PIERRE VON WARTBURG AND BENDICHT WERMUTH 506

86. 2-Ketoaldehyde Dehydrogenase from Rat Liver — DAVID L. VANDER JAGT 513

87. Diacetyl Reductase — ROBERTO MARTÍN SARMIENTO AND JUSTINO BURGOS 516

88. L-Glycol Dehydrogenase from Hen Muscle — JUSTINO BURGOS AND ROBERTO MARTÍN SARMIENTO 523

89. Formaldehyde Dehydrogenase from *Candida boidinii* — HORST SCHÜTTE, MARIA-REGINA KULA, AND HERMANN SAHM 527

90. Formate Dehydrogenase from *Pseudomonas oxalaticus* — TH. HÖPNER, U. RUSCHIG, U. MÜLLER, AND P. WILLNOW 531

91. Formate Dehydrogenase from *Escherichia coli* — HARRY G. ENOCH AND ROBERT L. LESTER 537

Section V. Isomerases, Epimerases, and Mutases

92. D-Ribose Isomerase — ALAN D. ELBEIN AND KEN IZUMORI 547

93. Glucosephosphate Isomerase from Catfish Muscle and Liver and from Mammalian Tissues — ROBERT W. GRACY 550

94. Phosphoglucose Isomerase from Mouse and *Drosophila melanogaster* — CHI-YU LEE 559

95. D-Galactose-6-phosphate Isomerase — RICHARD L. ANDERSON AND DONALD L. BISSETT 562

96. 6-Phospho-3-Ketohexulose Isomerase from *Methylomonas (Methylococcus) capsulatus* — J. RODNEY QUAYLE 566

97. Bacterial Ribosephosphate Isomerase — R. D. MACELROY AND C. R. MIDDAUGH 571

98. Triosephosphate Isomerase from Chicken and Rabbit Muscle — M. P. ESNOUF, R. P. HARRIS, AND J. D. MCVITTIE 579

99. Uridine Diphosphate Glucose-4-epimerase and Galactose-1-phosphate Uridylyltransferase from *Saccharomyces cerevisiae* — T. FUKASAWA, T. SEGAWA, AND Y. NOGI 584

100. Inositol Epimerase-Inosose Reductase from Bovine Brain — PAUL P. HIPPS, KAREN E. ACKERMANN, AND WILLIAM R. SHERMAN 593

101. Phosphoglucomutase from Yeast — J. G. JOSHI 599

VOLUME 90

CARBOHYDRATE METABOLISM (PART E)

Section I. Kinases

1. Hexokinase from Rabbit Red Blood Cells — GIORGIO FORNAINI, MARINA DACHÀ, MAURO MAGNANI, AND VILBERTO STOCCHI 3

2. Hexokinase Type II from Rat Skeletal Muscle	John S. Easterby and S. Saleheen Qadri	11
3. Hexokinases from *Drosophila melanogaster*	Chi-Yu Lee	16
4. Hexokinase from *Ascaris suum* Muscle	Scott C. Supowit and Ben G. Harris	21
5. Glucokinase from *Streptococcus mutans*	E. Victoria Porter and Bruce M. Chassy	25
6. Galactokinase from *Saccharomyces cerevisiae*	David B. Wilson and Mark A. Schell	30
7. Phosphofructokinase from Ehrlich Ascites Tumor	Michio Ui and Takahiko Sumi	35
8. Phosphofructokinase from Oyster Adductor Muscle	Kenneth B. Storey	39
9. Phosphofructokinase from *Ascaris suum* Muscle	Ben G. Harris, Jane A. Starling, and H. Werner Hofer	44
10. Phosphofructokinase from Yeast	Eberhard Hofmann and Gerhard Kopperschläger	49
11. Phosphofructokinases from *Escherichia coli*	Denise Kotlarz and Henri Buc	60
12. Phosphofructokinase from *Bacillus licheniformis*	Charles K. Marschke and Robert W. Bernlohr	70
13. Phosphofructokinase from *Streptococcus lactis*	Alison M. Fordyce, C. H. Moore, and G. G. Pritchard	77
14. Phosphofructokinase from *Clostridium pasteurianum*	Gerhard Gottschalk and Hasso v. Hugo	82
15. D-Tagatose-6-phosphate Kinase from *Staphylococcus aureus*	Richard L. Anderson and Donald L. Bissett	87
16. Inorganic Pyrophosphate : D-Fructose-6-phosphate 1-Phosphotransferase from Mung Bean	Richard L. Anderson and Dario C. Sabularse	91
17. 6-Phosphofructokinase (Pyrophosphate) from *Entamoeba histolytica*	Richard E. Reeves, Patricia Lobelle-Rich, and William B. Eubank	97
18. Phosphoglycerate Kinase from Animal Tissue	Günter W. K. Kuntz and Wolfgang K. G. Krietsch	103
19. Phosphoglycerate Kinase from Spinach, Blue-Green Algae, and Yeast	Günter W. K. Kuntz and Wolfgang K. G. Krietsch	110

20. 3-Phosphoglycerate Kinase from Bovine Liver and Yeast — KLAUS D. KULBE AND MARINA BOJANOVSKI — 115

21. 3-Phosphoglycerate Kinase Isozymes and Genetic Variants from Mouse — CHI-YU LEE — 121

22. Phosphoglycerate Kinase from *Bacillus stearothermophilus* — KOICHI SUZUKI AND KAZUTOMO IMAHORI — 126

23. Pyruvate Kinases from Human Erythrocytes and Liver — AXEL KAHN AND JOËLLE MARIE — 131

24. Pyruvate Kinase from Bovine Muscle and Liver — JANET M. CARDENAS — 140

25. Pyruvate Kinase Isozymes from Rat — KIICHI IMAMURA AND TAKEHIKO TANAKA — 150

26. Pyruvate Kinase from *Streptococcus lactis* — V. L. CROW AND G. G. PRITCHARD — 165

27. AMP- and Fructose 1,6-Bisphosphate-Activated Pyruvate Kinases from *Escherichia coli* — MASSIMO MALCOVATI AND GIOVANNA VALENTINI — 170

28. Acetate Kinase from *Bacillus stearothermophilus* — KOICHI SUZUKI, HIROSHI NAKAJIMA, AND KAZUTOMO IMAHORI — 179

29. Purification of Human and Canine Creatine Kinase Isozymes — ROBERT ROBERTS — 185

30. Pyruvate Dehydrogenase Kinase from Bovine Kidney — FLORA H. PETTIT, STEPHEN J. YEAMAN, AND LESTER J. REED — 195

31. cAMP-Dependent Protein Kinase, Soluble and Particulate, from Swine Kidney — JOSEPH MENDICINO, K. MUNIYAPPA, AND FREDRICH H. LEIBACH — 200

Section II. Aldolases and Transketolases

32. Transketolase from Yeast, Rat Liver, and Pig Liver — G. A. KOCHETOV — 209

33. Transketolase from Human Red Blood Cells — G. D. SCHELLENBERG N. M. WILSON, B. R. COPELAND, AND CLEMENT E. FURLONG — 223

34. D-Tagatose-1,6-bisphosphate Aldolase (Class I) from *Staphylococcus aureus* — RICHARD L. ANDERSON AND DONALD L. BISSETT — 228

35. D-Tagatose-1,6-bisphosphate Aldolase (Class II) from *Klebsiella pneumoniae* — RICHARD L. ANDERSON AND JOHN P. MARKWELL — 232

36. Fructose-1,6-bisphosphate Aldolase from *Bacillus subtilis* — SUSUMU UJITA AND KINUKO KIMURA — 235

37. Fructose-bisphosphate Aldolases from Mycobacteria — N. JAYANTHI BAI, M. RAMACHANDRA PAI, P. SURYANARAYANA MURTHY, AND T. A. VENKITASUBRAMANIAN — 241

38. Fructose-bisphosphate Aldolase from Human Erythrocytes — DON R. YELTMAN AND BEN G. HARRIS — 251

39. Fructose-bisphosphate Aldolase from *Ascaris suum* — MARIAN KOCHMAN AND DANUTA KWIATKOWSKA — 254

40. Fructose-bisphosphate Aldolase from *Helix pomatia* — MARIAN KOCHMAN, PAUL A. HARGRAVE, AND JANINA BUCZŁKO — 259

41. 2-Keto-3-deoxy-galactonate-6-phosphate Aldolase from *Pseudomonas saccharophila* — H. PAUL MELOCHE AND EDWARD L. O'CONNELL — 263

42. 2-Keto-3-deoxy-D-xylonate Aldolase (3-Deoxy-D-pentulosonic Acid Aldolase) — A. STEPHEN DAHMS AND ALAN DONALD — 269

43. 4-Hydroxy-4-methyl-2-ketoglutarate Aldolase from *Pseudomonas putida* — STANLEY DAGLEY — 272

44. 4-Hydroxy-2-ketopimelate Aldolase — STANLEY DAGLEY — 277

Section III. Dehydratases

45. D-Gluconate Dehydratase from *Clostridium pasteurianum* — GERHARD GOTTSCHALK AND RUDOLF BENDER — 283

46. D-Mannonate and D-Altronate Dehydratases of *Escherichia coli* K12 — J. ROBERT-BAUDOUY, J. JIMENO-ABENDANO, AND F. STOEBER — 288

47. D-Galactonate Dehydratase — A. STEPHEN DAHMS, DAVID SIBLEY, WILLIAM HUISMAN, AND ALAN DONALD — 294

48. D-Galactonate/D-Fuconate Dehydratase — RICHARD L. ANDERSON AND CHARLES L. HAUSWALD — 299

49. D-*xylo*-Aldonate Dehydratase — A. STEPHEN DAHMS AND ALAN DONALD — 302

Section IV. Synthases

50. L-*myo*-Inositol-1-phosphate Synthase from Bovine Testis — Yun-Hua H. Wong, Linda A. Mauck, and William R. Sherman 309

51. 3-Hexulose-6-phosphate Synthase from *Methylomonas* (*Methylococcus*) *capsulatus* — J. Rodney Quayle 314

52. 3-Hexulose-phosphate Synthase from *Methylomonas* M15 — Hermann Sahm, Horst Schütte, and Maria-Regina Kula 319

Section V. Phosphatases

53. Fructose-bisphosphatase, Zinc-Free, from Rabbit Liver — Margaret M. DeMaine, Carol A. Caperelli, and Stephen J. Benkovic 327

54. Fructose-bisphosphatase from Ox Liver — H. G. Nimmo and K. F. Tipton 330

55. Fructose-1,6-bisphosphatase from Turkey Liver — Peter F. Han and Joe Johnson, Jr. 334

56. Fructose-1,6-bisphosphatase from Chicken and Rabbit Muscle — John S. MacGregor, A. E. Annamalai, A. van Tol, W. J. Black, and B. L. Horecker 340

57. Fructose-1,6-bisphosphatase from Rabbit Liver (Neutral Form) — B. L. Horecker, W. C. McGregor, S. Traniello, E. Melloni, and S. Pontremoli 345

58. Fructose-1,6-bisphosphatase from Snake Muscle (*Zaocys dhumnades*) — Gen-jun Xu, Jian-ping Shi, and Ying-lai Wang 349

59. Fructose-1,6-bisphosphatase from Rat Liver — Frank Marcus, Judith Rittenhouse, Tapati Chatterjee, and M. Marlene Hosey 352

60. Fructose-1,6-bisphosphatase from Mouse and Rabbit Intestinal Mucosa — Yohtalou Tashima and Hideo Mizunuma 357

61. Fructose-1,6-bisphosphatase from Bumblebee Flight Muscle — Kenneth B. Storey 366

62. Fructose-bisphosphatase from Spinach Leaf Chloroplast and Cytoplasm — GRAHAME J. KELLY, GERHARD ZIMMERMANN, AND ERWIN LATZKO 371

63. Fructose-bisphosphatase from *Rhodopseudomonas palustris* — CLARK F. SPRINGGATE AND CHESTER S. STACHOW 378

64. Fructose-1,6-bisphosphatase from *Bacillus licheniformis* — DENNIS J. OPHEIM AND ROBERT W. BERNLOHR 384

65. Sedoheptulose-1,7-bisphosphatase from Wheat Chloroplasts — IAN E. WOODROW 392

66. Glucose-6-phosphatase from Cerebrum — MANFRED L. KARNOVSKY, JAMES MICHAEL ANCHORS, AND MICHAEL A. ZOCCOLI 396

67. Pyruvate Dehydrogenase Phosphatase from Bovine Heart — FLORA H. PETTIT, W. MARTIN TEAGUE, AND LESTER J. REED 402

68. Phosphoprotein Phosphatase from Swine Kidney — K. MUNIYAPPA, FREDRICH H. LEIBACH, AND JOSEPH MENDICINO 407

Section VI. Phosphoenolpyruvate : Glycose Phosphotransferase System

69. General Description and Assay Principles — SAUL ROSEMAN, NORMAN D. MEADOW, AND MARIA A. KUKURUZINSKA 417

70. Assays for the Phosphotransferase System from *Salmonella typhimurium* — E. BRUCE WAYGOOD AND NORMAN D. MEADOW 423

71. Enzyme I from *Salmonella typhimurium* — MARIA A. KUKURUZINSKA, NANCY WEIGEL, AND E. BRUCE WAYGOOD 431

72. HPr from *Salmonella typhimurium* — ATSUSHI NAKAZAWA AND NANCY WEIGEL 436

73. III^{Glc} from *Salmonella typhimurium* — NORMAN D. MEADOW 439

74. Phosphotransferase System from *Staphylococcus aureus* — JOHN B. HAYS AND ROBERT D. SIMONI 447

Section VII. Mono- and Disaccharide-Binding Proteins

75. Maltose-Binding Protein from *Escherichia coli* — ODILE KELLERMANN AND THOMAS FERENCI 459

76. L-Arabinose- and D-Galactose-Binding Proteins from *Escherichia coli* — ROBERT W. HOGG 463

77. Ribose-Binding Protein from *Escherichia coli* — Clement E. Furlong — 467

78. D-Xylose-Binding Protein (Periplasmic) from *Escherichia coli* — A. Stephen Dahms, William Huisman, Gerald Neslund, and Clarence Ahlem — 473

Section VIII. Procedures Yielding Several Enzymes

79. Purification of All Glycolytic Enzymes from One Muscle Extract — Robert K. Scopes and Ann Stoter — 479

80. Isolation of Several Abundant Muscle Enzymes — James K. Petell, Robert B. Killion, and Herbert G. Lebherz — 490

81. Simultaneous Purification of Glyceraldehyde-3-phosphate Dehydrogenase, 3-Phosphoglycerate Kinase, and Phosphoglycerate Mutase from Pig Liver and Muscle — Klaus D. Kulbe, Harald Foellmer, and Joachim Fuchs — 498

Section IX. Carboxylases and Decarboxylases

82. Ribulose-1,5-bisphosphate Carboxylase/Oxygenase from Spinach, Tomato, or Tobacco Leaves — Stephen D. McCurry, Robert Gee, and N. E. Tolbert — 515

83. Ribulosebisphosphate Carboxylase/Oxygenase from *Rhodospirillum rubrum* — John V. Schloss, E. F. Phares, Mary V. Long, I. Lucile Norton, Claude D. Stringer, and Fred C. Hartman — 522

84. Pyruvate Decarboxylase from Sweet Potato Roots — Kazuko Ôba and Ikuzo Uritani — 528

Section X. Miscellaneous Enzymes

85. Glyoxalase I from Human Erythrocytes — Bengt Mannervik, Anne-Charlotte Aronsson, and Gudrun Tibbelin — 535

86. Glyoxalase I from Mouse Liver — Bedii Oray and Scott J. Norton — 542

87. Glyoxalase II from Mouse Liver — Bedii Oray and Scott J. Norton — 547

88. Galactose-1-phosphate Uridylyltransferase from *Entamoeba histolytica* — PATRICIA LOBELLE-RICH AND RICHARD E. REEVES 552

89. Sucrose-6-phosphate Hydrolase from *Streptococcus mutans* — BRUCE M. CHASSY AND E. VICTORIA PORTER 556

VOLUME 91

ENZYME STRUCTURE (PART I)

Section I. Amino Acid Analysis and Related Procedures

1. A Buffer System for Amino Acid Analyzers with Automatic Integrators — C. H. W. HIRS 3

2. Analysis for Methylated Amino Acids in Proteins — MARSHALL ELZINGA AND NICHOLAS ALONZO 8

3. Analysis for γ-Carboxyglutamic Acid — PAUL A. PRICE 13

4. Analysis for Imino Acids with *o*-Phthalaldehyde — PETER BÖHLEN 17

5. Single Hydrolysis Method for All Amino Acids, Including Cysteine and Tryptophan — A. S. INGLIS 26

6. Amino Acid Composition of Proteins Eluted from Polyacrylamide Gels — WILLIAM E. BROWN AND GARY C. HOWARD 36

7. Amino Acid Analysis in the Picomole Range by Precolumn Derivatization and High-Performance Liquid Chromatography — JUI-YOA CHANG, RENÉ KNECHT, AND DIETMAR G. BRAUN 41

8. Reassessment of Ellman's Reagent — PETER W. RIDDLES, ROBERT L. BLAKELEY, AND BURT ZERNER 49

9. Relating Proteins by Amino Acid Composition — ATHEL CORNISH-BOWDEN 60

Section II. End-Group Methods

10. Amino-Terminal Analysis with Dimethylaminoazobenzene Isothiocyanate — JUI-YOA CHANG 79

11. Micro-Identification of Amino-Terminal Acetylamino Acids in Proteins by Using High-Performance Liquid Chromatography — SUSUMU TSUNASAWA AND KOZO NARITA 84

Section III. Chain Separation

12. Determination of Total Protein — GARY L. PETERSON 95

13. Analysis of Metal Ions in Proteins by X-Ray Fluorescence Spectrometry — SERGE N. VINOGRADOV AND MICHAEL J. ROKOSZ — 119

14. High-Performance Liquid Chromatography of Proteins — FRED E. REGNIER — 137

15. Very-High-Resolution Two-Dimensional Electrophoretic Separation of Proteins on Giant Gels — DONALD A. YOUNG, BRUCE P. VORIS, EDWARD V. MAYTIN, AND ROBERT A. COLBERT — 190

16. Preparative Polyacrylamide Gel Electrophoresis — STEVEN SPIKER AND IRVIN ISENBERG — 214

17. Isolation of Microgram Quantities of Proteins from Polyacrylamide Gels for Amino Acid Sequence Analysis — MICHAEL W. HUNKAPILLER, EVA LUJAN, FRANK OSTRANDER, AND LEROY E. HOOD — 227

18. Staining of Proteins on Gels: Comparisons of Dyes and Procedures — CURTIS M. WILSON — 236

19. Sensitive Detection of Proteins and Peptides in Polyacrylamide Gels after Formaldehyde Fixation — ROBERT R. BÜRK, MARGRET ESCHENBRUCH, PETER LEUTHARD, AND GERMAINE STECK — 247

20. Removal of Sodium Dodecyl Sulfate from Proteins by Ion-Pair Extraction — WILLIAM H. KONIGSBERG AND LOU HENDERSON — 254

21. Removal of Sodium Dodecyl Sulfate from Proteins by Ion-Retardation Chromatography — SERGE N. VINOGRADOV AND OSCAR H. KAPP — 259

22. Recovery of Functional Proteins in Sodium Dodecyl Sulfate Gels — AD SPANOS AND ULRICH HÜBSCHER — 263

23. Desalting Protein Solutions in a Centrifuge Column — RICHARD I. CHRISTOPHERSON — 278

24. Synthesis of Carrier Ampholytes for Isoelectric Focusing — WILHELM W. JUST — 281

25. Use of Noncommercial Carrier Ampholytes for Chromatofocusing — WILHELM W. JUST — 299

Section IV. Specific Cleavage of Peptide Chains

26. Cleavage at Tryptophan with *o*-Iodosobenzoic Acid — ANGELO FONTANA, DANIELE DALZOPPO, CLAUDIO GRANDI, AND MARCELLO ZAMBONIN — 311

27. Cleavage at Tryptophanyl Residues with Dimethyl Sulfoxide–Hydrochloric Acid and Cyanogen Bromide — HENRY V. HUANG, MARTHA W. BOND, MICHAEL W. HUNKAPILLER, AND LEROY E. HOOD — 318

28. Cleavage at Aspartic Acid — A. S. INGLIS — 324

Section V. Separation of Peptides

29. Countercurrent Chromatography — YOICHIRO ITO — 335

30. Separation of Peptides by Reversed-Phase High-Performance Liquid Chromatography — MARK HERMODSON AND WALTER C. MAHONEY — 352

31. Analyzer for Microscale Peptide Separations — HEINZ NIKA AND TORE HULTIN — 359

32. Purification of Peptides that Contain Methionine Residues — JACK KYTE, JAY DEGEN, AND RICHARD N. HARKINS — 367

33. Immobilized Anhydrotrypsin as a Specific Affinity Adsorbent for Tryptic Peptides — SHIN-ICHI ISHII, HIDEYOSHI YOKOSAWA, TAKASHI KUMAZAKI, AND IZUMI NAKAMURA — 378

34. High-Resolution Analytical and Preparative Peptide Mapping by a Combination of Ion-Exchange Chromatography and Thin-Layer Chromatography — DEBRA K. AROMATORIO, JUDY PARKER, AND WILLIAM E. BROWN — 384

35. Isolation of Sulfhydryl Peptides Alkylated with *N*-Ethylmaleimide by Diagonal Electrophoresis — HEINZ GEHRING AND PHILIPP CHRISTEN — 392

Section VI. Sequence Determination

36. High-Sensitivity Sequencing with a Gas-Phase Sequenator — MICHAEL W. HUNKAPILLER, RODNEY M. HEWICK, WILLIAM J. DREYER, AND LEROY E. HOOD — 399

37. Radiochemical Sequence Analysis of Biosynthetically Labeled Proteins — JOHN E. COLIGAN, FREDERICK T. GATES III, EDWARD S. KIMBALL, AND W. LEE MALOY — 413

38. Improvement of Efficiency of Commercial Spinning-Cup Sequenators — AJIT S. BHOWN AND J. CLAUDE BENNETT — 434

39. Modified Extraction Procedure in Spinning-Cup Sequenators — A. S. INGLIS — 443

40. High-Sensitivity Sequence Analysis of Proteins Recovered from Sodium Dodecyl Sulfate Gels — AJIT S. BHOWN AND J. CLAUDE BENNETT 450

41. Manual Micro-Sequence Analysis of Polypeptides Using Dimethylaminoazobenzene Isothiocyanate — JUI-YOA CHANG 455

42. Thin-Layer Peptide Mapping with Sequencing at the Nanomole Level — DENNIS A. POWERS, JAMES C. FISHBEIN, AND ALLEN R. PLACE 466

43. Analysis of Phenylthiohydantoins by Ultrasensitive Gradient High-Performance Liquid Chromatography — MICHAEL W. HUNKAPILLER AND LEROY E. HOOD 486

44. High-Performance Liquid Chromatography of Side Chain-Protected Phenylthiohydantoins: Application to Solid-Phase Peptide Synthesis — DAVID H. SCHLESINGER 494

45. Chemical Ionization Mass Spectral Analysis of Phenylthiohydantoin Derivatives — THOMAS FAIRWELL 502

46. Polypeptide Sequencing with Dipeptidyl Peptidases — HENRY C. KRUTZSCH 511

47. Establishing Homologies in Protein Sequences — MARGARET O. DAYHOFF, WINONA C. BARKER, AND LOIS T. HUNT 524

Section VII. Chemical Modification

48. Reduction of Sulfoxides in Peptides and Proteins — RICHARD A. HOUGHTEN AND CHOH HAO LI 549

49. Amidination — JOHN K. INMAN, RICHARD N. PERHAM, GARRETT C. DUBOIS, AND ETTORE APPELLA 559

50. Protein Labeling by Reductive Alkylation — NEIL JENTOFT AND DORR G. DEARBORN 570

51. Bifunctional Reagents — TAE H. JI 580

Section VIII. Active-Site Labeling

52. Photoaffinity Labeling of Nucleotide Binding Sites with 8-Azidopurine Analogs: Techniques and Applications — ROBERT L. POTTER AND BOYD E. HALEY 613

53. Labeling Acyl-CoA Binding Sites with Photolabile Analogs — ROLAND E. BARDEN, FIDELIS M. ACHENJANG, AND CHRISTOPHER M. ADAMS 633

VOLUME 92

Immunochemical Techniques (Part E: Monoclonal Antibodies and General Immunoassay Methods)

Section I. Hybridoma Technology

A. Production of Monoclonal Antibodies with Selected Applications

1. Human–Human Monoclonal Antibody-Producing Hybridomas: Technical Aspects — Lennart Olsson and Henry S. Kaplan — 3

2. Methods of Enhancing the Frequency of Antigen-Specific Hybridomas — Reuben P. Siraganian, Philip C. Fox, and Elsa H. Berenstein — 17

3. Spleen Cell Analysis and Optimal Immunization for High-Frequency Production of Specific Hybridomas — C. Stähli, Th. Staehelin, and V. Miggiano — 26

4. Estimation of the Number of Monoclonal Hybridomas in a Cell-Fusion Experiment — Angel L. De Blas, Makarand V. Ratnaparkhi, and James E. Mosimann — 36

5. Use of Human Endothelial Culture Supernatant (HECS) as a Growth Factor for Hybridomas — Giulia C. B. Astaldi — 39

6. Production, Purification, and Characterization of Antigen-Specific Murine Monoclonal Antibodies of IgE Class — Amar S. Tung — 47

7. Production and Properties of Monoclonal Antibodies to Guinea Pig Ia Antigens — Steven E. Zweig and Ethan M. Shevach — 66

8. Purification of Murine MHC Antigens by Monoclonal Antibody Affinity Chromatography — Matthew F. Mescher, Kathryn C. Stallcup, Cathleen P. Sullivan, Aaron P. Turkewitz, and Steven H. Herrmann — 86

9. Monoclonal Antibodies against HLA Products and Their Use in Immunoaffinity Production — Peter Parham — 110

10. Immunoassay for Sequence-Specific DNA–Protein Interactions — Ronald McKay — 138

B. Detection and Assessment of Monoclonal Antibodies

11. Quantitation of Hybridoma Immunoglobulins and Selection of Light-Chain Loss Variants — Timothy A. Springer — 147

12. Indirect ^{125}I-Labeled Protein A Assay for Monoclonal Antibodies to Cell-Surface Antigens — JOSEPH P. BROWN, KARL ERIK HELLSTRÖM, AND INGEGERD HELLSTRÖM 160

13. Enzyme-Linked Immunosorbent Assay for Screening Monoclonal Antibody Production Using Enzyme-Labeled Second Antibody — J. Y. DOUILLARD AND T. HOFFMAN 168

14. Rosette-Forming Cell Assay for Detection of Antibody-Synthesizing Hybridomas — PIERRE LEGRAIN, DOMINIQUE JUY, AND GÉRARD BUTTIN 175

15. Rapid Screening and Replica Plating of Hybridomas for the Production and Characterization of Monoclonal Antibodies — RICHARD B. BANKERT 182

16. Use of High-Resolution Two-Dimensional Gel Electrophoresis for Analysis of Monoclonal Antibodies and Their Specific Antigens — TERRY W. PEARSON AND N. LEIGH ANDERSON 196

17. Screening of Monoclonal Immunoglobulins by Immunofixation on Cellulose Acetate — M. A. PIZZOLATO 220

18. A Solid-Phase Immunofluorescence Assay (SIFA) for Detection and Characterization of Monoclonal Antibodies against Soluble Antigens — BURKHARD MICHEEL, HELMAR FIEBACH, AND UWE KARSTEN 227

19. Identification and Characterization of Lymphocyte Hybridomas by Electrophoresis of Glucose-6-phosphate Isomerase Isozymes — THOMAS J. ROGERS AND KATHLEEN O'DAY 237

20. Distinction of Epitopes by Monoclonal Antibodies — C. STÄHLI, V. MIGGIANO, J. STOCKER, TH. STAEHELIN, P. HÄRING, AND B. TAKÁCS 242

Section II. Immunoassay of Antigens and Antibodies

A. Labeling of Antigens and Antibodies

21. Methods for the Attachment of Haptens and Proteins to Erythrocytes — YI-HER JOU, PAUL K. MAZZAFERRO, GEORGE L. MAYERS, AND RICHARD B. BANKERT 257

22. Iodine Monochloride (ICl) Iodination Techniques — M. ANGELES CONTRERAS, WILLIAM F. BALE, AND IRVING L. SPAR 277

23. Application of High-Performance Liquid Chromatography to Characterize Radiolabeled Peptides for Radioimmunoassay, Biosynthesis, and Microsequence Studies of Polypeptide Hormones — N. G. SEIDAH AND M. CHRÉTIEN 292

24. Preparation of Stable Radioiodinated Polypeptide Hormones and Proteins Using Polyacrylamide Gel Electrophoresis — SUSANNE LINDE, BRUNO HANSEN, AND ÅKE LERNMARK — 309

B. Separation Methods in Immunoassay

25. Noncentrifugation Immunoassays: Novel Systems — MICHAEL CAIS — 336

26. Use of Activated Thiol-Sepharose in a Separation Method for Enzyme Immunoassay — KANEFUSA KATO — 345

27. Affinity Exclusion: A New Method for the Separation of Free and Bound Fractions in Enzyme Immunoassay — B. TEROUANNE AND J. C. NICOLAS — 359

28. Use of Chromatography Tubes in the Separation of Bound and Free Fractions in Radioimmunoassay — D. B. WAGNER, J. FEINGERS, D. INBAR, A. J. PICK, Y. TAMIR, AND O. WEISS — 366

C. Immunoassay Methods

29. Enzyme-Linked Immunoelectrotransfer Blot Techniques (EITB) for Studying the Specificities of Antigens and Antibodies Separated by Gel Electrophoresis — VICTOR C. W. TSANG, JOSE M. PERALTA, AND A. RAY SIMONS — 377

30. Quantitative, Single-Tube, Kinetic-Dependent Enzyme-Linked Immunosorbent Assay (k-ELISA) — VICTOR C. W. TSANG, BRITT C. WILSON, AND JOSE M. PERALTA — 391

31. Immunoassay Using Antigen-Coated Plastic Tubes and Radiolabeled or Enzyme-Labeled Protein A — ADRIAN P. GEE AND JOHN J. LANGONE — 403

32. Colorimetric Immunoassays Using Flavin Adenine Dinucleotide as Label — DAVID L. MORRIS AND ROBERT T. BUCKLER — 413

33. Bioluminescent Immunoassays — JON WANNLUND AND MARLENE DELUCA — 426

34. Immunoassay by Electrochemical Techniques — KENNETH R. WEHMEYER, MATTHEW J. DOYLE, H. BRIAN HALSALL, AND WILLIAM R. HEINEMAN — 432

35. Metalloimmunoassay: Principles and Practice — MICHAEL CAIS — 445

36. Fluorescence Fluctuation Immunoassay — V. B. ELINGS, D. F. NICOLI, AND J. BRIGGS — 458

37. Methods and Application of Hapten-Sandwich Labeling — LEON WOFSY — 472

38. Use of Lectin–Antibody Conjugates for Quantitation and Titration of Antigens and Antibodies — JEAN-LUC GUÉSDON AND STRATIS AVRAMEAS — 489

39. Partition Affinity Ligand Assay (PALA) for Quantifying Haptens, Macromolecules, and Whole Cells — BO MATTIASSON — 498

40. Preparation and Applications of Multivalent Antibodies with Dual Specificity — V. GHEŢIE AND I. MORARU — 523

D. Data Analysis

41. LIGAND: A Computerized Analysis of Ligand Binding Data — PETER J. MUNSON — 543

42. Two Simple Programs for the Analysis of Data from Enzyme-Linked Immunosorbent Assays (ELISA) on a Programmable Desk-Top Calculator — D. G. RITCHIE, J. M. NICKERSON, AND G. M. FULLER — 577

43. Determination of Affinity and Specificity of Anti-Hapten Antibodies by Competitive Radioimmunoassay — ROLF MÜLLER — 589

44. Methods of Measuring Confidence Limits in Radioimmunoassay — ROBERT C. BAXTER — 601

VOLUME 93

IMMUNOCHEMICAL TECHNIQUES (PART F: CONVENTIONAL ANTIBODIES, FC RECEPTORS, AND CYTOTOXICITY)

Section I. Production and Assessment of Conventional Antibodies

1. Production of Antibodies by Inoculation into Lymph Nodes — MORTON B. SIGEL, Y. N. SINHA, AND W. P. VANDERLAAN — 3

2. Production of Large Amounts of Antibodies, Nonspecific Immunoglobulins, and Other Serum Proteins in Ascitic Fluids of Individual Mice and Guinea Pigs — AMAR S. TUNG — 12

3. Preparation of Vaccines by the Action of Glutaraldehyde on Toxins, Bacteria, Viruses, Allergens, and Cells — E. H. RELYVELD AND S. BEN-EFRAIM — 24

4. Methods for the Analysis of Antibody Responses to Vaccines or Other Immune Stimuli — GEORGE R. SIBER AND BERNARD J. RANSIL — 60

5. Preparation of Polyclonal Antisera to Tumor-Associated Antigens Using Agarose-Entrapped Immune Complexes as Immunogens — PAUL J. HIGGINS — 78

6. Use of Liposomes as Biodegradable and Harmless Adjuvants — NICO VAN ROOIJEN AND RIA VAN NIEUWMEGEN 83

7. Role of Diffusion Regulation in Receptor–Ligand Interactions — CHARLES DELISI 95

Section II. Fc Receptors

8. Detection of Fc Receptors — ROBERT S. KERBEL AND BRUCE E. ELLIOTT 113

9. Quantitative Fluorometric Assay for Detection and Characterization of Fc Receptors — ALAIN B. SCHREIBER AND JOSEPH HAIMOVICH 147

10. Use of Solid-Phase Radioimmunoassay Specific for Human IgG or Human C3b to Detect Fcγ or C3b Receptors on Human Lymphoblastoid Cell Surfaces — RAYMOND FRADE, MONIQUE BAREL, AND CHRISTIANE CHARRIAUT 155

11. Demonstration of Lymphocyte Surface Markers Using Alkaline Phosphatase-Labeled Reagents. Application to the Enumeration of Lymphocyte Populations in Whole Peripheral Blood — ELIZABETH O. PEPYS AND M. B. PEPYS 164

12. Purification of Fc_ε Receptors and Fc_γ Receptors — ANTHONY KULCZYCKI, JR. 178

13. The Fc Receptor of the Fetal Rabbit Yolk Sac Membrane — MAX SCHLAMOWITZ AND ANITA R. SHAW 190

14. Isolation and Characterization of Biologically Active Fcγ Receptors of Human B Lymphocytes — TSUNEO SUZUKI 219

Section III. Cytotoxicity Tests and Cytotoxic Agents

15. Use of Dyes and Radioisotopic Markers in Cytotoxicity Tests — SEYMOUR I. SCHLAGER AND ANDREA C. ADAMS 233

16. Use of Antibody-Dependent Cell-Mediated Cytotoxicity (ADCC) Assay in Basic and Clinical Immunology — HILLEL S. KOREN 245

17. Quantitation of Cell Surface Antigens by Antibody-Complement-Mediated Cytotoxicity — JOHN E. T. MOEN AND SVEN O. WARNAAR 253

18. Poly-L-lysine-Induced Immobilization of Tumor Cells to Plastic for the Study of Cellular Metabolic Responses to Attack by Cytotoxic T Lymphocytes — CARMEN A. SEPULVEDA AND SEYMOUR I. SCHLAGER 260

19. Frequency Determination of Killer Cells by a Single-Cell Cytotoxic Assay — BENJAMIN BONAVIDA, THOMAS P. BRADLEY, AND ELIZABETH A. GRIMM 270

20. Preparation of Antibody-Linked Cytotoxic Agents — Tarun I. Ghose, A. Huntley Blair, and Padmaja N. Kulkarni — 280

21. Preparation and Application of Antibodies Coupled to the A Chain of Ricin — K. A. Krolick, J. W. Uhr, and E. S. Vitetta — 333

22. Molecular Titration of Components of the Classical Complement Pathway — Adrian P. Gee — 339

23. Purification and Quantitation of the Components of the Alternative Complement Pathway — Joe O. Minta and Adrian P. Gee — 375

24. Preparation and Isolation of Specific Antibodies to Complement Components — Sucharit Bhakdi, Marion Muhly, and Margit Roth — 409

VOLUME 94

Polyamines

Section I. Analytical Methods for Amines

1. Thin-Layer Chromatography and Thin-Layer Electrophoresis of Polyamines and Their Derivatives — Nikolaus Seiler — 3

2. Liquid Chromatographic Methods for Assaying Polyamines Using Prechromatographic Derivatization — Nikolaus Seiler — 10

3. Ion-Pair Partition Chromatographic Separation of Polyamines and Their Monoacetyl Derivatives — Nikolaus Seiler — 25

4. Quantitative Determination of Naturally Occurring Aliphatic Diamines and Polyamines by an Automated Liquid Chromatography Procedure — Celia White Tabor and Herbert Tabor — 29

5. A Double-Isotope Derivative Assay for Polyamines — Thomas J. Paulus and Rowland H. Davis — 36

6. A Simple Method for the Determination of Polyamines and Histamine and Its Application to the Assay of Ornithine and Histidine Decarboxylase Activities — Yasuo Endo — 42

7. Gas Chromatographic–Mass Spectrometric Analysis of Polyamines and Polyamine Conjugates — G. Doyle Daves, Jr., Ronald G. Smith, and Cornelius A. Valkenburg — 48

Section II. Analytical and Preparative Methods for Adenosylmethionine, Decarboxylated Adenosylmethionine, and Related Compounds

8. High-Performance Liquid Chromatographic Analysis of Adenosyl-Sulfur Compounds Related to Polyamine Biosynthesis — Vincenzo Zappia, Maria Cartenì-Farina, Patrizia Galletti, Fulvio Della Ragione, and Giovanna Cacciapuoti 57

9. Determination of Tissue *S*-Adenosylmethionine by Radioenzymatic Assay — Peter H. Yu 66

10. Determination of Cellular Decarboxylated *S*-Adenosylmethionine Content — Anthony E. Pegg and R. A. Bennett 69

11. Methods for the Preparation and Assay of *S*-Adenosyl-(5′)-3-methylthiopropylamine — Vincenzo Zappia, Patrizia Galletti, Adriana Oliva, and Marina Porcelli 73

Section III. Genetic Techniques

A. Mutants in Biosynthetic Pathways

12. Mass Screening for Mutants in the Biosynthetic Pathway for Polyamines in *Escherichia coli:* A General Method for Mutants in Enzymatic Reactions Producing CO_2 — Celia White Tabor, Herbert Tabor, and Edmund W. Hafner 83

13. Localized Mutagenesis of Any Specific Region of the *Escherichia coli* Chromosome with Bacteriophage Mu — Herbert Tabor, Edmund W. Hafner, and Celia White Tabor 91

14. Mass Screening for Mutants in the Polyamine Biosynthetic Pathway in *Saccharomyces cerevisiae* — Herbert Tabor, Celia White Tabor, and Murray S. Cohn 104

15. Selection of Ornithine Decarboxylase-Deficient Mutants of Chinese Hamster Ovary Cells — C. Steglich and I. E. Scheffler 108

16. Uses of Arginaseless Cells in the Study of Polyamine Metabolism (*Neurospora crassa*) — Rowland H. Davis and Thomas J. Paulus 112

B. Preparation of Plasmids for Overproduction of Enzymes in Biosynthetic Pathway (*Escherichia coli*)

17. Cloning of the *Escherichia coli* Genes for the Biosynthetic Enzymes for Polyamines — Celia White Tabor, Herbert Tabor, Edmund W. Hafner, George D. Markham, and Stephen M. Boyle 117

Section IV. Ornithine Decarboxylase, Arginine Decarboxylase, Lysine Decarboxylase

A. Enzyme Assays and Preparations

18. Biosynthetic and Biodegradative Ornithine and Arginine Decarboxylases from *Escherichia coli* — DAVID R. MORRIS AND ELIZABETH A. BOEKER — 125

19. Ornithine Decarboxylase (*Saccharomyces cerevisiae*) — ANIL K. TYAGI, CELIA WHITE TABOR, AND HERBERT TABOR — 135

20. Ornithine Decarboxylase and the Ornithine Decarboxylase-Modifying Protein of *Physarum polycephalum* — JOHN L. A. MITCHELL — 140

21. Polyamine-Dependent Protein Kinase from the Slime Mold *Physarum polycephalum* — GLENN D. KUEHN, VALERIE J. ATMAR, AND GARY R. DANIELS — 147

22. Ornithine Decarboxylase (Rat Liver) — SHIN-ICHI HAYASHI AND TAKAAKI KAMEJI — 154

23. Ornithine Decarboxylase (Mouse Kidney) — JAMES E. SEELY AND ANTHONY E. PEGG — 158

24. Ornithine Decarboxylase (Germinated Barley Seeds) — DIMITRIOS A. KYRIAKIDIS, CHRISTOS A. PANAGIOTIDIS, AND JOHN G. GEORGATSOS — 162

25. Immunocytochemical Localization of Ornithine Decarboxylase — LO PERSSON, ELSA ROSENGREN, FRANK SUNDLER, AND ROLF UDDMAN — 166

26. Autoradiographic Localization of Ornithine Decarboxylase — IAN S. ZAGON, JAMES E. SEELY, AND ANTHONY E. PEGG — 169

27. Arginine Decarboxylase (Oat Seedlings) — TERENCE A. SMITH — 176

28. Lysine Decarboxylase (*Escherichia coli* B) — ELIZABETH A. BOEKER AND EDMOND H. FISCHER — 180

29. Antizyme and Antizyme Inhibitor of Ornithine Decarboxylase (Rat Liver) — SHIN-ICHI HAYASHI AND KAZUNOBU FUJITA — 185

30. Purification of Ornithine Decarboxylase Antizymes (*Escherichia coli*) — DIMITRIOS A. KYRIAKIDIS, JOHN S. HELLER, AND EVANGELOS S. CANELLAKIS — 193

B. Enzyme Inhibitors

31. Synthesis of Irreversible Inhibitors of Polyamine Biosynthesis — PHILIPPE BEY, PATRICK CASARA, JEAN-PAUL VEVERT, AND BRIAN METCALF 199

32. Labeling and Quantitation of Ornithine Decarboxylase Protein by Reaction with α-[5-^{14}C]Difluoromethylornithine — JAMES E. SEELY, HANNU PÖSÖ, AND ANTHONY E. PEGG 206

33. Methods for the Study of the Treatment of Protozoan Diseases by Inhibitors of Ornithine Decarboxylase — PETER P. MCCANN, CYRUS J. BACCHI, WILLIAM L. HANSON, HENRY C. NATHAN, SEYMOUR H. HUTNER, AND ALBERT SJOERDSMA 209

34. Methods for the Study of the Experimental Interruption of Pregnancy by Ornithine Decarboxylase Inhibitors: Effects of DL-α-Difluoromethylornithine — JOHN R. FOZARD AND MARIE-LOUISE PART 213

Section V. Adenosylmethionine Synthetase (Methionine Adenosyltransferase) and Adenosylmethionine Decarboxylase

A. Enzyme Assays and Preparations

35. *S*-Adenosylmethionine Synthetase (Methionine Adenosyltransferase) (*Escherichia coli*) — GEORGE D. MARKHAM, EDMUND W. HAFNER, CELIA WHITE TABOR, AND HERBERT TABOR 219

36. Fractionation of Methionine Adenosyltransferase Isozymes (Rat Liver) — JERALD L. HOFFMAN 223

37. *S*-Adenosylmethionine Decarboxylase (*Escherichia coli*) — GEORGE D. MARKHAM, CELIA WHITE TABOR, AND HERBERT TABOR 228

38. *S*-Adenosylmethionine Decarboxylase (*Saccharomyces cerevisiae*) — MURRAY S. COHN, CELIA WHITE TABOR, AND HERBERT TABOR 231

39. *S*-Adenosylmethionine Decarboxylase (Rat Liver) — ANTHONY E. PEGG AND HANNU PÖSÖ 234

B. Enzyme Inhibitors

40. Inhibitors of *S*-Adenosylmethionine Decarboxylase — ANTHONY E. PEGG 239

41. Two Enzyme Inhibition Assays for Methylglyoxal Bis(guanylhydrazone) — P. SEPPÄNEN, L. ALHONEN-HONGISTO, K. KÄPYAHO, AND J. JÄNNE 247

Section VI. Putrescine Aminopropyltransferase (Spermidine Synthase) and Spermidine Aminopropyltransferase (Spermine Synthase)

A. Enzyme Assays and Preparations

42. Rapid Assays for Putrescine Aminopropyltransferase (Spermidine Synthase) and Spermidine Aminopropyltransferase (Spermine Synthase) — AARNE RAINA, TERHO ELORANTA, AND RAIJA-LEENA PAJULA 257

43. Assay of Aminopropyltransferases — ANTHONY E. PEGG 260

44. Putrescine Aminopropyltransferase (*Escherichia coli*) — CELIA WHITE TABOR AND HERBERT TABOR 265

45. Purification of Putrescine Aminopropyltransferase (Spermidine Synthase) from Eukaryotic Tissues — KEIJIRO SAMEJIMA, AARNE RAINA, BANRI YAMANOHA, AND TERHO ELORANTA 270

46. Purification of Spermidine Aminopropyltransferase (Spermine Synthase) from Bovine Brain — AARNE RAINA, RAIJA-LEENA PAJULA, AND TERHO ELORANTA 276

47. Putrescine Aminopropyltransferase (Spermidine Synthase) of Chinese Cabbage — RAM K. SINDHU AND SEYMOUR S. COHEN 279

B. Enzyme Inhibitors

48. Aminopropyltransferase Substrates and Inhibitors — JAMES K. COWARD, GARY L. ANDERSON, AND KUO-CHANG TANG 286

49. Inhibition of Aminopropyltransferases — ANTHONY E. PEGG 294

Section VII. Amine Oxidases and Dehydrogenases

50. Purification of Putrescine Oxidase from *Micrococcus rubens* by Affinity Chromatography — MASATO OKADA, SEIICHI KAWASHIMA, AND KAZUTOMO IMAHORI 301

51. Purification of Spermidine Dehydrogenase from *Serratia marcescens* by Affinity Chromatography — MASATO OKADA, SEIICHI KAWASHIMA, AND KAZUTOMO IMAHORI 303

52. Polyamine Oxidase (Rat Liver) — ERKKI HÖLTTÄ — 306

53. Polyamine Oxidase (Oat Seedlings) — TERENCE A. SMITH — 311

54. Purification of Bovine Plasma Amine Oxidase — B. MONDOVÌ, P. TURINI, O. BEFANI, AND S. SABATINI — 314

Section VIII. Spermidine Acetylation and Deacetylation

55. Spermidine N^1-Acetyltransferase — FULVIO DELLA RAGIONE AND ANTHONY E. PEGG — 321

56. Purification of Two Spermidine *N*-Acetyltransferases (Histone *N*-Acetyltransferases) from Calf Liver Nuclei — PAUL R. LIBBY — 325

57. Acetylspermidine Deacetylase (Rat Liver) — PAUL R. LIBBY — 329

Section IX. Other Enzymes Involved in Polyamine Synthesis and Metabolism

58. Putrescine Synthase from *Lathyrus sativus* (Grass Pea) Seedlings — K. S. SRIVENUGOPAL AND P. R. ADIGA — 335

59. Putrescine Carbamoyltransferase (*Streptococcus faecalis*) — VICTOR STALON — 339

60. Agmatine Coumaroyltransferase (Barley Seedlings) — COLIN R. BIRD AND TERENCE A. SMITH — 344

61. γ-Glutamylamine Cyclotransferase (Rabbit Kidney) — MARY LYNN FINK AND J. E. FOLK — 347

Section X. Metabolism of 5′-Methylthioadenosine and 5-Methylthioribose

62. Purification and Properties of 5′-Methylthioadenosine Phosphorylase from *Caldariella acidophila* — VINCENZO ZAPPIA, MARIA CARTENÌ-FARINA, GIOVANNA ROMEO, MARIO DE ROSA, AND AGATA GAMBACORTA — 355

63. 5-Methylthioribose Kinase (*Enterobacter aerogenes*) — ADOLPH J. FERRO AND KEVIN S. MARCHITTO — 361

64. 5′-Methylthioadenosine Nucleosidase (*Lupinus luteus* Seeds) — ANDRZEJ B. GURANOWSKI, PETER K. CHIANG, AND GIULIO L. CANTONI — 365

Section XI. Methods for the Study of Polyamines in Lymphocytes and Mammary Gland

65. Methods for the Study of the Physiological Effects of Inhibitors of Polyamine Biosynthesis in Mitogen-Activated Lymphocytes — CHRISTINE E. SEYFRIED AND DAVID R. MORRIS — 373

66. Use of Mammary Gland Tissue for the Study of Polyamine Metabolism and Function — TAKAMI OKA AND JOHN W. PERRY — 389

67. Ornithine Decarboxylase Assay Permitting Early Determination of Histocompatibility in the Mixed Lymphocyte Reaction — ACHILLES A. DEMETRIOU, CELIA WHITE TABOR, AND HERBERT TABOR — 396

Section XII. Analogs and Derivatives

68. Novel Polyamines in *Thermus thermophilus:* Isolation, Identification, and Chemical Synthesis — TAIRO OSHIMA — 401

69. Analogs of Spermine and Spermidine — MERVYN ISRAEL AND EDWARD J. MODEST — 411

70. Preparation of Thermospermine — BRUCE GANEM AND KAN CHANTRAPROMMA — 416

71. Synthesis of Putreanine and of Spermic Acid — HERBERT TABOR AND CELIA WHITE TABOR — 418

72. Synthesis of Acetylated Derivatives of Polyamines — HERBERT TABOR AND CELIA WHITE TABOR — 420

73. α-Putrescinylthymine — A. M. B. KROPINSKI, K. L. MALTMAN, AND R. A. J. WARREN — 422

74. Preparation and Purification of *N*-Carbamoylputrescine and [*ureido*-^{14}C]-*N*-Carbamoylputrescine — K. S. SRIVENUGOPAL AND P. R. ADIGA — 429

75. Hydroxyputrescine: 1,4-Diaminobutan-2-ol — JIRO TOBARI AND T. T. TCHEN — 431

76. Glutathionylspermidine — HERBERT TABOR AND CELIA WHITE TABOR — 434

77. Isolation and Assay of 2,3-Dihydroxybenzoyl Derivatives of Polyamines: The Siderophores Agrobactin and Parabactin from *Agrobacterium tumefaciens* and *Paracoccus denitrificans* — J. B. NEILANDS — 437

78. Edeine A, Edeine B, and Guanidospermidine — Z. KURYLO-BOROWSKA AND J. HEANEY-KIERAS — 441

79. Synthesis of N^1-(γ-Glutamyl)spermidine, N^8-(γ-Glutamyl)spermidine, N^1,N^8-Bis(γ-glutamyl)-spermidine, N^1-(γ-Glutamyl)spermine, N^1,N^{12}-Bis(γ-glutamyl)spermine, and N^1,N^4-Bis(γ-Glutamyl)putrescine	J. E. Folk	451
80. Chromatographic Identification of Hypusine[N^1-(4-amino-2-hydroxybutyl)lysine] and Deoxyhypusine [N^1-(4-aminobutyl)lysine]	Myung Hee Park, Herbert L. Cooper, and J. E. Folk	458
81. Isolation of *S*-Adenosyl-3-thiopropylamine	Shosuke Ito	463

VOLUME 96

Biomembranes [Part J: Membrane Biogenesis: Assembly and Targeting (General Methods, Eukaryotes)]

Section I. Biogenesis and Assembly of Membrane Proteins

A. General Methods

1. Preparation of Rough Microsomes and Membrane-Bound Polysomes That Are Active in Protein Synthesis *in Vitro*	Sancia Gaetani, Julia A. Smith, Ricardo A. Feldman, and Takashi Morimoto	3
2. Methods for the Preparation of Messenger RNA	Paul M. Lizardi	24
3. Cell-Free Translation of Messenger RNA in a Wheat Germ System	Ann. H. Erickson and Günter Blobel	38
4. Preparation and Use of Nuclease-Treated Rabbit Reticulocyte Lysates for the Translation of Eukaryotic Messenger RNA	Richard J. Jackson and Tim Hunt	50
5. *In Vivo* and *in Vitro* Systems for Studying Bacterial Membrane Biogenesis	Pamela Green and Masayori Inouye	74
6. Preparation of Microsomal Membranes for Cotranslational Protein Translocation	Peter Walter and Günter Blobel	84
7. Methods for the Study of Protein Translocation across the RER Membrane Using the Reticulocyte Lysate Translation System and Canine Pancreatic Microsomal Membranes	George Scheele	94
8. Immunoprecipitation of Proteins from Cell-Free Translations	David J. Anderson and Günter Blobel	111

9. Use of Proteases for the Study of Membrane Insertion — Takashi Morimoto, Monique Arpin, and Sancia Gaetani — 121

10. Identifying Primary Translation Products: Use of *N*-Formylmethionyl-tRNA and Prevention of NH_2-Terminal Acetylation — Richard D. Palmiter — 150

11. Selection of Lectin-Resistant Mutants of Animal Cells — Pamela Stanley — 157

12. Two-Dimensional Polyacrylamide Gel Electrophoresis of Membrane Proteins — R. W. Rubin and C. L. Leonardi — 184

13. Immunochemical Identification of Membrane Proteins after Sodium Dodecyl Sulfate–Polyacrylamide Gel Electrophoresis — Albert Haid and Mordechai Suissa — 192

14. Immunologic Detection of Specific Proteins in Cell Extracts by Fractionation in Gels and Transfer to Paper — Jakob Reiser and George R. Stark — 205

15. Use of Fluorography for Sensitive Isotope Detection in Polyacrylamide Gel Electrophoresis and Related Techniques — William M. Bonner — 215

16. Peptide Mapping in One Dimension by Limited Proteolysis of Sodium Dodecyl Sulfate-Solubilized Proteins — Don W. Cleveland — 222

17. Silver Staining Methods for Polyacrylamide Gel Electrophoresis — Carl R. Merril, David Goldman, and Margaret L. Van Keuren — 230

18. Slab Gel System for the Resolution of Oligopeptides below Molecular Weight of 10,000 — F. A. Burr and B. Burr — 239

19. Two-Dimensional Immunoelectrophoresis of Membrane Antigens — Carolyn A. Converse and David S. Papermaster — 244

B. Eukaryotic Membranes

Plasma Membrane

20. Methods for Study of the Synthesis and Maturation of the Erythrocyte Anion Transport Protein — Harvey F. Lodish and William A. Braell — 257

21. Glycophorins: Isolation, Orientation, and Localization of Specific Domains — Heinz Furthmayr and Vincent T. Marchesi — 268

22. Glycophorin A: *In Vitro* Biogenesis and Processing — Carl G. Gahmberg, Mikko Jokinen, Kimmo K. Karhi, Olle Kämpe, Per A. Peterson, and Leif C. Andersson — 281

23. Erythrocyte Membrane Proteins: Detection of Spectrin Oligomers by Gel Electrophoresis — Jon S. Morrow and Wallace B. Haigh, Jr. — 298

24. Isolation of the Chemical Domains of Human Erythrocyte Spectrin — William J. Knowles and Marcia L. Bologna — 305

25. Proteins Involved in Membrane–Cytoskeleton Association in Human Erythrocytes: Spectrin, Ankyrin, and Band 3 — Vann Bennett — 313

26. Assembly of Histocompatibility Antigens — Bernhard Dobberstein and Sune Kvist — 325

27. Biosynthesis and Intracellular Transport of Acetylcholine Receptors — Douglas M. Fambrough — 331

28. Acetylcholinesterase Biosynthesis and Transport in Tissue Culture — Richard L. Rotundo — 353

29. Biosynthesis of Acetylcholine Receptor *in Vitro* — David J. Anderson and Günter Blobel — 367

30. Biosynthesis of Myelin-Specific Proteins — David R. Colman, Gert Kreibich, and David D. Sabatini — 378

31. Sucrase–Isomaltase of the Small-Intestinal Brush Border Membrane: Assembly and Biosynthesis — Josef Brunner, Hans Wacker, and Giorgio Semenza — 386

32. Structure and Biosynthesis of Aminopeptidases — Suzanne Maroux and Hélène Feracci — 406

Enveloped Viruses

33. Use of the Heavy-Isotope Density-Shift Method to Investigate Insulin Receptor Synthesis, Turnover, and Processing — Michael N. Krupp, Victoria P. Knutson, Gabriele V. Ronnett, and M. Daniel Lane — 423

34. Biosynthesis of Myxovirus Glycoproteins with Special Emphasis on Mutants Defective in Glycoprotein Processing — Hans-Dieter Klenk — 434

35. Methods for Assay of Cellular Receptors for Picornaviruses — RICHARD L. CROWELL, DAVID L. KRAH, JOHN MAPOLES, AND BURTON J. LANDAU — 443

36. Transport of Virus Membrane Glycoproteins, Use of Temperature-Sensitive Mutants and Organelle-Specific Lectins — LEEVI KÄÄRIÄINEN, ISMO VIRTANEN, JAAKKO SARASTE, AND SIRKKA KERÄNEN — 453

37. Immunoelectron Microscopy Using Thin, Frozen Sections: Application to Studies of the Intracellular Transport of Semliki Forest Virus Spike Glycoproteins — G. GRIFFITHS, K. SIMONS, G. WARREN, AND K. T. TOKUYASU — 466

38. Immunocytochemistry of Retinal Membrane Protein Biosynthesis at the Electron Microscopic Level by the Albumin Embedding Technique — BARBARA G. SCHNEIDER AND DAVID S. PAPERMASTER — 485

39. Expression of Viral Membrane Proteins from Cloned cDNA by Microinjection into Eukaryotic Cell Nuclei — BEATE TIMM, CLAUDIA KONDOR-KOCH, HANS LEHRACH, HEIMO RIEDEL, JAN-ERIK EDSTRÖM, AND HENRIK GAROFF — 496

40. Biosynthesis of Sindbis Virus Membrane Glycoproteins *in Vitro* — STEFANO BONATTI — 512

Endoplasmic Reticulum

41. Ribophorins I and II: Membrane Proteins Characteristic of the Rough Endoplasmic Reticulum — GERT KREIBICH, EUGENE E. MARCANTONIO, AND DAVID D. SABATINI — 520

42. Biosynthesis of Hepatocyte Endoplasmic Reticulum Proteins — GERT KREIBICH, DAVID D. SABATINI, AND MILTON ADESNIK — 530

43. Membrane Induction by Drugs — GUSTAV DALLNER AND JOSEPH W. DEPIERRE — 542

44. Preparation of Microsomal β-Glucuronidase and Its Membrane Anchor Protein Egasyn — ALDONS J. LUSIS — 557

Sarcoplasmic Reticulum

45. Biosynthesis of Sarcoplasmic Reticulum Proteins — DAVID H. MACLENNAN AND STELLA DE LEON — 570

Plant Vacuoles

46. Plant Vacuoles — CLARENCE A. RYAN AND MARY WALKER-SIMMONS — 580

Nuclear Membrane

47. Preparation of a Nuclear Matrix–Pore Complex–Lamina Fraction from Embryos of *Drosophila melanogaster* — PAUL A. FISHER AND GÜNTER BLOBEL — 589

48. Proteins of Pore Complex–Lamina Structures from Nuclei and Nuclear Membranes — GEORG KROHNE AND WERNER W. FRANKE — 597

Other Specialized Systems

49. Subcellular Fractionation and Immunochemical Analysis of Membrane Biosynthesis of Photoreceptor Proteins — DAVID S. PAPERMASTER — 609

50. Transducin and the Cyclic GMP Phosphodiesterase of Retinal Rod Outer Segments — LUBERT STRYER, JAMES B. HURLEY, AND BERNARD K.-K. FUNG — 617

51. Avian Salt Gland: A Model for the Study of Membrane Biogenesis — RUSSELL J. BARRNETT, JOSEPH E. MAZURKIEWICZ, AND JOHN S. ADDIS — 627

Section II. Targeting: Selected Techniques to Study Transfer of Newly Synthesized Proteins into or across Membranes (Eukaryotic Cells)

52. Control of Intracellular Protein Traffic — GÜNTER BLOBEL — 663

53. Signal Recognition Particle: A Ribonucleoprotein Required for Cotranslational Translocation of Proteins, Isolation and Properties — PETER WALTER AND GÜNTER BLOBEL — 682

54. Proteins Mediating Vectorial Translocation: Purification of the Active Domain of the Endoplasmic Reticulum Docking Protein — DAVID I. MEYER AND BERNHARD DOBBERSTEIN — 692

55. Biosynthesis of Glyoxysomal Proteins — H. KINDL AND C. KRUSE — 700

56. *In Vitro* Processing of Plant Preproteins — F. A. BURR AND B. BURR — 716

57. Biogenesis of Peroxisomal Content Proteins: *In Vivo* and *in Vitro* Studies — PAUL B. LAZAROW — 721

58. Studies of Lysosomal Enzyme Biosynthesis in Cultured Cells — RACHEL MYEROWITZ, APRIL R. ROBBINS, RICHARD L. PROIA, G. GARY SAHAGIAN, CHRISTINA M. PUCHALSKI, AND ELIZABETH F. NEUFELD 729

59. Inhibitors of Lysosomal Function — PER O. SEGLEN 737

60. Biosynthesis of Lysosomal Enzymes — MELVIN G. ROSENFELD, GERT KREIBICH, DAVID D. SABATINI, AND KEITARO KATO 764

61. Applications of Amino Acid Analogs for Studying Co- and Posttranslational Modifications of Proteins — GLEN HORTIN AND IRVING BOIME 777

62. Quantitative Assay for Signal Peptidase — ROBERT C. JACKSON 784

63. Fatty Acid Acylation of Eukaryotic Cell Proteins — MILTON J. SCHLESINGER 795

64. Yeast Secretory Mutants: Isolation and Characterization — RANDY SCHEKMAN, BRENT ESMON, SUSAN FERRO-NOVICK, CHARLES FIELD, AND PETER NOVICK 802

65. Secretory Mutants in the Cellular Slime Mold *Dictyostelium discoideum* — RANDALL L. DIMOND, DAVID A. KNECHT, KEVIN B. JORDAN, ROBERT A. BURNS, AND GEORGE P. LIVI 815

VOLUME 97

BIOMEMBRANES [PART K: MEMBRANE BIOGENESIS: ASSEMBLY AND TARGETING (PROKARYOTES, MITOCHONDRIA, AND CHLOROPLASTS)]

Section I. Prokaryotic Membranes

A. General Methods

1. Genetic Analysis of Protein Export in *Escherichia coli* — JONATHAN BECKWITH AND THOMAS J. SILHAVY 3

2. Isolation and Characterization of Mutants of *Escherichia coli* K12 Affected in Protein Localization — THOMAS J. SILHAVY AND JONATHAN BECKWITH 11

3. Purification and Characterization of Leader Peptidase from *Escherichia coli*	P. B. WOLFE, C. ZWIZINSKI, AND WILLIAM WICKNER	40
4. Molecular Genetics of *Escherichia coli* Leader Peptidase	TAKAYASU DATE, PAMELA SILVER, AND WILLIAM WICKNER	46
5. Pulse-Labeling Studies of Membrane Assembly and Protein Secretion in Intact Cells: M13 Coat Protein	WILLIAM WICKNER, TAKAYASU DATE, RICHARD ZIMMERMANN, AND KOREAKI ITO	57
6. Synthesis of Proteins by Membrane-Associated Polysomes and Free Polysomes	PNANG C. TAI, MICHAEL P. CAULFIELD, AND BERNARD D. DAVIS	62
7. Preparation of Free and Membrane-Bound Polysomes from *Escherichia coli*	LINDA L. RANDALL AND SIMON J. S. HARDY	70
8. Analysis of Cotranslational Proteolytic Processing of Nascent Chains Using Two-Dimensional Gel Electrophoresis	LARS-GÖRAN JOSEFSSON AND LINDA L. RANDALL	77

B. Outer Membrane

9. Proteins Forming Large Channels from Bacterial and Mitochondrial Outer Membranes: Porins and Phage Lambda Receptor Protein	HIROSHI NIKAIDO	85
10. Phage λ Receptor (LamB Protein) in *Escherichia coli*	MAXIME SCHWARTZ	100
11. Synthesis and Assembly of the Outer Membrane Proteins OmpA and OmpF of *Escherichia coli*	IAN CROWLESMITH AND KONRAD GAMON	112
12. Isolation of Mutants of the Major Outer Membrane Lipoprotein of *Escherichia coli* for the Study of Its Assembly	JACK COLEMAN, SUMIKO INOUYE, AND MASAYORI INOUYE	124

C. Inner Membrane

13. Analysis of M13 Procoat Assembly into Membranes *in Vitro*	COLIN WATTS, JOEL M. GOODMAN, PAMELA SILVER, AND WILLIAM WICKNER	130
14. Insertion of Proteins into Bacterial Membranes	PETER MODEL AND MARJORIE RUSSEL	138
15. Influence of Membrane Potential on the Insertion and Transport of Proteins in Bacterial Membranes	ROBERT C. LANDICK, CHARLES J. DANIELS, AND DALE L. OXENDER	146
16. Penicillinase Secretion *in Vivo* and *in Vitro*	JENNIFER B. K. NIELSEN	153

17. Lactose Permease of *Escherichia coli* — J. K. Wright, R. M. Teather, and P. Overath — 158

18. Cloning of the Structural Genes of the *Escherichia coli* Adenosinetriphosphatase Complex — David A. Jans and Frank Gibson — 176

19. Biogenesis of an Oligomeric Membrane Protein Complex: The Proton Translocating ATPase of *Escherichia coli* — William S. A. Brusilow, Robert P. Gunsalus, and Robert D. Simoni — 188

20. Analysis of *Escherichia coli* ATP Synthase Subunits by DNA and Protein Sequencing — John E. Walker and Nicholas J. Gay — 195

21. Biogenesis of Purple Membrane in Halobacteria — Dorothea-Ch. Neugebauer, Horst-Peter Zingsheim, and Dieter Oesterhelt — 218

22. Isolation of the Bacterioopsin Gene by Colony Hybridization — Heike Vogelsang, Wolfgang Oertel, and Dieter Oesterhelt — 226

Section II. Mitochondria

23. Assessing Import of Proteins into Mitochondria: An Overview — Susan M. Gasser and Rick Hay — 245

24. Molecular Cloning of Middle-Abundant mRNAs from *Neurospora crassa* — Adelheid Viebrock, Angela Perz, and Walter Sebald — 254

25. Biogenesis of Cytochrome *c* in *Neurospora crassa* — Bernd Hennig and Walter Neupert — 261

26. Biosynthesis and Assembly of Nuclear-Coded Mitochondrial Membrane Proteins in *Neurospora crassa* — Richard Zimmermann and Walter Neupert — 275

27. Isolation and Properties of the Porin of the Outer Mitochondrial Membrane from *Neurospora crassa* — Helmut Freitag, Roland Benz, and Walter Neupert — 286

28. Synthesis and Assembly of Subunit 6 of the Mitochondrial ATPase in Yeast — Sangkot Marzuki and Anthony W. Linnane — 294

29. Preparation and Use of Antibodies against Insoluble Membrane Proteins — Mordechai Suissa and Graeme A. Reid — 305

30. Processing of Mitochondrial Polypeptide Precursors in Yeast — Peter C. Böhni and Günther Daum — 311

31. Pulse Labeling of Yeast Cells and Spheroplasts — Graeme A. Reid — 324

32. Import of Polypeptides into Isolated Yeast Mitochondria — SUSAN M. GASSER — 329

33. A Yeast Mitochondrial Chelator-Sensitive Protease That Processes Cytoplasmically Synthesized Protein Precursors: Isolation from Yeast and Assay — PHYLLIS C. MCADA AND MICHAEL G. DOUGLAS — 337

34. Selection and Characterization of Nuclear Genes Coding Mitochondrial Proteins: Genetic Complementation of Yeast *pet* Mutants — KAREN O'MALLEY AND MICHAEL G. DOUGLAS — 344

35. Transformation of Nuclear Respiratory Deficient Mutants of Yeast — CAROL L. DIECKMANN AND ALEXANDER TZAGOLOFF — 355

36. Analysis of Yeast Mitochondrial Genes — CAROL L. DIECKMANN AND ALEXANDER TZAGOLOFF — 361

37. Genetics and Biogenesis of Cytochrome *b* — PHILIP S. PERLMAN AND HENRY R. MAHLER — 374

38. Synthesis and Intracellular Transport of Mitochondrial Matrix Proteins in Rat Liver: Studies *in Vivo* and *in Vitro* — GORDON C. SHORE, RICHARD A. RACHUBINSKI, CAROLE ARGAN, RIMA ROZEN, MARCEL POUCHELET, CAROL J. LUSTY, AND YVES RAYMOND — 396

39. Biosynthesis of Cytochrome *c* and Its Posttranslational Transfer into Mitochondria — TAKASHI MORIMOTO, SHIRO MATSUURA, AND MONIQUE ARPIN — 408

40. Isolation of Mammalian Mitochondrial DNA and RNA and Cloning of the Mitochondrial Genome — DOUGLAS P. TAPPER, RICHARD A. VAN ETTEN, AND DAVID A. CLAYTON — 426

41. Analysis of Human Mitochondrial RNA — GIUSEPPE ATTARDI AND JULIO MONTOYA — 435

42. Isolation of a Hexokinase Binding Protein from the Outer Mitochondrial Membrane — JOHN E. WILSON, JANICE L. MESSER, AND PHILIP L. FELGNER — 469

43. Protein Synthesis by Isolated Plant Mitochondria — C. J. LEAVER, E. HACK, AND B. G. FORDE — 476

Section III. Chloroplasts

44. Synthesis and Assembly of Thylakoid Membrane Proteins in Isolated Pea Chloroplasts — RICHARD S. WILLIAMS AND JOHN BENNETT — 487

45. *In Vitro* Reconstitution of Synthesis, Uptake, and Assembly of Cytoplasmically Synthesized Chloroplast Proteins — JOHN E. MULLET AND NAM-HAI CHUA — 502

46. Structure and Synthesis of Chloroplast ATPase — NATHAN NELSON — 510

47. Cloning in Physical Mapping of Maize Plastid Genes — LAWRENCE BOGORAD, EARL J. GUBBINS, ENNO KREBBERS, IGNACIO M. LARRINUA, BERNARD J. MULLIGAN, KAREN M. T. MUSKAVITCH, ELIZABETH A. ORR, STEVEN R. RODERMEL, RUDI SCHANTZ, ANDRE A. STEINMETZ, GUIDO DE VOS, AND YUKUN K. YE — 524

48. Role of Thylakoid Polypeptide Phosphorylation and Turnover in the Assembly and Function of Photosystem II — M. WETTERN, J. C. OWENS, AND I. OHAD — 554

Section IV. Summary of Membrane Proteins

49. Membrane Proteins: A Summary of Known Structural Information — DAVID R. NELSON AND NEAL C. ROBINSON — 571

Addendum

Addendum to Article [39] — TAKASHI MORIMOTO, SHIRO MATSUURA, AND MONIQUE ARPIN — 621

VOLUME 98

BIOMEMBRANES [PART L: MEMBRANE BIOGENESIS (PROCESSING AND RECYCLING)]

1. Intracellular Membrane Traffic: Pathways, Carriers, and Sorting Devices — MARILYN GIST FARQUHAR — 1

Section I. Specialized Methods

2. Pancreatic Lobules in the *in Vitro* Study of Pancreatic Acinar Cell Function — GEORGE SCHEELE — 17

3. High-Voltage Techniques for Gaining Access to the Interior of Cells: Application to the Study of Exocytosis and Membrane Turnover — P. F. BAKER AND D. E. KNIGHT — 28

4. Transport-Coupled Oligosaccharide Processing in a Cell-Free System — WILLIAM E. BALCH, ERIK FRIES, WILLIAM G. DUNPHY, LENORE J. URBANI, AND JAMES E. ROTHMAN — 37

5. Perturbation of the Structure and Function of the Golgi Complex by Monovalent Carboxylic Ionophores — ALAN MICHAEL TARTAKOFF 47

6. Isolation of Golgi Apparatus from Rat Liver Using D_2O–Sucrose Gradients — BECCA FLEISCHER 60

7. Membranes of Pancreatic Zymogen Granules — JACOPO MELDOLESI 67

8. Secretory Membranes of the Rat Parotid Gland: Preparation and Comparative Characterization — PETER ARVAN, RICHARD S. CAMERON, AND J. DAVID CASTLE 75

Section II. Importance of Glycosylation and Trimming; Targeting

9. Overview: Role of Intracellular Membrane Systems in Glycosylation of Proteins — WILLIAM LENNARZ 91

10. Glycosyltransferases Involved in Elongation of N-Glycosidically Linked Oligosaccharides of the Complex or *N*-Acetyllactosamine Type — HARRY SCHACHTER, SAROJA NARASIMHAN, PAUL GLEESON, AND GEORGE VELLA 98

11. Inhibitors of Glycoprotein Synthesis — ALAN D. ELBEIN 135

12. Preparation of Glycoprotein and Oligosaccharide Probes for Receptor Studies — JACQUES U. BAENZIGER 154

13. Immunological Recognition of Modifications on Functionally Related Proteins — DAVID A. KNECHT, ROBERT C. MIERENDORF, JR., AND RANDALL L. DIMOND 159

Section III. Exocytosis

14. Role of Cytoskeleton in Liver: *In Vivo* Effect of Colchicine on Hepatic Protein Secretion — COLVIN M. REDMAN 169

15. Role of the Cytoskeleton in Secretory Processes: Lacrimal and Salivary Glands — B. ROSSIGNOL, A. M. CHAMBAUT-GUÉRIN, AND P. MULLER 175

16. A Subcellular Fractionation Approach for Studying Insulin Release Mechanisms and Calcium Metabolism in Islets of Langerhans — MICHAEL L. MCDANIEL, JERRY R. COLCA, NIRMALA KOTAGAL, AND PAUL E. LACY 182

Section IV. Internalization of Plasma Membrane Components during Absorptive Endocytosis

17. Use of Electron-Opaque Tracers for Studies on Endocytosis and Membrane Recycling — VOLKER HERZOG AND MARILYN GIST FARQUHAR 203

A. Receptor-Mediated Uptake

18. Use of Isolated, Perfused Liver in Studies of Receptor-Mediated Endocytosis — William A. Dunn, Doris A. Wall, and Ann L. Hubbard 225

19. Receptor-Mediated Endocytosis of Low-Density Lipoprotein in Cultured Cells — Joseph L. Goldstein, Sandip K. Basu, and Michael S. Brown 241

20. Binding, Endocytosis, and Degradation of Enveloped Animal Viruses — Mark Marsh, Ari Helenius, Karl Matlin, and Kai Simons 260

21. Image Intensification Techniques for Detection of Proteins in Cultured Cells — Mark C. Willingham and Ira H. Pastan 266

22. Receptor-Mediated Endocytosis of Epidermal Growth Factor — Harry T. Haigler 283

23. Determinants in the Uptake of Lysosomal Enzymes by Cultured Fibroblasts — Kim E. Creek, H. David Fischer, and William S. Sly 290

24. Uptake of Glycoproteins and Glycoconjugates by Macrophages — M. Konish, V. Shepherd, G. Holt, and P. Stahl 301

25. Isolation of a Phosphomannosyl Receptor from Bovine Liver Membranes — Diane C. Mitchell, G. Gary Sahagian, Jack J. Distler, Renate M. Wagner, and George W. Jourdian 304

26. Purification of Rat Liver Fucose Binding Protein — Mark A. Lehrman and Robert L. Hill 309

B. Coated Vesicles

27. Isolation of Coated Vesicles — B. M. F. Pearse 320

28. Isolation and Characterization of Coated Vesicles from Rat Liver — F. Anthony Simion, David Winek, Enrique Brandan, Becca Fleischer, and Sidney Fleischer 326

29. Isolation of Coated Vesicles: Comparative Studies — John L. Daiss and Thomas F. Roth 337

30. Dissociation and Reassociation of Clathrin — W. Schook and S. Puszkin 350

31. Preparation of Antibodies to Clathrin and Use in Cytochemical Localization — JAMES H. KEEN — 359

32. Structural Investigations on the Role of Microfilaments in Ligand Translocation — JEFFREY L. SALISBURY AND GILBERT A. KELLER — 368

Section V. Recycling of Plasma Membrane Proteins

33. Use of Immunocytochemical Techniques in Studying the Biogenesis of Cell Surfaces in Polarized Epithelia — HUBERT REGGIO, PAUL WEBSTER, AND DANIEL LOUVARD — 379

34. Use of Antibody to 5′-Nucleotidase as a Marker to Study Membrane Flow during Pinocytosis — CHRISTOPHER C. WIDNELL — 395

35. Intracellular Iodination of Lysosome Membrane for Studies of Membrane Composition and Recycling — WILLIAM A. MULLER, RALPH M. STEINMAN, AND ZANVIL A. COHN — 404

36. Labeling of Plasma Membrane Glycoconjugates by Terminal Glycosylation (Galactosyltransferase and Glycosidase) — LUTZ THILO — 415

37. *Dictyostelium discoideum* as a Model System to Study Recognition Mechanisms in Phagocytosis — GÜNTER VOGEL — 421

38. Recycling of Insulin-Sensitive Glucose Transporter in Rat Adipocytes — TETSURO KONO — 431

Section VI. Transcellular Transport

39. Preparation of Inside-Out Thyroid Follicles for Studies on Transcellular Transport (Transcytosis) — VOLKER HERZOG — 447

40. Biosynthesis, Processing, and Function of Secretory Component — KEITH E. MOSTOV AND GÜNTER BLOBEL — 458

41. Transcellular Transport of Proteins *in Vivo* — GERHARD ROHR AND GEORGE SCHEELE — 466

Section VII. Plasma Membrane Differentiation

42. Polarized Multicellular Structures Designed for the *in Vitro* Study of Thyroid Cell Function and Polarization — JEAN MAUCHAMP, MARIANNE CHAMBARD, JACQUELINE GABRION, AND BERNARD VERRIER — 477

43. Polarized Assembly of Enveloped Viruses from Cultured Epithelial Cells — ENRIQUE RODRIGUEZ-BOULAN — 486

44. Isolation and Characterization of Liver Gap Junctions — ELLIOT L. HERTZBERG — 501

45. Techniques for Studying the Cell-Free Synthesis of the Major Lens Fiber Membrane Polypeptide — EDWARD H. WILLIAMS, NALIN M. KUMAR, AND NORTON B. GILULA — 510

46. Gap Junctions in Liver: Isolation, Morphological Analysis, and Quantitation — BRUCE J. NICHOLSON AND JEAN-PAUL REVEL — 519

47. Studies on the Biogenesis of Cell–Cell Channels — GERHARD DAHL, RUDOLF WERNER, AND ROOBIK AZARNIA — 537

48. Selective Labeling and Quantitative Analysis of Internalized Plasma Membrane — IRA MELLMAN AND CYNTHIA J. GALLOWAY — 545

49. Spontaneous Transfer of Exogenous Epidermal Growth Factor Receptors into Receptor-Negative Mutant Cells — MANJUSRI DAS, JEFFREY FEINMAN, MARK PITTENGER, HERMAN MICHAEL, AND SUBAL BISHAYEE — 555

Section VIII. Transfer of Phospholipids between Membranes

50. Lipid Transfer Proteins: Overview and Applications — DONALD B. ZILVERSMIT — 565

51. Nonspecific Lipid Transfer Protein from Rat and Beef Liver: Purification and Properties — BERNABÉ BLOJ AND DONALD B. ZILVERSMIT — 574

52. Phosphatidylcholine Transfer Protein from Bovine Liver — JAN WESTERMAN, H. H. KAMP, AND K. W. A. WIRTZ — 581

53. Phosphatidylcholine Transfer Protein from Rat Liver: Purification and Radioimmunoassay — T. TEERLINK, B. J. H. M. POORTHUIS, AND K. W. A. WIRTZ — 586

54. Nonspecific Phospholipid Transfer Protein from Rat Liver — B. J. H. M. POORTHUIS AND K. W. A. WIRTZ — 592

55. Phospholipid Exchange Protein-Dependent Synthesis of Sphingomyelin — DENNIS R. VOELKER AND EUGENE P. KENNEDY — 596

56. Use of a Nonspecific Lipid Transfer Protein to Modify the Cholesterol Content of Synaptic Membranes — PAULA NORTH AND SIDNEY FLEISCHER — 599

57. Cerebroside Transfer Protein — NORMAN S. RADIN AND RAYMOND J. METZ — 613

58. Synthesis of Labeled Phospholipids in High Yield — HANSJORG EIBL, J. OLIVER MCINTYRE, EDUARD A. M. FLEER, AND SIDNEY FLEISCHER — 623

Addendum

Addendum to Article [21] — MARK C. WILLINGHAM AND IRA H. PASTAN — 635

VOLUME 99

HORMONE ACTION (PART F: PROTEIN KINASES)

Section I. General Methodology

1. Assays of Protein Kinase — ROBERT ROSKOSKI, JR. — 3
2. Measurement of Chemical Phosphate in Proteins — JANICE E. BUSS AND JAMES T. STULL — 7
3. Removal of Phosphate from Proteins by the Reverse Reaction — DAVID A. FLOCKHART — 14
4. Measurement of Hormone-Stimulated Phosphorylation in Intact Cells — JAMES C. GARRISON — 20
5. Peptide Mapping and Purification of Phosphopeptides Using High-Performance Liquid Chromatography — HENNING JUHL AND THOMAS R. SODERLING — 37

Section II. Purification and Properties of Specific Protein Kinases

A. Cyclic Nucleotide-Dependent Protein Kinases

6. Catalytic Subunit of cAMP-Dependent Protein Kinase — ERWIN M. REIMANN AND RICHARD A. BEHAM — 51
7. Regulatory Subunits of Bovine Heart and Rabbit Skeletal Muscle cAMP-Dependent Protein Kinase Isozymes — STEPHEN R. RANNELS, ALFREDA BEASLEY, AND JACKIE D. CORBIN — 55
8. cGMP-Dependent Protein Kinase — THOMAS M. LINCOLN — 62
9. Insect (cAMP–cGMP)-Dependent Protein Kinase — ALEXANDER VARDANIS — 71
10. Preparation of Partially Purified Protein Kinase Inhibitor — KEITH K. SCHLENDER, JENNIFER L. TYMA, AND ERWIN M. REIMANN — 77
11. Inhibitor Protein of the cAMP-Dependent Protein Kinase: Characteristics and Purification — SUSAN WHITEHOUSE AND DONAL A. WALSH — 80
12. Use of NMR and EPR to Study cAMP-Dependent Protein Kinase — ALBERT S. MILDVAN, PAUL R. ROSEVEAR, JOSEPH GRANOT, CATHERINE A. O'BRIAN, H. NEAL BRAMSON, AND E. T. KAISER — 93

13. Synthesis of Oligopeptides for the Study of Cyclic Nucleotide-Dependent Protein Kinases — DAVID B. GLASS — 119

14. Affinity Labeling of cAMP-Dependent Protein Kinases — SUSAN S. TAYLOR, ANTHONY R. KERLAVAGE, AND MARK J. ZOLLER — 140

15. Photoaffinity Labeling of the Regulatory Subunit of cAMP-Dependent Protein Kinase — ULRICH WALTER AND PAUL GREENGARD — 154

16. Use of 1,N^6-Etheno-cAMP as a Fluorescent Probe to Study cAMP-Dependent Protein Kinase — HILLARY D. WHITE, STEPHEN B. SMITH, AND EDWIN G. KREBS — 162

17. Using Analogs to Study Selectivity and Cooperativity of Cyclic Nucleotide Binding Sites — STEPHEN R. RANNELS AND JACKIE D. CORBIN — 168

18. Reversible Autophosphorylation of Type II cAMP-Dependent Protein Kinase: Distinction between Intramolecular and Intermolecular Reactions — JACK ERLICHMAN, RAPHAEL RANGEL-ALDAO, AND ORA M. ROSEN — 176

19. Use of Immunological Approaches to Identify a Brain Protein Kinase Isozyme — DWIJEN SARKAR, JACK ERLICHMAN, NORMAN FLEISCHER, AND CHARLES S. RUBIN — 187

20. Using Mutants to Study cAMP-Dependent Protein Kinase — MICHAEL GOTTESMAN — 197

21. Protein Modulation of Cyclic Nucleotide-Dependent Protein Kinases — GORDON M. WALTON AND GORDON N. GILL — 206

22. Substrate-Directed Regulation of cAMP-Dependent Phosphorylation — M. R. EL-MAGHRABI, T. H. CLAUS, AND S. J. PILKIS — 212

23. Use of Microinjection Techniques to Study Protein Kinases and Protein Phosphorylation in Amphibian Oocytes — JAMES L. MALLER — 219

24. Determination of the cAMP-Dependent Protein Kinase Activity Ratio in Intact Tissues — JACKIE D. CORBIN — 227

25. Radiolabeling and Detection Methods for Studying Metabolism of Regulatory Subunit of cAMP-Dependent Protein Kinase I in Intact Cultured Cells — ROBERT A. STEINBERG — 233

B. Calcium-Dependent Protein Kinases

26. Phosphorylase Kinase from Rabbit Skeletal Muscle — PHILIP COHEN — 243

27. Cardiac Phosphorylase Kinase: Preparation and Properties — HEI SOOK SUL, BERNADETTE DIRDEN, KAREN L. ANGELOS, PATRICK HALLENBECK, AND DONAL A. WALSH 250

28. Separation of the Subunits of Muscle Phosphorylase Kinase — K.-F. JESSE CHAN AND DONALD J. GRAVES 259

29. Use of Peptide Substrates to Study the Specificity of Phosphorylase Kinase Phosphorylation — DONALD J. GRAVES 268

30. Smooth Muscle Myosin Light Chain Kinase — MICHAEL P. WALSH, SUSAN HINKINS, RENATA DABROWSKA, AND DAVID J. HARTSHORNE 279

31. Calcium-Activated, Phospholipid-Dependent Protein Kinase (Protein Kinase C) from Rat Brain — USHIO KIKKAWA, RYOJI MINAKUCHI, YOSHIMI TAKAI, AND YASUTOMI NISHIZUKA 288

32. Liver Calmodulin-Dependent Glycogen Synthase Kinase — M. ELIZABETH PAYNE AND THOMAS R. SODERLING 299

C. Cyclic Nucleotide and Calcium-Independent Protein Kinases

33. Casein Kinase I — GARY M. HATHAWAY, POLYGENA T. TUAZON, AND JOLINDA A. TRAUGH 308

34. Casein Kinase II — GARY M. HATHAWAY AND JOLINDA A. TRAUGH 317

35. Pyruvate Dehydrogenase Kinase from Bovine Kidney — FLORA H. PETTIT, STEPHEN J. YEAMAN, AND LESTER J. REED 331

36. Glycogen Synthase Kinase-3 from Rabbit Skeletal Muscle — BRIAN A. HEMMINGS AND PHILIP COHEN 337

37. Double-Stranded RNA-Dependent eIF-2α Protein Kinase — RAY PETRYSHYN, DANIEL H. LEVIN, AND IRVING M. LONDON 346

38. Rhodopsin Kinase — HITOSHI SHICHI, ROBERT L. SOMERS, AND KATSUHIKO YAMAMOTO 362

39. Polyamine-Dependent Protein Kinase and Phosphorylation of Ornithine Decarboxylase in *Physarum polycephalum* — VALERIE J. ATMAR AND GLENN D. KUEHN — 366

D. Tyrosine-Specific Protein Kinases

40. Characterization of the Abelson Murine Leukemia Virus-Encoded Tyrosine-Specific Protein Kinase — JEAN YIN JEN WANG AND DAVID BALTIMORE — 373

41. Purification of the Receptor for Epidermal Growth Factor from A-431 Cells: Its Function as a Tyrosyl Kinase — STANLEY COHEN — 379

42. Detection and Quantification of Phosphotyrosine in Proteins — JONATHAN A. COOPER, BARTHOLOMEW M. SEFTON, AND TONY HUNTER — 387

43. Base Hydrolysis and Amino Acid Analysis for Phosphotyrosine in Proteins — TODD M. MARTENSEN AND RODNEY L. LEVINE — 402

VOLUME 100

RECOMBINANT DNA (PART B)

Section I. Use of Enzymes in Recombinant DNA Research

1. Guide to the Use of Type II Restriction Endonucleases — ROY FUCHS AND ROBERT BLAKESLEY — 3

2. Using T4 RNA Ligase with DNA Substrates — CATHERINE A. BRENNAN, ALICE E. MANTHEY, AND RICHARD I. GUMPORT — 38

3. Joining of RNA Molecules with RNA Ligase — PAUL J. ROMANIUK AND OLKE C. UHLENBECK — 52

4. Exonuclease III: Use for DNA Sequence Analysis and in Specific Deletions of Nucleotides — LI-HE GUO AND RAY WU — 60

5. Terminal Transferase: Use in the Tailing of DNA and for *in Vitro* Mutagenesis — GUO-REN DENG AND RAY WU — 96

6. Generation of Overlapping Labeled DNA Fragments for DNA Sequencing: A Mercury–Thiol Affinity System — JAMES L. HARTLEY AND JOHN E. DONELSON — 116

Section II. Enzymes Affecting the Gross Morphology of DNA

A. Topoisomerases Type I

7. HeLa Topoisomerase — Leroy F. Liu — 133

8. Multiple Forms of Rat Liver Type I Topoisomerase — Susan R. Martin, William K. McCoubrey, Jr., Betty L. McConaughy, Lisa S. Young, Michael D. Been, Bonita J. Brewer, and James J. Champoux — 137

B. Topoisomerases Type II

9. *Escherichia coli* Phage T4 Topoisomerase — Kenneth N. Kreuzer and Cornelis Victor Jongeneel — 144

10. Purification and Properties of Type II DNA Topoisomerase from Embryos of *Drosophila melanogaster* — Tao-shih Hsieh — 161

11. *Escherichia coli* DNA Gyrase — Richard Otter and Nicholas R. Cozzarelli — 171

Section III. Proteins with Specialized Functions Acting at Specific Loci

12. The Bacteriophage Lambda Terminase Enzyme — M. Gold, A. Becker, and W. Parris — 183

13. The Resolvase Protein of the Transposon $\gamma\delta$ — Randall R. Reed — 191

14. Purification of *recA* Protein from *Escherichia coli* — Takehiko Shibata, Lynn Osber, and Charles M. Radding — 197

15. Purification and Properties of the Bacteriophage Lambda Int Protein — Howard A. Nash — 210

16. Analysis of the ϕX174 Gene *A* Protein Using *in Vitro* DNA Replication Systems — David R. Brown, Danny Reinberg, Thomas Schmidt-Glenewinkel, Stephen L. Zipursky, and Jerard Hurwitz — 217

Section IV. New Methods for DNA Isolation, Hybridization, and Cloning

17. A Rapid Alkaline Extraction Method for the Isolation of Plasmid DNA — H. C. BIRNBOIM — 243

18. Hybridization of Denatured RNA Transferred or Dotted to Nitrocellulose Paper — PATRICIA S. THOMAS — 255

19. Isolation of Multigene Families and Determination of Homologies by Filter Hybridization Methods — GERALD A. BELTZ, KENNETH A. JACOBS, THOMAS H. EICKBUSH, PETER T. CHERBAS, AND FOTIS C. KAFATOS — 266

20. Synthesis of ds-cDNA Involving Addition of dCMP Tails to Allow Cloning of 5′-Terminal mRNA Sequences — HARTMUT LAND, MANUEL GREZ, HANSJÖRG HAUSER, WERNER LINDENMAIER, AND GÜNTHER SCHÜTZ — 285

21. β-Galactosidase Gene Fusions for Analyzing Gene Expression in *Escherichia coli* and Yeast — MALCOLM J. CASADABAN, ALFONSO MARTINEZ-ARIAS, STUART K. SHAPIRA, AND JOANY CHOU — 293

22. Efficient Transfer of Small DNA Fragments from Polyacrylamide Gels to Diazo or Nitrocellulose Paper and Hybridization — ERICH FREI, ABRAHAM LEVY, PETER GOWLAND, AND MARKUS NOLL — 309

23. Electrophoretic Transfer of DNA, RNA, and Protein onto Diazobenzyloxymethyl Paper — ALBERT E. DAHLBERG AND EDMUND J. STELLWAG — 326

24. Plasmid Screening at High Colony Density — DOUGLAS HANAHAN AND MATTHEW MESELSON — 333

25. Chromogenic Method to Screen Very Large Populations of Bacteriophage Plaques for the Presence of Specific Antigen — DONALD A. KAPLAN, LAWRENCE GREENFIELD, AND R. JOHN COLLIER — 342

26. Purification of Nucleic Acids by RPC-5 ANALOG Chromatography: Peristaltic and Gravity-Flow Applications — J. A. THOMPSON, R. W. BLAKESLEY, K. DORAN, C. J. HOUGH, AND ROBERT D. WELLS — 368

27. Hybrid Selection of Specific DNAs Using DNA Covalently Coupled to Macroporous Supports — HANS BÜNEMANN AND PETER WESTHOFF — 400

Section V. Analytical Methods for Gene Products

28. Quantitative Two-Dimensional Gel Electrophoresis of Proteins — JAMES I. GARRELS — 411

29. Peptide Mapping in Gels — STUART G. FISCHER — 424

30. Methods of RNA Sequence Analysis — ALAN DIAMOND AND BERNARD DUDOCK — 431

Section VI. Mutagenesis: *In Vitro* and *in Vivo*

31. Directed Mutagenesis with Sodium Bisulfite — DAVID SHORTLE AND DAVID BOTSTEIN — 457

32. Oligonucleotide-Directed Mutagenesis of DNA Fragments Cloned into M13 Vectors — MARK J. ZOLLER AND MICHAEL SMITH — 468

33. Mud(Ap, *lac*)-Generated Fusions in Studies of Gene Expression — JUDY H. KRUEGER AND GRAHAM C. WALKER — 501

VOLUME 101

RECOMBINANT DNA (PART C)

Section I. New Vectors for Cloning Genes

1. New Bacteriophage Lambda Vectors with Positive Selection for Cloned Inserts — JONATHAN KARN, SYDNEY BRENNER, AND LESLIE BARNETT — 3

2. New M13 Vectors for Cloning — JOACHIM MESSING — 20

3. An Integrated and Simplified Approach to Cloning into Plasmids and Single-Stranded Phages — GRAY F. CROUSE ANNEMARIE FRISCHAUF, AND HANS LEHRACH — 78

4. Cutting of M13mp7 Phage DNA and Excision of Cloned Single-Stranded Sequences by Restriction Endonucleases — MICHAEL D. BEEN AND JAMES J. CHAMPOUX — 90

5. Kilo-Sequencing: Creation of an Ordered Nest of Asymmetric Deletions across a Large Target Sequence Carried on Phage M13 — WAYNE M. BARNES, MICHAEL BEVAN, AND PAULA H. SON — 98

6. The Use of pKC30 and Its Derivatives for Controlled Expression of Genes — MARTIN ROSENBERG, YEN-SEN HO, AND ALLAN SHATZMAN — 123

7. Amplification of DNA Repair Genes Using Plamid pKC30 — GEORGE H. YOAKUM — 138

8. Plasmids Containing the *trp* Promoters of *Escherichia coli* and *Serratia marcescens* and Their Use in Expressing Cloned Genes — BRIAN P. NICHOLS AND CHARLES YANOFSKY — 155

Section II. Cloning of Genes into Yeast Cells

9. Construction and Use of Gene Fusions to *lacZ* (β-Galactosidase) Which Are Expressed in Yeast — MARK ROSE AND DAVID BOTSTEIN — 167

10. Yeast Promoters and *lacZ* Fusions Designed to Study Expression of Cloned Genes in Yeast — LEONARD GUARENTE — 181

11. Expression of Genes in Yeast Using the *ADCI* Promoter — GUSTAV AMMERER — 192

12. One-Step Gene Disruption in Yeast — RODNEY J. ROTHSTEIN — 202

13. Eviction and Transplacement of Mutant Genes in Yeast — FRED WINSTON, FORREST CHUMLEY, AND GERALD R. FINK — 211

14. Genetic Applications of Yeast Transformation with Linear and Gapped Plasmids — TERRY L. ORR-WEAVER, JACK W. SZOSTAK, AND RODNEY J. ROTHSTEIN — 228

15. A Rapid Procedure for the Construction of Linear Yeast Plasmids — J. W. SZOSTAK — 245

16. Cloning Regulated Yeast Genes from a Pool of *lacZ* Fusions — STEPHANIE W. RUBY, JACK W. SZOSTAK, AND ANDREW W. MURRAY — 253

17. Construction of Specific Chromosomal Rearrangements in Yeast — NEAL SUGAWARA AND JACK W. SZOSTAK — 269

18. Yeast Vectors with Negative Selection — PATRICIA A. BROWN AND JACK W. SZOSTAK — 278

19. Use of Integrative Transformation of Yeast in the Cloning of Mutant Genes and Large Segments of Contiguous Chromosomal Sequences — JOHN I. STILES — 290

20. Selection Procedure for Isolation of Centromere DNAs from *Saccharomyces cerevisiae* — LOUISE CLARKE, CHU-LAI HSIAO, AND JOHN CARBON — 300

21. Construction of High Copy Yeast Vectors Using 2-μm Circle Sequences — JAMES R. BROACH — 307

22. Cloning of Yeast *STE* Genes in 2 μm Vectors — VIVIAN L. MACKAY — 325

Section III. Systems for Monitoring Cloned Gene Expression
A. Intact Cell Systems

23. Analysis of Recombinant DNA Using *Escherichia coli* Minicells — JOSEPHINE E. CLARK-CURTISS AND ROY CURTISS III — 347

24. Uses of Transposon $\gamma\delta$ in the Analysis of Cloned Genes — MARK S. GUYER — 362

25. The Use of *Xenopus* Oocytes for the Expression of Cloned Genes — J. B. GURDON AND M. P. WICKENS — 370

B. Introduction of Genes into Mammalian Cells

26. Eukaryotic Cloning Vectors Derived from Bovine Papillomavirus DNA — PETER M. HOWLEY, NAVA SARVER, AND MING-FAN LAW — 387

27. High-Efficiency Transfer of DNA into Eukaryotic Cells by Protoplast Fusion — ROZANNE M. SANDRI-GOLDIN, ALAN L. GOLDIN, MYRON LEVINE, AND JOSEPH GLORIOSO — 402

28. Gene Transfer into Mouse Embryos: Production of Transgenic Mice by Pronuclear Injection — JON W. GORDON AND FRANK H. RUDDLE — 411

29. Introduction of Exogenous DNA into Cotton Embryos — GUANG-YU ZHOU, JIAN WENG, YISHEN ZENG, JUNGI HUANG, SIYING QIAN, AND GUILING LIU — 433

30. Microinjection of Tissue Culture Cells — M. GRAESSMANN AND A. GRAESSMANN — 482

31. Fusogenic Reconstituted Sendai Virus Envelopes as a Vehicle for Introducing DNA into Viable Mammalian Cells — A. VAINSTEIN, A. RAZIN, A. GRAESSMANN, AND A. LOYTER — 492

32. Liposomes as Carriers for Intracellular Delivery of Nucleic Acids — ROBERT M. STRAUBINGER AND DEMETRIOS PAPAHADJOPOULOS — 512

33. *Agrobacterium* Ti Plasmids as Vectors for Plant Genetic Engineering — KENNETH A. BARTON AND MARY-DELL CHILTON — 527

C. Cell-Free Systems; Transcription

34. Isolation of Bacterial and Bacteriophage RNA Polymerases and Their Use in Synthesis of RNA *in Vitro* — M. CHAMBERLIN, R. KINGSTON, M. GILMAN, J. WIGGS, AND A. DEVERA — 540

35. *In Vitro* Transcriptions: Whole-Cell Extract — JAMES L. MANLEY, ANDREW FIRE, MARK SAMUELS, AND PHILLIP A. SHARP 568

36. Eukaryotic Gene Transcription with Purified Components — JOHN D. DIGNAM, PAUL L. MARTIN, BARKUR S. SHASTRY, AND ROBERT G. ROEDER 582

D. Cell-Free Systems; Translation

37. Bacterial *in Vitro* Protein-Synthesizing Systems — GLENN H. CHAMBLISS, TINA M. HENKIN, AND JUDITH M. LEVENTHAL 598

38. Translation of Exogenous mRNAs in Reticulocyte Lysates — WILLIAM C. MERRICK 606

39. Translational Systems Prepared from the Ehrlich Ascites Tumor Cell — EDGAR C. HENSHAW AND RICHARD PANNIERS 616

40. Preparation of a Cell-Free System from Chinese Hamster Ovary Cells That Translates Natural and Synthetic Messenger Ribonucleic Acid Templates — KIVIE MOLDAVE AND ITZHAK FISCHER 629

41. Preparation of a Cell-Free Protein-Synthesizing System from Wheat Germ — CARL W. ANDERSON, J. WILLIAM STRAUS, AND BERNARD S. DUDOCK 635

42. Preparation of a Cell-Free System from *Saccharomyces cerevisiae* That Translates Exogenous Messenger Ribonucleic Acids — KIVIE MOLDAVE AND EUGENIUSZ GASIOR 644

43. Methods Utilizing Cell-Free Protein-Synthesizing Systems for the Identification of Recombinant DNA Molecules — JACQUELINE S. MILLER, BRUCE M. PATERSON, ROBERT P. RICCIARDI, LAWRENCE COHEN, AND BRYAN E. ROBERTS 650

44. Prokaryotic Coupled Transcription–Translation — HUI-ZHU CHEN AND GEOFFREY ZUBAY 674

45. A Coupled DNA-Directed *in Vitro* System to Study Gene Expression Based on Di- and Tripeptide Formation — NIKOLAOS ROBAKIS, YVES CENATIEMPO, LUIS MEZA-BASSO, NATHAN BROT, AND HERBERT WEISSBACH 690

METHODS IN ENZYMOLOGY

EDITED BY

Sidney P. Colowick and Nathan O. Kaplan

VANDERBILT UNIVERSITY
SCHOOL OF MEDICINE
NASHVILLE, TENNESSEE

DEPARTMENT OF CHEMISTRY
UNIVERSITY OF CALIFORNIA
AT SAN DIEGO
LA JOLLA, CALIFORNIA

I. Preparation and Assay of Enzymes
II. Preparation and Assay of Enzymes
III. Preparation and Assay of Substrates
IV. Special Techniques for the Enzymologist
V. Preparation and Assay of Enzymes
VI. Preparation and Assay of Enzymes (*Continued*)
Preparation and Assay of Substrates
Special Techniques
VII. Cumulative Subject Index

METHODS IN ENZYMOLOGY

EDITORS-IN-CHIEF

Sidney P. Colowick and Nathan O. Kaplan

VOLUME VIII. Complex Carbohydrates
Edited by ELIZABETH F. NEUFELD AND VICTOR GINSBURG

VOLUME IX. Carbohydrate Metabolism
Edited by WILLIS A. WOOD

VOLUME X. Oxidation and Phosphorylation
Edited by RONALD W. ESTABROOK AND MAYNARD E. PULLMAN

VOLUME XI. Enzyme Structure
Edited by C. H. W. HIRS

VOLUME XII. Nucleic Acids (Parts A and B)
Edited by LAWRENCE GROSSMAN AND KIVIE MOLDAVE

VOLUME XIII. Citric Acid Cycle
Edited by J. M. LOWENSTEIN

VOLUME XIV. Lipids
Edited by J. M. LOWENSTEIN

VOLUME XV. Steroids and Terpenoids
Edited by RAYMOND B. CLAYTON

VOLUME XVI. Fast Reactions
Edited by KENNETH KUSTIN

VOLUME XVII. Metabolism of Amino Acids and Amines (Parts A and B)
Edited by HERBERT TABOR AND CELIA WHITE TABOR

VOLUME XVIII. Vitamins and Coenzymes (Parts A, B, and C)
Edited by DONALD B. MCCORMICK AND LEMUEL D. WRIGHT

VOLUME XIX. Proteolytic Enzymes
Edited by GERTRUDE E. PERLMANN AND LASZLO LORAND

VOLUME XX. Nucleic Acids and Protein Synthesis (Part C)
Edited by KIVIE MOLDAVE AND LAWRENCE GROSSMAN

VOLUME XXI. Nucleic Acids (Part D)
Edited by LAWRENCE GROSSMAN AND KIVIE MOLDAVE

VOLUME XXII. Enzyme Purification and Related Techniques
Edited by WILLIAM B. JAKOBY

VOLUME XXIII. Photosynthesis (Part A)
Edited by ANTHONY SAN PIETRO

VOLUME XXIV. Photosynthesis and Nitrogen Fixation (Part B)
Edited by ANTHONY SAN PIETRO

VOLUME XXV. Enzyme Structure (Part B)
Edited by C. H. W. HIRS AND SERGE N. TIMASHEFF

VOLUME XXVI. Enzyme Structure (Part C)
Edited by C. H. W. HIRS AND SERGE N. TIMASHEFF

VOLUME XXVII. Enzyme Structure (Part D)
Edited by C. H. W. HIRS AND SERGE N. TIMASHEFF

VOLUME XXVIII. Complex Carbohydrates (Part B)
Edited by VICTOR GINSBURG

VOLUME XXIX. Nucleic Acids and Protein Synthesis (Part E)
Edited by LAWRENCE GROSSMAN AND KIVIE MOLDAVE

VOLUME XXX. Nucleic Acids and Protein Synthesis (Part F)
Edited by KIVIE MOLDAVE AND LAWRENCE GROSSMAN

VOLUME XXXI. Biomembranes (Part A)
Edited by SIDNEY FLEISCHER AND LESTER PACKER

VOLUME XXXII. Biomembranes (Part B)
Edited by SIDNEY FLEISCHER AND LESTER PACKER

VOLUME XXXIII. Cumulative Subject Index Volumes I–XXX
Edited by MARTHA G. DENNIS AND EDWARD A. DENNIS

VOLUME XXXIV. Affinity Techniques (Enzyme Purification: Part B)
Edited by WILLIAM B. JAKOBY AND MEIR WILCHEK

VOLUME XXXV. Lipids (Part B)
Edited by JOHN M. LOWENSTEIN

VOLUME XXXVI. Hormone Action (Part A: Steroid Hormones)
Edited by BERT W. O'MALLEY AND JOEL G. HARDMAN

VOLUME XXXVII. Hormone Action (Part B: Peptide Hormones)
Edited by BERT W. O'MALLEY AND JOEL G. HARDMAN

VOLUME XXXVIII. Hormone Action (Part C: Cyclic Nucleotides)
Edited by JOEL G. HARDMAN AND BERT W. O'MALLEY

VOLUME XXXIX. Hormone Action (Part D: Isolated Cells, Tissues, and Organ Systems)
Edited by JOEL G. HARDMAN AND BERT W. O'MALLEY

VOLUME XL. Hormone Action (Part E: Nuclear Structure and Function)
Edited by BERT W. O'MALLEY AND JOEL G. HARDMAN

VOLUME XLI. Carbohydrate Metabolism (Part B)
Edited by W. A. WOOD

VOLUME XLII. Carbohydrate Metabolism (Part C)
Edited by W. A. WOOD

VOLUME XLIII. Antibiotics
Edited by JOHN H. HASH

VOLUME XLIV. Immobilized Enzymes
Edited by KLAUS MOSBACH

VOLUME XLV. Proteolytic Enzymes (Part B)
Edited by LASZLO LORAND

VOLUME XLVI. Affinity Labeling
Edited by WILLIAM B. JAKOBY AND MEIR WILCHEK

VOLUME XLVII. Enzyme Structure (Part E)
Edited by C. H. W. HIRS AND SERGE N. TIMASHEFF

VOLUME XLVIII. Enzyme Structure (Part F)
Edited by C. H. W. HIRS AND SERGE N. TIMASHEFF

VOLUME XLIX. Enzyme Structure (Part G)
Edited by C. H. W. HIRS AND SERGE N. TIMASHEFF

VOLUME L. Complex Carbohydrates (Part C)
Edited by VICTOR GINSBURG

VOLUME LI. Purine and Pyrimidine Nucleotide Metabolism
Edited by PATRICIA A. HOFFEE AND MARY ELLEN JONES

VOLUME LII. Biomembranes (Part C: Biological Oxidations)
Edited by SIDNEY FLEISCHER AND LESTER PACKER

VOLUME LIII. Biomembranes (Part D: Biological Oxidations)
Edited by SIDNEY FLEISCHER AND LESTER PACKER

VOLUME LIV. Biomembranes (Part E: Biological Oxidations)
Edited by SIDNEY FLEISCHER AND LESTER PACKER

VOLUME LV. Biomembranes (Part F: Bioenergetics)
Edited by SIDNEY FLEISCHER AND LESTER PACKER

VOLUME LVI. Biomembranes (Part G: Bioenergetics)
Edited by SIDNEY FLEISCHER AND LESTER PACKER

VOLUME LVII. Bioluminescence and Chemiluminescence
Edited by MARLENE A. DELUCA

VOLUME LVIII. Cell Culture
Edited by WILLIAM B. JAKOBY AND IRA PASTAN

VOLUME LIX. Nucleic Acids and Protein Synthesis (Part G)
Edited by KIVIE MOLDAVE AND LAWRENCE GROSSMAN

VOLUME LX. Nucleic Acids and Protein Synthesis (Part H)
Edited by KIVIE MOLDAVE AND LAWRENCE GROSSMAN

VOLUME 61. Enzyme Structure (Part H)
Edited by C. H. W. HIRS AND SERGE N. TIMASHEFF

VOLUME 62. Vitamins and Coenzymes (Part D)
Edited by DONALD B. MCCORMICK AND LEMUEL D. WRIGHT

VOLUME 63. Enzyme Kinetics and Mechanism (Part A: Initial Rate and Inhibitor Methods)
Edited by DANIEL L. PURICH

VOLUME 64. Enzyme Kinetics and Mechanism (Part B: Isotopic Probes and Complex Enzyme Systems)
Edited by DANIEL L. PURICH

VOLUME 65. Nucleic Acids (Part I)
Edited by LAWRENCE GROSSMAN AND KIVIE MOLDAVE

VOLUME 66. Vitamins and Coenzymes (Part E)
Edited by DONALD B. MCCORMICK AND LEMUEL D. WRIGHT

VOLUME 67. Vitamins and Coenzymes (Part F)
Edited by DONALD B. MCCORMICK AND LEMUEL D. WRIGHT

VOLUME 68. Recombinant DNA
Edited by RAY WU

VOLUME 69. Photosynthesis and Nitrogen Fixation (Part C)
Edited by ANTHONY SAN PIETRO

VOLUME 70. Immunochemical Techniques (Part A)
Edited by HELEN VAN VUNAKIS AND JOHN J. LANGONE

VOLUME 71. Lipids (Part C)
Edited by JOHN M. LOWENSTEIN

VOLUME 72. Lipids (Part D)
Edited by JOHN M. LOWENSTEIN

VOLUME 73. Immunochemical Techniques (Part B)
Edited by JOHN J. LANGONE AND HELEN VAN VUNAKIS

VOLUME 74. Immunochemical Techniques (Part C)
Edited by JOHN J. LANGONE AND HELEN VAN VUNAKIS

VOLUME 75. Cumulative Subject Index Volumes XXXI, XXXII, and XXXIV–LX
Edited by EDWARD A. DENNIS AND MARTHA G. DENNIS

VOLUME 76. Hemoglobins
Edited by ERALDO ANTONINI, LUIGI ROSSI-BERNARDI, AND EMILIA CHIANCONE

VOLUME 77. Detoxication and Drug Metabolism
Edited by WILLIAM B. JAKOBY

VOLUME 78. Interferons (Part A)
Edited by SIDNEY PESTKA

VOLUME 79. Interferons (Part B)
Edited by SIDNEY PESTKA

VOLUME 80. Proteolytic Enzymes (Part C)
Edited by LASZLO LORAND

VOLUME 81. Biomembranes (Part H: Visual Pigments and Purple Membranes, I)
Edited by LESTER PACKER

VOLUME 82. Structural and Contractile Proteins (Part A: Extracellular Matrix)
Edited by LEON W. CUNNINGHAM AND DIXIE W. FREDERIKSEN

VOLUME 83. Complex Carbohydrates (Part D)
Edited by VICTOR GINSBURG

VOLUME 84. Immunochemical Techniques (Part D: Selected Immunoassays)
Edited by JOHN J. LANGONE AND HELEN VAN VUNAKIS

VOLUME 85. Structural and Contractile Proteins (Part B: The Contractile Apparatus and the Cytoskeleton)
Edited by DIXIE W. FREDERIKSEN AND LEON W. CUNNINGHAM

VOLUME 86. Prostaglandins and Arachidonate Metabolites
Edited by WILLIAM E. M. LANDS AND WILLIAM L. SMITH

VOLUME 87. Enzyme Kinetics and Mechanism (Part C: Intermediates, Stereochemistry, and Rate Studies)
Edited by DANIEL L. PURICH

VOLUME 88. Biomembranes (Part I: Visual Pigments and Purple Membranes, II)
Edited by LESTER PACKER

VOLUME 89. Carbohydrate Metabolism (Part D)
Edited by WILLIS A. WOOD

VOLUME 90. Carbohydrate Metabolism (Part E)
Edited by WILLIS A. WOOD

VOLUME 91. Enzyme Structure (Part I)
Edited by C. H. W. HIRS AND SERGE N. TIMASHEFF

VOLUME 92. Immunochemical Techniques (Part E: Monoclonal Antibodies and General Immunoassay Methods)
Edited by JOHN J. LANGONE AND HELEN VAN VUNAKIS

VOLUME 93. Immunochemical Techniques (Part F: Conventional Antibodies, Fc Receptors, and Cytotoxicity)
Edited by JOHN J. LANGONE AND HELEN VAN VUNAKIS

VOLUME 94. Polyamines
Edited by HERBERT TABOR AND CELIA WHITE TABOR

VOLUME 95. Cumulative Subject Index Volumes 61–74 and 76–80
Edited by EDWARD A. DENNIS AND MARTHA G. DENNIS

VOLUME 96. Biomembranes [Part J: Membrane Biogenesis: Assembly and Targeting (General Methods; Eukaryotes)]
Edited by SIDNEY FLEISCHER AND BECCA FLEISCHER

VOLUME 97. Biomembranes [Part K: Membrane Biogenesis: Assembly and Targeting (Prokaryotes, Mitochondria, and Chloroplasts)]
Edited by SIDNEY FLEISCHER AND BECCA FLEISCHER

VOLUME 98. Biomembranes [Part L: Membrane Biogenesis (Processing and Recycling)]
Edited by SIDNEY FLEISCHER AND BECCA FLEISCHER

VOLUME 99. Hormone Action (Part F: Protein Kinases)
Edited by JACKIE D. CORBIN AND JOEL G. HARDMAN

VOLUME 100. Recombinant DNA (Part B)
Edited by RAY WU, LAWRENCE GROSSMAN, AND KIVIE MOLDAVE

VOLUME 101. Recombinant DNA (Part C)
Edited by RAY WU, LAWRENCE GROSSMAN, AND KIVIE MOLDAVE

VOLUME 102. Hormone Action (Part G: Calmodulin and Calcium-Binding Proteins)
Edited by ANTHONY R. MEANS AND BERT W. O'MALLEY

VOLUME 103. Hormone Action (Part H: Neuroendocrine Peptides)
Edited by P. MICHAEL CONN

VOLUME 104. Enzyme Purification and Related Techniques (Part C)
Edited by WILLIAM B. JAKOBY

VOLUME 105. Oxygen Radicals in Biological Systems
Edited by LESTER PACKER

VOLUME 106. Posttranslational Modifications (Part A)
Edited by FINN WOLD AND KIVIE MOLDAVE

VOLUME 107. Posttranslational Modifications (Part B)
Edited by FINN WOLD AND KIVIE MOLDAVE

VOLUME 108. Immunochemical Techniques (Part G: Separation and Characterization of Lymphoid Cells)
Edited by GIOVANNI DI SABATO, JOHN J. LANGONE, AND HELEN VAN VUNAKIS

VOLUME 109. Hormone Action (Part I: Peptide Hormones)
Edited by LUTZ BIRNBAUMER AND BERT W. O'MALLEY

VOLUME 110. Steroids and Isoprenoids (Part A)
Edited by JOHN H. LAW AND HANS C. RILLING

VOLUME 111. Steroids and Isoprenoids (Part B)
Edited by JOHN H. LAW AND HANS C. RILLING

VOLUME 112. Drug and Enzyme Targeting (Part A)
Edited by KENNETH J. WIDDER AND RALPH GREEN

VOLUME 113. Glutamate, Glutamine, Glutathione, and Related Compounds
Edited by ALTON MEISTER

VOLUME 114. Diffraction Methods for Biological Macromolecules (Part A)
Edited by HAROLD W. WYCKOFF, C. H. W. HIRS, AND SERGE N. TIMASHEFF

VOLUME 115. Diffraction Methods for Biological Macromolecules (Part B)
Edited by HAROLD W. WYCKOFF, C. H. W. HIRS, AND SERGE N. TIMASHEFF

VOLUME 116. Immunochemical Techniques (Part H: Effectors and Mediators of Lymphoid Cell Functions)
Edited by GIOVANNI DI SABATO, JOHN J. LANGONE, AND HELEN VAN VUNAKIS

VOLUME 117. Enzyme Structure (Part J)
Edited by C. H. W. HIRS AND SERGE N. TIMASHEFF

VOLUME 118. Plant Molecular Biology
Edited by ARTHUR WEISSBACH AND HERBERT WEISSBACH

VOLUME 119. Interferons (Part C)
Edited by SIDNEY PESTKA

VOLUME 120. Cumulative Subject Index Volumes 81–94, 96–101

VOLUME 121. Immunochemical Techniques (Part I: Hybridoma Technology and Monoclonal Antibodies)
Edited by JOHN J. LANGONE AND HELEN VAN VUNAKIS

VOLUME 122. Vitamins and Coenzymes (Part G)
Edited by FRANK CHYTIL AND DONALD B. MCCORMICK

VOLUME 123. Vitamins and Coenzymes (Part H)
Edited by FRANK CHYTIL AND DONALD B. MCCORMICK

VOLUME 124. Hormone Action (Part J: Neuroendocrine Peptides)
Edited by P. MICHAEL CONN

VOLUME 125. Biomembranes (Part M: Transport in Bacteria, Mitochondria, and Chloroplasts: General Approaches and Transport Systems)
Edited by SIDNEY FLEISCHER AND BECCA FLEISCHER

VOLUME 126. Biomembranes (Part N: Transport in Bacteria, Mitochondria, and Chloroplasts: Protonmotive Force)
Edited by SIDNEY FLEISCHER AND BECCA FLEISCHER

VOLUME 127. Biomembranes (Part O: Protons and Water: Structure and Translocation)
Edited by LESTER PACKER

VOLUME 128. Plasma Lipoproteins (Part A: Preparation, Structure, and Molecular Biology)
Edited by JERE P. SEGREST AND JOHN J. ALBERS

VOLUME 129. Plasma Lipoproteins (Part B: Characterization, Cell Biology, and Metabolism)
Edited by JOHN J. ALBERS AND JERE P. SEGREST

VOLUME 130. Enzyme Structure (Part K)
Edited by C. H. W. HIRS AND SERGE N. TIMASHEFF

VOLUME 131. Enzyme Structure (Part L)
Edited by C. H. W. HIRS AND SERGE N. TIMASHEFF

VOLUME 132. Immunochemical Techniques (Part J: Phagocytosis and Cell-Mediated Cytotoxicity) (in preparation)
Edited by GIOVANNI DI SABATO AND JOHANNES EVERSE

VOLUME 133. Bioluminescence and Chemiluminescence (Part B) (in preparation)
Edited by MARLENE DELUCA AND WILLIAM D. MCELROY

VOLUME 134. Structural and Contractile Proteins (Part C: The Contractile Apparatus and the Cytoskeleton) (in preparation)
Edited by RICHARD B. VALLEE

VOLUME 135. Immobilized Enzymes and Cells (Part B) (in preparation)
Edited by KLAUS MOSBACH

VOLUME 136. Immobilized Enzymes and Cells (Part C) (in preparation)
Edited by KLAUS MOSBACH

VOLUME 137. Immobilized Enzymes and Cells (Part D) (in preparation)
Edited by KLAUS MOSBACH

VOLUME 138. Complex Carbohydrates (Part E) (in preparation)
Edited by VICTOR GINSBURG

Subject Index

Boldface numerals indicate volume number.

A

Abelson murine leukemia virus
  encoded protein-tyrosine kinase
    assay, **99**, 374
    characterization, **99**, 373
    purification, **99**, 376
Acanthamoeba castellanii
  myosin isozymes
    assays, **85**, 357
    properties, **85**, 360, 363
    purification, **85**, 359, 361
Acetate kinase
  Bacillus stearothermophilus
    assay, **90**, 179
    properties, **90**, 183
    purification, **90**, 180
Acetobacter
  A. aceti
    cultivation, **89**, 451
    purification of membrane-bound dehydrogenases, **89**, 454, 494
  A. rancens, purification of 2-keto-D-gluconate reductase, **89**, 206
Acetoin dehydrogenase
  assay, **89**, 516
  disc gel electrophoresis, **89**, 517
  properties, **89**, 521
  purification from
    bovine liver, **89**, 517
    Escherichia coli, **89**, 519
    pigeon liver, **89**, 518
Acetylamino acids
  amino-terminal, identification by HPLC, **91**, 84
2-Acetylaminofluorene, *see* *N*-2-Fluorenylacetamide
Acetylation
  protein amino-terminus, prevention, **96**, 156
Acetylcholine
  arylsulfatase effects, **86**, 20
  radioimmunoassay, **84**, 555
Acetylcholine receptors
  skeletal muscle
    biosynthesis, **96**, 341
      early events, **96**, 348
    degradation
      process, **96**, 350
      regulation, **96**, 352
    intracellular transport, **96**, 341, 350
    measurement, **96**, 332
    metabolic labeling with heavy isotopes, **96**, 335
    role in membrane recycling, **96**, 348
  subunits, *Torpedo californica* electroplax
    cell-free synthesis, **96**, 369
    mRNA isolation, **96**, 368
    in vitro synthesized
      electrophoresis, **96**, 371
      immunoprecipitation, **96**, 370
      integration into microsomal membranes, **96**, 374
Acetylcholinesterase
  Electrophorus electricus
    intact tail subunits, isolation, **82**, 335
    pepsin-resistant fragments, isolation, **82**, 332
    purification, **82**, 327
  muscle
    intracellular transport and fate, **96**, 363
    molecular forms, localization and assembly, **96**, 365
    subcellular distribution, **96**, 354
    synthesis, **96**, 361
Acetyl-CoA:α-glucosaminide *N*-acetyltransferase, *see* Glucosamine acetyltransferase
N-Acetylgalactosamine-4-sulfatase
  human placental
    effect on
      spasmogenic mediators, **86**, 20
      synthetic leukotrienes, **86**, 28
    inactivation of slow-reacting substance, **86**, 20
    purification, **86**, 18
N-Acetylgalactosamine-6-sulfatase
  deficiency in mucopolysaccharidoses, assay, **83**, 563
  radioactive substrates, preparation, **83**, 570

- α-*N*-Acetylgalactosaminide α2→6-sialyltransferase
  - porcine submaxillary gland
    - assay, **83**, 480
    - properties, **83**, 486
    - purification, **83**, 481
- *N*-Acetylglucosamine-6-sulfatase (heparan sulfate-degrading)
  - deficiency in mucopolysaccharidoses, assay, **83**, 562
  - radioactive substrates, preparation, **83**, 568
- *N*-Acetyl-D-glucosamine transferases
  - acceptor preparation, **98**, 105
  - assays, **98**, 111, 116, 117, 120
  - properties, **98**, 113, 116, 119, 120
  - purification from
    - bovine colostrum, **98**, 112, 116
    - porcine liver, **98**, 113
    - rabbit liver, **98**, 113
  - substrate preparation, **98**, 102
- *N*-Acetylglucosaminide α1→3-fucosyltransferase
  - human milk
    - assay, **83**, 501
    - properties, **83**, 505
    - purification, **83**, 503
- *N*-Acetylglucosaminide α1→4-fucosyltransferase
  - human milk
    - assay, **83**, 501
    - properties, **83**, 505
    - purification, **83**, 503
- β-*N*-Acetylhexosaminidase B
  - pinocytosis measurements in Sandhoff disease fibroblasts, **98**, 292
  - purification from Tay–Sachs disease fibroblast secretions, **98**, 291
- *N*′-Acetylnornicotine analogs
  - radioimmunoassay, **84**, 641
    - sensitivity and specificity, **84**, 646
  - synthesis, **84**, 642
- Acetylspermidine deacetylase
  - rat liver
    - assay, **94**, 329
    - properties, **94**, 331
    - purification, **94**, 329
- Acid hydrolysis
  - γ-carboxyglutamic acid in proteins, **91**, 14
  - phosphotyrosine in proteins, **99**, 396
  - single method for all amino acids, **91**, 26
- Acinar cells
  - pancreatic, *in vitro* functional analysis, **98**, 17
- *Acinetobacter*
  - 4-hydroxy-2-ketopimelate aldolase
    - purification, **90**, 278
- ω-Acrylamidoalkyl glycosides
  - copolymerized, in polyacrylamide gels, preparation, **83**, 294
- Actin
  - binding protein, macrophage
    - purification, **85**, 483
    - sedimentation assay for crosslinking activity, **85**, 481
  - cytoplasmic
    - assay, **85**, 371
    - properties, **85**, 372
    - purification from porcine brain, **85**, 371
  - filaments
    - equilibrium binding to
      - heavy meromyosin, **85**, 714
      - myosin single-headed fragment, **85**, 710
    - network characterization by
      - falling-ball assay, **85**, 216
      - Fukada viscoelastometry, **85**, 231
      - gelmeter, **85**, 221
      - low-speed sedimentation, **85**, 215
      - micromagnetorheometry, **85**, 226
      - rotational viscometry, **85**, 223
      - test tube inversion, **85**, 213
      - Zaner viscoelastometry, **85**, 228
    - properties, **85**, 182
    - role in ligand translocation, **98**, 368
    - steady-state binding to myosin single-headed fragment, **85**, 717
  - interaction with myosin, EPR, **85**, 613
  - polymerization
    - assay by
      - capillary viscometry, **85**, 185
      - DNase I inhibition, **85**, 204
      - electron microscopy, **85**, 200
      - flow birefringence, **85**, 192
      - fluorescence, **85**, 195
      - light scattering, **85**, 198

Millipore filtration, **85**, 207
ultracentrifugation, **85**, 206
uv absorption, **85**, 190
mechanism, **85**, 182
properties
breaking and annealing, **85**, 209
critical concentration, **85**, 209
elongation rates, **85**, 208
filament length distribution, **85**, 210
nucleation rate, **85**, 208
purification from rabbit skeletal muscle, **85**, 164
regulated
equilibrium binding to myosin single-headed fragment, **85**, 715
steady-state binding to myosin single-headed fragment, **85**, 722
removal from crude myosin preparation, **85**, 134
saturation transfer studies, **85**, 612
spin labeling, **85**, 595, 608
selectivity and stoichiometry, **85**, 596
structure–function relationship, optical activity measurements, **85**, 684
α-Actinin
preparation from smooth muscle, **85**, 316
Actinomycin D, *see* Dactinomycin
Actomyosin
ATP hydrolysis
steady-state rate measurements, **85**, 698
transient kinetic measurements, **85**, 701
interaction analysis, **85**, 723
Acylation
fatty acids in eukaryotic cell proteins
detection, **96**, 797
distribution, **96**, 796
site and mechanism, **96**, 799
Acyl CoA
photolabile analogs, synthesis, **91**, 635
radiolabeled unsaturated, preparation, **86**, 371
Acyl-CoA receptors
photoaffinity labeling, **91**, 639
2-Acylglycerophosphocholine
preparation, **86**, 374
2-Acylglycerophosphoethanolamine
preparation, **86**, 374
Adenosine 5′-monophosphate
ribose 5-phosphate preparation from, **89**, 118
Adenosinetriphosphatase, H^+-transporting, *see* ATP synthase (H^+-transporting)
Adenosine 5′-triphosphate
hydrolysis by actomyosin and myosin
steady-state rate measurements, **85**, 698
transient kinetic measurements, **85**, 701
S-Adenosyl-1,8-diamino-3-thiooctane
synthesis, **94**, 289
S-Adenosylmethionine
decarboxylated, *see* *S*-Adenosyl-(5′)-3-methylthiopropylamine
HPLC, **94**, 58
metabolites
analysis in biological samples, **94**, 61
HPLC, **94**, 58
preparation, **94**, 57
separation from biological samples, **94**, 61
preparation, **94**, 57
related nucleosides, inhibition of
S-adenosylmethionine decarboxylase, **94**, 245
aminopropyltransferases, **94**, 296
tissue, content determination by radioenzymatic assay, **94**, 66
S-Adenosylmethionine decarboxylase
Escherichia coli
assay, **94**, 228
deficient mutants, screening, **94**, 84
gene cloning, **94**, 121
properties, **94**, 230
purification, **94**, 229
in inhibition assay for methylglyoxal bis(guanylhydrazone), **94**, 247
rat liver
antibody preparation, **94**, 238
assay, **94**, 234
properties, **94**, 237
purification, **94**, 235
Saccharomyces cerevisiae
assay, **94**, 231

deficient mutants, mass screening, **94**, 107
properties, **94**, 233
purification, **94**, 232
S-Adenosylmethionine decarboxylase inhibitors
synthesis and characterization, **94**, 239
S-Adenosylmethionine synthetase, *see* Methionine adenosyltransferase
S-Adenosyl-(5′)-3-methylthiopropylamine
cellular, content determination, **94**, 69
HPLC, **94**, 58
preparation, **94**, 57, 73, 263
purification, **94**, 74
purity testing, **94**, 77
radiolabeled, preparation, **94**, 293
Adenosyl-sulfur compounds
related to polyamine biosynthesis, HPLC, **94**, 57
S-Adenosyl-3-thiopropylamine
isolation, **94**, 463
properties, **94**, 464
Adenylate cyclase
retinal, characterization and localization in
microdissected sample, **81**, 518
tissue homogenates, **81**, 517
Adipocytes
rat
glucose transporter solubilization and reconstitution, **98**, 438
isolation, **98**, 432
reconstituted, glucose transport assay, **98**, 441
subcellular fractions, isolation, **98**, 434
Adrenal medulla
cellular exocytosis and endocytosis, high-voltage analytical techniques, **98**, 33
Affinity chromatography
acetylcholinesterase, **82**, 327
alcohol dehydrogenase, **89**, 446
on 5′-AMP–Sepharose 4B, **89**, 436
on magnetic 5′-AMP–Sepharose 4B, **89**, 442
calmodulin-dependent glycogen synthase kinase, **99**, 299
carbohydrates on lectin–Sepharose, **83**, 127
casein kinase I, **99**, 311
cGMP-dependent protein kinase, **99**, 64
cyclic nucleotide-dependent protein kinase, **99**, 73
dsRNA-dependent eIF-2α protein kinase, **99**, 351
epidermal growth factor-related tyrosyl kinase, **99**, 379
Fc receptors, **93**, 178, 219
free and bound fractions in enzyme immunoassay, **92**, 359
glycogen synthase kinase-3, **99**, 340
glycophorin A on lectin–Sepharose, **96**, 286
heat-stable cAMP-dependent protein kinase inhibitor, **99**, 78, 84
HLA products with monoclonal antibodies, **92**, 128
affinity considerations, **92**, 137
immune sera to clathrin, **98**, 362
lens mRNA, **98**, 514
murine major histocompatibility complex antigens with monoclonal antibodies, **92**, 86
myosin light-chain kinase, **99**, 281
nonmuscle myosin, **85**, 342
putrescine oxidase, **94**, 301
pyruvate dehydrogenase complex mutants and components on ethanol–Sepharose 2B, **89**, 398
retinal rod and cone pigments, **81**, 160
rhodopsin on concanavalin A–Sepharose, **81**, 150, 169, 197
rhodopsin kinase, **99**, 364
Sindbis virus membrane glycoproteins on concanavalin A–Sepharose, **96**, 519
spermidine dehydrogenase, **94**, 303
tryptic peptides with immobilized anhydrotrypsin as adsorbent, **91**, 381
type-I cAMP-dependent protein kinase regulatory subunit, **99**, 57
type-II cAMP-dependent protein kinase regulatory subunit, **99**, 56
Affinity labeling
application to analysis of enzyme structure and function
active-site topography, **87**, 471
catalytic mechanisms, **87**, 478

essential amino acid residues, **87**, 470, 472
cAMP-dependent protein kinase
ATP-binding site, **99**, 141
cAMP-binding site, **99**, 147
regulatory subunit, **99**, 154
reagents, active site-directed
design considerations, **87**, 481
evaluation, **87**, 486
reaction facilitation, **87**, 495
Agarose gels
isotope detection by fluorography, **96**, 220
mRNA recovery, **96**, 35
Agmatine coumaroyltransferase
barley seedling
assay, **94**, 345
properties, **94**, 347
purification, **94**, 346
substrate synthesis, **94**, 344
Agmatine ureohydrolase
Escherichia coli
deficient mutants, screening, **94**, 90
gene cloning, **94**, 118
Agrobacterium tumefaciens
agrobactin isolation, **94**, 438
Ti plasmids, as vectors for plant genetic engineering, **101**, 527
Agrobactin
Agrobacterium tumefaciens
assay, **94**, 439
isolation, **94**, 438
properties, **94**, 439
Alanine dehydrogenase
isotope exchange study, fingerprint kinetic patterns for modifier action, **87**, 652
Albumin
bovine serum, *see* Bovine serum albumin
human
heterogeneous immunoassay with differential pulse anodic stripping voltammetry, **92**, 437
stable radioiodinated, preparation by gel electrophoresis, **92**, 335
Alcohol
microdetermination with membrane-bound dehydrogenase, **89**, 28
Alcohol dehydrogenase
Acetobacter aceti membrane-bound
assay, **89**, 450
properties, **89**, 455
purification, **89**, 454
active-center overlapping subsites, analysis by Yonetani–Theorell graphical method, **87**, 504
Drosophila melanogaster
assay, **89**, 446
purification, **89**, 446
sequencing at nanomole level, **91**, 470
thin-layer peptide mapping, **91**, 466
equine liver
assay, **89**, 436
–NAD complexes
assays, **89**, 461
properties, **89**, 464
synthesis, **89**, 459
purification, **89**, 436
steroid-active SS isozyme
assays, **89**, 430
catalytic properties, **89**, 431
gel electrophoresis, **89**, 431
purification, **89**, 432, 439
Gluconobacter suboxydans membrane-bound
assay, **89**, 450
properties, **89**, 455
purification, **89**, 452
membrane-bound, microdetermination of alcohol, **89**, 28
Alcohol dehydrogenase ($NADP^+$)
human liver
assay, **89**, 507
properties, **89**, 510
purification, **89**, 508
Alcohol oxidase
Candida boidinii
assay, **89**, 424
production, **89**, 425
properties, **89**, 427
purification, **89**, 426
Aldehyde dehydrogenase
Acetobacter aceti membrane-bound
assay, **89**, 491
properties, **89**, 495
purification, **89**, 494

Gluconobacter suboxydans membrane-bound
assay, **89**, 491
properties, **89**, 495
purification, **89**, 493
membrane-bound, microdetermination of aldehydes, **89**, 27
Aldehyde dehydrogenase (NAD^+)
bovine liver
assay, **89**, 497
properties, **89**, 499
purification, **89**, 498
isozymes, equine liver
assay, **89**, 474
properties, **89**, 477
purification, **89**, 475
Pseudomonas aeruginosa
assay, **89**, 484
properties, **89**, 488
purification, **89**, 485
Aldehyde dehydrogenase ($NADP^+$)
Proteus vulgaris
assay, **89**, 480
properties, **89**, 483
purification, **89**, 480
Pseudomonas aeruginosa
assay, **89**, 489
properties, **89**, 490
purification, **89**, 489
Aldehyde dehydrogenase ($NAD(P)^+$)
yeast
assay, **89**, 469
properties, **89**, 472
purification, **89**, 469
Aldehyde reductase, *see also* Alcohol dehydrogenase ($NADP^+$); Glucuronate reductase
human brain
assay, **89**, 182
properties, **89**, 184
purification, **89**, 183
Aldehydes
microdetermination with membrane-bound dehydrogenase, **89**, 27
protein labeling by reductive alkylation, **91**, 575
ω-Aldehydoalkyl 1-thioglycosides
in neoglycoprotein preparation, **83**, 289
Alditols, oligosaccharide, *see* Oligosaccharide alditols
Aldolase, *see* Fructose-bisphosphate aldolase
D-*xylo*-Aldonate dehydratase
pseudomonad MSU-1
assay, **90**, 302
properties, **90**, 304
purification, **90**, 303
Aldononitriles
enriched with carbon isotopes, chemical synthesis, **89**, 64
Aldose phosphate esters
enriched with carbon isotopes, enzymatic synthesis, **89**, 83
Aldose phosphates
enriched with carbon isotopes
chemical synthesis, **89**, 73
epimeric mixtures, separation, **89**, 76
Aldose reductase, *see* Aldehyde reductase
Aldoses
enriched with
carbon isotopes
chemical synthesis, **89**, 64
enzymatic synthesis, **89**, 83
hydrogen and oxygen isotopes, chemical synthesis, **89**, 79
molybdate-catalyzed interconversion, **89**, 72
Aleuria aurantia
lectin purification, **83**, 375
Algae
blue-green, phosphoglycerate kinase purification, **90**, 111
Alkaline phosphatase
functional, recovery from SDS gels, **91**, 274
linked immunoassay for prostaglandin $F_{2\alpha}$, **86**, 272
Alkylamines
weak-base
cellular uptake, **96**, 738
entry into lysosomes, **96**, 742
inhibitory effects on lysosomes, **96**, 743
Almond emulsin
α-L-fucosidase isolation, **83**, 625
Altronate dehydratase
Escherichia coli K12
assay, **90**, 288
properties, **90**, 291
purification, **90**, 289

D-Altronate-NAD$^+$ dehydrogenase, *see* Tagaturonate reductase
Amido black
procurement, **91**, 239
–protein complexes in gels
absorption spectra, **91**, 245
metachromasia, **91**, 245
photography, **91**, 246
scanning, **91**, 246
protein staining, **91**, 237
Amine oxidase (copper-containing)
in inhibition assay for methylglyoxal bis(guanylhydrazone), **94**, 250
Amine oxidase (flavin-containing)
bovine plasma
assays, **94**, 315
–polyamine system, structure–activity relationship, **94**, 416
properties, **94**, 316
purification, **94**, 315
Amines
weak-base
cellular uptake, **96**, 738
entry into lysosomes, **96**, 742
inhibitory effects on lysosomes, **96**, 743
Amino acid analogs
effects on protein processing, **96**, 783
incorporation into proteins, **96**, 778
Amino acid analyzers
with automated integrators, buffer system preparation, **91**, 3
for methylated amino acids in proteins, **91**, 8
Amino acid composition
comparison indexes
definitions, **91**, 61
effect of experimental error, **91**, 64
theoretical properties, **91**, 64
proteins eluted from polyacrylamide gels, determination, **91**, 36
in testing protein relatedness, **91**, 65
combination with sequence data, **91**, 73
phylogenetic tree construction, **91**, 72
Amino acids, *see also specific amino acids*
amino-terminal, determination with dimethylaminoazobenzene isothiocyanate, **91**, 79
analysis in picomole range by derivatization with
dimethylaminoazobenzenesulfonyl chloride, **91**, 43
o-phthalaldehyde, **91**, 46
HPLC, **91**, 44
biosynthetic interconversions, **91**, 420
methylated, resolution, **91**, 8
protein, sequence homologies
alignment scores: ALIGN program, **91**, 537
scoring matrices, **91**, 544
segment comparison scores, RELATE program, **91**, 540
selection by
computer database, **91**, 528
printed tabulations, **91**, 527
separation and purification by countercurrent chromatography, **91**, 341
single hydrolysis, **91**, 26
Amino acid sequencing
bacteriorhodopsin, **88**, 56
peptides
with commercial spinning-cup sequenator, **91**, 434, 443
with gas-phase sequenator, **91**, 399
at nanomole level, **91**, 470
polypeptides
with dipeptidyl peptidases, **91**, 511
manually with dimethylaminoazobenzene isothiocyanate, **91**, 455
proteins
with commercial spinning-cup sequenator, **91**, 434, 443
with gas-phase sequenator, **91**, 399
in microgram quantities, preparative isolation from polyacrylamide gels, **91**, 227
recovered from SDS gels, **91**, 450
radiochemical
advantages and disadvantages, **91**, 433
biosynthetically labeled proteins, **91**, 413
side chain-protected phenylthiohydantoin decapeptides, **91**, 500
2-Aminoethylalkylenediamine
derivative synthesis, **94**, 415

Amino groups
peptide, amidination, **91**, 559
reversal, **91**, 566
protein
amidination, **91**, 559
reversal, **91**, 566
reductive alkylation, **91**, 570
specific crosslinking reagents, **91**, 591
Aminopeptidase A
rabbit intestinal brush-border membrane
mode of integration, relation to biosynthesis, **96**, 417
structure, relation to mode of integration, **96**, 407
Aminopeptidase, microsomal, *see* Microsomal aminopeptidase
Aminophenyl thioether paper
preparation, **96**, 207
protein electrophoretic transfer, **96**, 209
Ammonyx LO
in extraction of visual pigments, **81**, 133
Amoeba proteus
nonmuscle myosin aggregate formation and assay, **85**, 38
AMP, *see* Adenosine 5′-monophosphate
Ampholytes
carrier, *see* Carrier ampholytes
Amphomycin
inhibitory effects on glycoprotein lipid-linked saccharide formation, **98**, 144
Amylose
chlorite-activated, inhibitory effects on lysosomes, **96**, 763
Anaphylaxis
determination of murine monoclonal antibodies of IgE class, **92**, 58
slow-reacting substance
guinea pig lung
extraction, **86**, 427
preparation, **86**, 426
purification, **86**, 428
structural analysis, **86**, 431
rat
arylsulfatase effects, **86**, 20
purification, **86**, 22
structural analysis, **86**, 22
Anazolene sodium
procurement, **91**, 239
–protein complexes in gels
absorption spectra, **91**, 245
metachromasia, **91**, 245
photography, **91**, 246
scanning, **91**, 246
in protein quantitation method, **91**, 108
staining, for protein resolution, **91**, 211, 237
Anhydrotrypsin
as affinity adsorbent for tryptic peptides, **91**, 381
immobilization on Sepharose, **91**, 381
preparation and purification, **91**, 379
Anileridine
radioimmunoassay, **84**, 516
applications, **84**, 523
Ankyrin
human erythrocyte
binding assay, **96**, 324
purification, **96**, 318
radiolabeling, **96**, 322
spectrin-binding site, purification, **96**, 320
Antibiotic 24010
inhibitory effects on glycoprotein lipid-linked saccharide formation, **98**, 138
Antibiotic MM 19290
inhibitory effects on glycoprotein lipid-linked saccharide formation, **98**, 138
Antibodies, *see also* Antiserum
S-adenosylmethionine decarboxylase, preparation, **94**, 238
alkaline phosphatase-conjugated, preparation, **84**, 194
bacterial lipopolysaccharide, preparation, **84**, 267
bleomycin, production, **84**, 463
n-butylbiguanide
production, **84**, 580
specificity, **84**, 582
calmodulin, purification, **84**, 142
cathepsin D, preparation, **96**, 770
cholera toxin
purification, **84**, 239, 241
radioimmunoassay, **84**, 251
radioiodination, **84**, 241

collagen
  applications, **82**, 487
  characterization, **82**, 479
  preparation, **82**, 472
  purification, **82**, 476
complement component
  assays, **93**, 415
  isolation, **93**, 412
  preparation, **93**, 411
conformation-specific, to vitamin K-dependent blood coagulation proteins
  interaction with protein antigens, **84**, 76
  preparation, **84**, 64, 68
  in radioimmunoassay, **84**, 76
conventional, production and assessment, **93**, 3
cotinine, production, **84**, 633
coupled to ricin A chain
  preparation, **93**, 334
  structural and functional analysis, **93**, 336
desmosine, preparation, **82**, 759
digoxin
  affinity assessment, **84**, 573
  detection, **84**, 567
  preparation, **84**, 564
  specificity determination, **84**, 571
5α,7α-dihydroxy-11-ketotetranorprostane-1,16-dioic acid
  production, **86**, 343
  properties, **86**, 344
3,3′-diiodothyronine, preparation, **84**, 280
DNA
  native and denatured, ELISA, **84**, 194
  passive hemagglutination and hemolysis tests, **84**, 201
    applications, **84**, 208
    characteristics, **84**, 207
DNA–RNA hybrid, radioimmunoassay, **84**, 210
ecdysone
  incubation and separation, **84**, 356
  preparation, **84**, 354
  specificity analysis, **84**, 355
Escherichia coli enterotoxin
  purification, **84**, 239, 241
  radioimmunoassay, **84**, 251
  in Fc receptor detection, **93**, 141
ferritin
  purification, **84**, 155
  radioiodination, **85**, 157
α-fetoprotein, preparation, **84**, 4
fibrinopeptide
  detection, **84**, 109
  specificity determination, **84**, 111
fibronectin, preparation, **82**, 816
β-glucuronidase, preparation, **96**, 770
haloperidol
  production, **84**, 535
  specificity and sensitivity, **84**, 536
hapten
  affinity and specificity
    determinants, **92**, 598
    determination by competitive radioimmunoassay, **92**, 590
  radioactivity, determination by competitive radioimmunoassay, **92**, 598
hepatocyte surface, effect on collagen adhesion, **82**, 525
12-hydroxyeicosatetraenoic acid
  preparation, **86**, 247
  serological specificity, **86**, 249
immunoassays, **92**, 377
  data analysis, **92**, 543
insoluble membrane protein
  immunoprecipitation, **97**, 310
  preparation, **97**, 306
  specificity test, **97**, 308
  titer determination, **97**, 309
insulin
  labeling with β-D-galactosidase, **92**, 357
  separation in enzyme immunoassay with activated thiol–Sepharose, **92**, 353
iodothyronine, preparation, **84**, 280
15-keto-13,14-dihydro-prostaglandin $F_{2\alpha}$
  preparation, **84**, 322
  properties, **84**, 323
labeling, **92**, 257
–lectin conjugates, preparation, **92**, 492
linked cytotoxic agents
  preparation
    modification procedures for disulfide, amide, and diazo linkage formation, **93**, 323

strategy, **93**, 281
structural considerations, **93**, 282, 285
techniques, **93**, 288
comparison, **93**, 331
via intermediaries and spacers, **93**, 320
purification and characterization, **93**, 328
metallothionein
detection, **84**, 130
production, **84**, 128
methotrexate, coupling to magnetizable particles, **84**, 430
monoclonal, *see* Monoclonal antibodies
multivalent, with dual specificity
applications
cell-surface antigen detection, **92**, 534
cytotoxic lymphocyte coating, **92**, 538
preparation, **92**, 523
properties, **92**, 528
rosette formation, **92**, 530
myelin-specific protein, preparation, **96**, 385
neocarzinostatin, preparation, **84**, 484
nicotine, production, **84**, 633
to *N*′-nitrosonornicotine and *N*′-acylnornicotine analogs, production, **84**, 642
5′-nucleotidase
binding assays, **98**, 400
labeling, **98**, 398
phosphatidylcholine transfer protein, preparation, **98**, 589
pituitary hormone, production by lymph node inoculation, **93**, 3
procollagen
applications, **82**, 487
characterization, **82**, 479
preparation, **82**, 472
purification, **82**, 476
production in ascitic fluids of mice and guinea pigs, **93**, 12
to pyrazolone derivatives, production, **84**, 597
γ-(3-pyridyl)-γ-oxo-*N*-methylbutyramide, production, **84**, 633
quantitation and titration with lectin–antibody conjugates, **92**, 489
retinal rod outer segment large protein
affinity purification, **81**, 244
assay, **81**, 245
preparation, **81**, 240
rhodopsin
affinity purification, **81**, 244
assay, **81**, 245
preparation, **81**, 240
ribophorin, preparation, **96**, 525
separated by gel electrophoresis, specificity analysis with enzyme-linked immunoelectrotransfer blot technique, **92**, 377
application notes, **92**, 387
sources of error, **92**, 388
separation in immunoassay, **92**, 336
trinitrotoluene, preparation, **92**, 429
tropoelastin, purification, **82**, 763
tumor-associated antigen, preparation, **93**, 325
Antibody responses
to polyvalent vaccines, **93**, 76
to vaccines and other immune stimuli
distribution testing and transformations to normal, **93**, 61
graphical methods for analysis, **93**, 71
mean antibody concentrations before and after immunization, **93**, 65
proportion of individuals achieving defined outcome, **93**, 70
Antigens
attachment to liposomes
by covalent coupling, **93**, 91
via inserted receptor, **93**, 92
in bacteriophage plaques
production, influencing factors, **100**, 354
screening, chromogenic technique, **100**, 342
influencing factors, **100**, 364
carcinoembryonic
antiserum preparation, **84**, 40
chemical and physical properties, **84**, 39
purification, **84**, 33

radioimmunoassay, **84**, 43
radioiodination, **84**, 42
cell-surface
detection by multivalent antibodies with dual specificity, **92**, 534
monoclonal antibodies
immunoglobulin isotype determination, **92**, 166
indirect assay with ^{125}I-labeled protein A, **92**, 160
quantitation by antibody–complement-mediated cytotoxicity assay, **93**, 253
coated plastic tubes and radiolabeled or enzyme-labeled protein A in
enzyme immunoassays, **92**, 411
hapten radioimmunoassay, **92**, 409
immunoglobulin radioimmunoassay, **92**, 404
entrapment in liposomes, **93**, 88
entrapped and exposed on liposome surface, immune responses to, **93**, 85
factor VIII, immunoradiometric assays, **84**, 51
guinea pig Ia, monoclonal antibodies
epitope binding assay, **92**, 77
in fluorescence-activated cell sorter analysis, **92**, 81
immunoprecipitation with, **92**, 72
production, **92**, 68
in T-lymphocyte proliferation inhibition, **92**, 82
high-resolution two-dimensional gel electrophoresis, **92**, 196
pattern interpretation, **92**, 202
histocompatibility, assembly, **96**, 325
HLA, monoclonal antibodies to products
characterization, **92**, 125
coupling to cyanogen bromide–Sepharose, **92**, 129
in immunoaffinity chromatography, **92**, 128
affinity considerations, **92**, 137
production, **92**, 114, 123, 136
purification, **92**, 123
HLA-DR
assembly *in vivo*, **96**, 328
immunoprecipitation, **96**, 328
mRNA isolation, **96**, 326
immunoassays, **92**, 377
data analysis, **92**, 543
intracellular, antibody labeling, **96**, 467
labeling, **92**, 257
membrane, two-dimensional immunoelectrophoresis, **96**, 244
murine
H-2
immunoprecipitation, **96**, 327
membrane-spanning segment, size determination, **96**, 330
mRNA isolation, **96**, 326
purification by monoclonal antibody affinity chromatography, **92**, 86
H-2K transplantation, radiosequencing, **91**, 413
Ia, purification by monoclonal antibody affinity chromatography, **92**, 86
nonspecific association with liposome surfaces, **93**, 90
quantitation and titration with lectin–antibody conjugates, **92**, 489
ragweed, polymerization by glutaraldehyde, application in immunotherapy, **93**, 50
separated by gel electrophoresis, specificity analysis with enzyme-linked immunoelectrotransfer blot technique, **92**, 377
application notes, **92**, 387
source of error, **92**, 388
separation in immunoassay, **92**, 336
soluble, monoclonal antibodies, solid-phase immunofluorescence assay, **92**, 227
tumor-associated, polyclonal antiserum preparation with agarose-entrapped immune complexes as immunogen, **93**, 78
Antihemophilic factor
immunoradiometric assay, **84**, 56
Antipain
inhibitory effects on lysosomes, **96**, 759

Antiserum, *see also* Antibodies
1-β-D-arabinofuranosylcytosine, production, **84**, 474
calmodulin, production, **84**, 141
carcinoembryonic antigen, preparation, **84**, 40
chlordiazepoxide, preparation, **84**, 498
chlorinated dibenzo-*p*-dioxin, production and screening, **84**, 623
cholera toxin, preparation, **84**, 238
clonazepam, preparation, **84**, 501
to *Clostridium botulinum* toxins, preparation, **84**, 244
cotinine nucleotide, preparation, **84**, 651
N-demethyldiazepam, production, **84**, 512
desmin
production, **85**, 495
specificity analysis, **85**, 497
diazepam, preparation, **84**, 492
egasyn, preparation, **96**, 566
to erythrocyte band-III proteolytic fragments, production, **96**, 263
Escherichia coli enterotoxin, preparation, **84**, 238
fentanyl, preparation, **84**, 527
ferritin, preparation, **84**, 153
fibrinopeptide
choice for radioimmunoassay, **84**, 113
preparation, **84**, 106
insoluble elastin
cross-reactivity, **82**, 758
preparation, **82**, 747
leukotriene D_4
preparation, **86**, 253
serological specificity, **86**, 254
methotrexate, preparation, **84**, 428
metyrapone
preparation, **84**, 589
specificity analysis, **84**, 591
motilin, production, **84**, 361
myoglobin, preparation, **84**, 173
nicotine nucleotide, preparation, **84**, 651
5′-nucleotidase, preparation, **98**, 396
pimozide, preparation, **84**, 544
platelet factor 4, preparation, **84**, 87
polyclonal, to tumor-associated antigens
preparation with agarose-entrapped immune complexes as immunogen, **93**, 78
specificity analysis, **93**, 81
prostaglandin E
binding properties, **86**, 260, 263
specificity problems, **86**, 258
prostaglandin H synthase
Fab fragment preparation, **86**, 215
IgG isolation, **86**, 215
in immunocytochemistry, **86**, 216
preparation, **86**, 214, 223
proteoglycan, preparation, **83**, 221
serum bile acid
characterization, **84**, 333
production, **84**, 323
soluble elastin
cross-reactivity, **82**, 758
preparation, **82**, 747
specific for modifications in functionally related proteins
cross-reactivity, **98**, 163
production, **98**, 161
β-thromboglobulin, preparation, **84**, 96
thromboxane B_2, preparation, **86**, 289
tropoelastin, preparation, **82**, 762
type-II cAMP-dependent protein kinase regulatory subunit, preparation, **99**, 190
vimentin
production, **85**, 495
specificity analysis, **85**, 497
Aorta
bovine, prostaglandin I_2 synthase purification, **86**, 240
chick
lathyrism induction, **82**, 667
tissue homogenization and tropoelastin isolation, **82**, 652
porcine
copper-deficient
harvesting, **82**, 661
preparation, **82**, 658
lathyrism induction, **82**, 667
vascular tissue, prostacyclin synthase preparation, **86**, 93
D-Apiitol dehydrogenase
micrococcal
assay, **89**, 229

properties, **89**, 231
purification, **89**, 229
α-D-Apiofuranosyl 1-phosphate isomers
properties, **89**, 124
synthesis, **89**, 121
α-Apiose 1,2-cyclic phosphate isomers
properties, **89**, 124
synthesis, **89**, 121
D-Apiose tetraacetate
synthesis, **89**, 122
Apocytochrome *c*
antibody preparation, **97**, 267
binding to mitochondrial receptor sites, **97**, 274
immunoprecipitation, **97**, 267
labeled, preparation by reductive methylation, **97**, 272
preparation, **97**, 266
synthesis in cell-free system, **97**, 269
transfer into mitochondria and conversion to holocytochrome *c*, **97**, 272
Apovitellenin I
goose, cleavage at aspartic acid residue with hydrochloric acid and formic acid, **91**, 324
1-β-D-Arabinofuranosylcytosine, *see* Cytarabine
1-β-D-Arabinofuranosyluracil
radioimmunoassay, **84**, 482
D-Arabinose
containing oxygen isotopes, chemical synthesis, **89**, 82
L-Arabinose
binding protein, *Escherichia coli*
assay, **90**, 464
purification, **90**, 465
Arachidonic acid
deuterated, preparation, **86**, 366
extraction with octadecylsilyl silica, **86**, 467
^{3}H-labeled, preparation, **86**, 101
metabolites
extraction, **86**, 477
with octadecylsilyl silica, **86**, 467
HPLC, **86**, 511
immunochemical assays, **86**, 246
^{18}O-labeled, preparation for gas chromatography–mass spectrometry, **86**, 547
open tubular glass capillary gas chromatography, **86**, 622
TLC, **86**, 484
related enzymes
assays, **86**, 3, 213
purification, **86**, 3
Arachidonic acid 12-lipoxygenase
bovine platelet
assays, **86**, 50
partial purification, **86**, 51
properties, **86**, 53
rabbit peritoneal polymorphonuclear leukocyte
assay, **86**, 46
properties, **86**, 48
purification, **86**, 46
Arginase
deficient mutants, *Neurospora crassa*, in analysis of polyamine metabolism, **94**, 112
Arginine decarboxylase
Escherichia coli
biosynthetic and biodegradative forms
assay, **94**, 125
properties, **94**, 133
purification, **94**, 127
deficient mutants, screening, **94**, 89
gene cloning, **94**, 118
oat seedling
assays, isotopic and spectrophotometric, **94**, 177
properties, **94**, 180
purification, **94**, 178
Arylsulfatase
limpet
effect on spasmogenic mediators, **86**, 20
inactivation of
slow-reacting substance of anaphylaxis, **86**, 20
synthetic leukotrienes, **86**, 28
Arylsulfatase A, *see* Cerebroside-sulfatase
Arylsulfatase B, *see* *N*-Acetylgalactosamine-4-sulfatase
Ascitic fluids
production of antibodies, nonspecific immunoglobulins, and serum proteins, **93**, 12

Asparagine
-linked sugar chains, hydrazinolysis, **83**, 263
Aspartate aminotransferase
sulfhydryl peptide alkylated with *N*-ethylmaleimide, isolation by diagonal electrophoresis, **91**, 394
Aspartate carbamoyltransferase
isotope exchange study, fingerprint kinetic patterns for modifier action, **87**, 651
Aspartic acid residues
protein, cleavage with hydrochloric acid and formic acid, **91**, 324
Aspergillus japonicus
glycerol oxidase purification, **89**, 244
Aspirin
^{3}H-labeled
applications, **86**, 397
assays, **86**, 396
preparation, **86**, 393
ATP, *see* Adenosine 5′-triphosphate
ATPase
Ca^{2+},Mg^{2+}-, pancreatic islet plasma membrane, assay, **98**, 192
dynein-associated, *see* Dynein ATPase
H^{+}-transporting, *see* ATP synthase (H^{+}-transporting)
Mg^{2+}-
light-activated, in amphibian rod outer segments
ADPase activity, **81**, 516
assay, **81**, 515
in vertebrate photoreceptor rod outer segments
assays, **81**, 509
ATP hydrolysis, **81**, 510
structural consequences, **81**, 511
microtubule-associated, assay, **85**, 430
myosin-associated, *see* Myosin ATPase
Na^{+},K^{+}-
avian salt gland
assay, **96**, 638
cytochemical localization with
antibodies, **96**, 646
ouabain, **96**, 643
various substrates, **96**, 640
methionine-containing peptide purification, **91**, 374
rat parotid gland, activity in plasmalemma and secretion granule membranes, **98**, 85
ATP synthase (H^{+}-transporting)
chloroplast
assay, **97**, 513
labeling *in vivo* in *Chlamydomonas reinhardi* cells, **97**, 520
polypeptide electrotransfer and immunodecoration, **97**, 514
purification, **97**, 510
subunit-specific antibodies, preparation, **97**, 515
synthesis assay, **97**, 518
Escherichia coli
DNA isolation and restriction mapping, **97**, 189
structural genes, cloning, **97**, 176
subunit analysis by
amino-terminal sequencing, **97**, 211
DNA sequencing, **97**, 196
in vitro transcription, **97**, 210
synthesis *in vitro*, **97**, 191
yeast mitochondrial
labeling, **97**, 296
subunit 6
immunoprecipitation, **97**, 302
one-dimensional gel electrophoresis, **97**, 297
source, **97**, 304
two-dimensional gel electrophoresis, **97**, 299
Autophosphorylation
casein kinase II, **99**, 325
cyclic nucleotide-dependent protein kinase, effects of poly(L-arginine) and core histones, **99**, 210
type-II cAMP-dependent protein kinase, **99**, 177
application to study of enzyme structure and regulation, **99**, 185
intra- and intermolecular, distinction, **99**, 179
reversal, **99**, 183
Autoradiography
erythrocyte band-III proteolytic fragments, **96**, 262
membrane proteins in polyacrylamide gels, **96**, 190

phosphorylated proteins in intact cells, **99**, 34
prostaglandins and related compounds, **86**, 498
in protein resolution, **91**, 213
Avian myeloblastosis virus
major envelope glycoprotein
isolation, **83**, 321
properties, **83**, 323
L-Azaserine
^{14}C-labeled, reaction with formylglycinamide ribonucleotide amidotransferase, **87**, 78
Azidoarylisothiocyanates
hydrophobic, bacteriorhodopsin heterobifunctional crosslinking, **88**, 207
p-Azidobenzoyl-CoA
acyl CoA-binding site photoaffinity labeling, **91**, 639
synthesis, **91**, 635
S-[12-*N*-(4-Azido-2-nitrophenyl)aminododecanoyl]-CoA
acyl CoA-binding site photoaffinity labeling, **91**, 639
synthesis, **91**, 635
S-[13-*N*-(4-Azido-2-nitrophenyl)amino-2-oxotridecyl]-CoA
acyl CoA-binding site photoaffinity labeling, **91**, 639
synthesis, **91**, 635
8-Azidopurine nucleotide cyclic monophosphates
nucleotide-binding protein
cellular localization, **91**, 629
photoaffinity labeling, **91**, 619
storage, **91**, 613
8-Azidopurine nucleotide triphosphates
nucleotide-binding protein
cellular localization, **91**, 629
photoaffinity labeling, **91**, 622
storage, **91**, 613

B

Bacillus
B. caldolyticus, ribose-5-phosphate isomerase purification, **89**, 574
B. licheniformis, purification of
fructose-bisphosphatase, **90**, 385
phosphofructokinase, **90**, 72
B. pumilus, β-D-xylosidase purification, **83**, 633
B. stearothermophilus, purification of
acetate kinase, **90**, 180
6-phospho-D-gluconate dehydrogenase, **89**, 283
phosphoglycerate kinase, **90**, 127
pyruvate dehydrogenase complex, **89**, 402
B. subtilis
cultivation and harvest, **89**, 258
protein-synthesizing systems
for DNA-dependent protein synthesis, **101**, 602
preparation, **101**, 599
for RNA-dependent protein synthesis, **101**, 601
purification of
fructose-bisphosphate aldolase, **90**, 237
pyruvate dehydrogenase complex, **89**, 402
vegetative and spore glucose-6-phosphate dehydrogenases, **89**, 259
B. symbiosus, pyruvate,orthophosphate dikinase purification, **87**, 55
Bacitracin
inhibition of glycoprotein lipid-linked saccharide formation, **98**, 146
Bacteria, *see also specific bacteria*
DNA high-efficiency transfer into eukaryotic cells by protoplast fusion, **101**, 402
methanogenic, electron transfer components, EPR detection, **88**, 412
Bacterio-opsin
conversion *in vivo* to bacteriorhodopsin, **88**, 394
crystalline white membranes, isolation from *Halobacterium halobium* R_1mW, **88**, 407
denaturation and refolding, **88**, 81
Bacteriophages
f1, coat protein insertion in bacterial membranes, **97**, 139, 143
lambda
DNA terminase
assay, **100**, 183
host factors, **100**, 189

properties, **100**, 189
purification, **100**, 185
Int protein
assay, **100**, 210
properties, **100**, 215
purification, **100**, 212
LamB receptor protein from *Escherichia coli*
assay, **97**, 101
exclusion limit evaluation, **97**, 91
interactions with phages, **97**, 107
penetration rate evaluation, **97**, 94
properties, **97**, 86, 99, 104
purification, **97**, 89, 103
reconstitution into planar films, **97**, 98
role in maltose and maltodextrin transport, **97**, 109
variants, **97**, 110
vectors with positive selection for cloned inserts
identification, **101**, 17
preparation, **101**, 9
structure, **101**, 7
M13
cloned DNA fragments, oligonucleotide-directed mutagenesis, **100**, 468
screening, **100**, 483
target specificity, preliminary *in vitro* tests, **100**, 477
cloning into, integrated and simplified approach, **101**, 78
coat protein
assembly in
Escherichia coli inner membrane, **97**, 130
intact cells, pulse labeling, **97**, 57
precursor isolation, **97**, 137
secretion in intact cells, pulse labeling, **97**, 57
transmembrane insertion criteria, **97**, 131
kilo-sequencing, **101**, 98
life cycle and cloning applications, **101**, 20
M13mp7 DNA, cutting and excision of cloned single-stranded sequences by restriction endonucleases, **101**, 90
vectors
cloning, **101**, 43
controls, **101**, 42
strategies, **101**, 30, 40
constant and variable primers, **101**, 25
preparation, **101**, 29, 38
recombinant identification, **101**, 31, 53
transformation, **101**, 51
Mu
localized mutagenesis of *Escherichia coli* chromosome
deletion methods for
speA, *speB*, and *speC*, **94**, 102
speD, **94**, 101
insertion methods, **94**, 95
Mud(Ap, *lac*)
generated *lac* fusions, stabilization, **100**, 505
insertion mapping, **100**, 504
isolation of regulatory mutants, **100**, 507
lysogen isolation, **100**, 504
preparation, **100**, 503
preparation, **94**, 94
ϕW-14, DNA
digestion to mononucleotides, **94**, 426
isolation of
α-putrescinyldeoxythymidine, **94**, 426
α-putrescinylthymine, **94**, 423
ϕX174, gene *A* protein
analysis with *in vitro* DNA replication systems, **100**, 219
properties, **100**, 217
purification, **100**, 223
recombinant plasmid construction, **100**, 230
plaques
antigen production, influencing factors, **100**, 354

chromogenic screening for specific antigen, **100**, 342
influencing factors, **100**, 364
T4, DNA topoisomerase II
assay, **100**, 150
comparison with other type-II topoisomerases, **100**, 159
properties, **100**, 154
purification, **100**, 145
T7, RNA polymerase
assay, **101**, 559
purification, **101**, 556
in *in vitro* RNA synthesis, **101**, 567
Bacteriorhodopsin
absorption spectrum, pressure effect, **88**, 719
biogenesis, genetics, and sources, **88**, 360
chemistry, spectroscopy, and photochemistry, **88**, 133
–chromophore interactions as spectroscopic and photochemical determinants, **88**, 462
-containing planar lipid membranes, assembly by sequential transfer of air–water interface films, **88**, 45
crystal forms
three-dimensional, preparation, **88**, 114
two-dimensional, preparation, **88**, 112
delipidation and renaturation, **88**, 74
deuterated, isolation from fully deuterated *Halobacterium halobium*, **88**, 3
discrimination from halorhodopsin
in cell suspensions by measuring light-induced pH changes, **88**, 388
by flash spectroscopy, **88**, 434
emission spectroscopy
low-temperature, **88**, 218
room-temperature, **88**, 221
strongly allowed transition, **88**, 225
temperature dependence, **88**, 227
energy storage mechanisms, **88**, 466
folding in membrane, **88**, 460
gene isolation by colony hybridization, **97**, 226
α-helix orientation in membrane, measurement by polarized infrared spectroscopy, **88**, 707
heterobifunctional crosslinking by hydrophobic azidoarylisothiocyanates, **88**, 207
incorporation into collodion film, **88**, 37
ion transport and physiology, **88**, 334
13-*cis* and *trans* isomers
assay, **88**, 173
flash spectroscopy, **88**, 170
preparation, **88**, 168
isotopic analogs, resonance Raman spectroscopy, **88**, 633
–lipid interactions, analysis with diffusible spin labels, **88**, 762
molecular structure, **88**, 56, 447
comparison with rhodopsin, **88**, 458
modeling techniques
AIMS approach, **88**, 775
classification, **88**, 773
first-guess, selection, **88**, 776
refinement method
graphic demonstration, **88**, 783
mathematical formulation, **88**, 778
X-ray diffraction analysis, **88**, 124
monomeric
endogenous phospholipids, removal, **88**, 9
preparation, **88**, 6
detergent choice and removal, **88**, 9
properties, **88**, 8
reconstituted into phospholipid vesicles
preparation, **88**, 31
properties and applications, **88**, 33
neutron scattering, **88**, 241
photocycle
comparison with rhodopsin, **88**, 470
later stages, **88**, 484
primary events
kinetic aspects, **88**, 470
molecular mechanisms, **88**, 476
viscosity effect, **88**, 297

photocycle intermediates
absorption spectra
low-temperature spectrophotometry, **88**, 163
pressure effect, **88**, 719
high-field solid-state NMR, **88**, 310
kinetic infrared spectroscopy, **88**, 729
kinetic resonance Raman spectroscopy with rotating sample, **88**, 643, 659
monitoring, **88**, 265
nanosecond X-ray diffraction, **88**, 276
rapid kinetic optical absorption spectroscopy, **88**, 133
time-resolved protein fluorescence, **88**, 306, 617
time-resolved resonance Raman spectroscopy, **88**, 617
X-ray diffraction, **88**, 282
photoelectric response, measurement in
collodion film, **88**, 37
natural membranes, **88**, 38
polypeptide chain
chemical modification, position determination, **88**, 89
helical segments, identification
modification systems for, **88**, 84
by proteolytic cleavage, **88**, 83
structural constraints, **88**, 82
by transfer free energy calculations, **88**, 85
primary photoproduct, resonance Raman spectroscopy with spinning sample at 77°K, **88**, 648
protonation state during photocycle, spectroscopic determination, **88**, 254
reconstitution
in *Halobacterium halobium*
apomembranes and white membranes, **88**, 16
intact cells, **88**, 14
retinal-deficient cells, **88**, 355
in vesicles, **88**, 14, 17, 78
resonance Raman spectroscopy, **88**, 561
retinal in
extraction techniques, **88**, 553
heavy atom labeling, **88**, 178
rotational motions, measurement, photodichroism relaxation techniques, **88**, 741
sequencing, **88**, 56
spin-labeled
preparation, **88**, 176
properties, **88**, 176
synthetic analogs
functionality tests, **88**, 520
preparation, **88**, 496, 516
precautions, **88**, 500
in thin layers, orientation and photocycle kinetics, **88**, 180
transient dichroism
electric field-induced, **88**, 253
measurement, **88**, 249
in purple membranes, **88**, 251
in reconstituted membranes, **88**, 252
Barley seeds
germinated, purification of
agmatine coumaroyltransferase, **94**, 346
ornithine decarboxylase, **94**, 163
Base hydrolysis
protein phosphotyrosines, **99**, 402
Bean sprouts
purification of
mitochondrial porin, **97**, 90
pyrophosphate–fructose-6-phosphate 1-phosphotransferase, **90**, 94
Benzodiazepines
radioimmunoassay, **84**, 490
Benzoyl chloride
prechromatographic derivatization of polyamines, **94**, 22
BHK-21 cells (baby hamster kidney)
Semliki Forest virus
binding, **98**, 261
degradation, **98**, 263
endocytosis, **98**, 262
nucleocapsid uncoating, **98**, 264
spike glycoprotein transport, immunoelectron microscopy, **96**, 466
Bile acids
serum, *see* Serum bile acids
Birefringence
and birefringence gradients in rod outer segments, measurement, **81**, 655

Bis(3-aminopropyl)amine
preparation, **94**, 414
N,N'-Bis(3-aminopropyl)pentane-1,5-diamine
preparation, **94**, 414
N,N'-Bis(2-cyanoethyl)decane-1,10-diamine
preparation, **94**, 413
N,N'-Bis(2-cyanoethyl)pentane-1,5-diamine
preparation, **94**, 413
Bis(2-ethylhexyl) phthalate
induction of mitochondria, **96**, 557
N^1,N^4-Bis(γ-glutamyl)putrescine
ion-exchange chromatography, **94**, 456
synthesis, **94**, 455
N^1,N^8-Bis(γ-glutamyl)spermidine
ion-exchange chromatography, **94**, 456
synthesis, **94**, 451
N^1,N^{12}-Bis(γ-glutamyl)spermine
ion-exchange chromatography, **94**, 456
synthesis, **94**, 451
2,3-Bisphosphoglycerate
inhibition of casein kinases, **99**, 314, 327
Bisphosphoglycerate mutase
phosphorylated intermediates
kinetic competence, **87**, 47
phosphoprotein bond characterization, **87**, 46
preparation and assay, **87**, 50
primary structure at active site, homology with phosphoglycerate mutase, **87**, 49
role in enzyme reaction, **87**, 43
Bisphosphoglycerate synthase, *see* Bisphosphoglycerate mutase
Bleomycin
antibody production, **84**, 463
iodination, **84**, 464
radioimmunoassay, **84**, 465
clinical applications, **84**, 469
validation, **84**, 465
Blood
coagulation proteins
and peptides, immunoassays, **84**, 51
vitamin K-dependent
antibodies
interaction with protein antigens, **84**, 76
preparation, **84**, 64
radioimmunoassay, **84**, 75
prostaglandin-F major metabolite, radioimmunoassay, **86**, 339
whole peripheral, lymphocyte population enumeration with alkaline phosphatase-labeled reagents, **93**, 164
Bones
murine femur, macrophage preparation, **98**, 301
Borane dimethylamine
reduction of retinal–opsin linkage in rhodopsin, **81**, 290
Bordetella pertussis
vaccine preparation with glutaraldehyde, **93**, 41
Bovine serum albumin
–mannose, internalization by macrophages, **98**, 304
tissue embedding, **96**, 486
advantages, **96**, 492
Brain
bovine, purification of
clathrin, **98**, 359
inositol epimerase–inosose reductase, **89**, 597
microtubule-associated protein, **85**, 409
putrescine aminopropyltransferase, **94**, 273
spermidine aminopropyltransferase, **94**, 277
tubulin by
assembly and disassembly cycles with chromatography on phosphocellulose, **85**, 376
modified Weisenberg procedure, **85**, 381, 394
type-II cAMP-dependent protein kinase and regulatory subunit, **99**, 189
canine, creatine kinase isozyme purification, **90**, 195
chick, nonmuscle myosin aggregate formation and assay, **85**, 40
human, purification of
aldose reductase, **89**, 183
creatine kinase isozymes, **90**, 187
glyceraldehyde-3-phosphate dehydrogenase, **89**, 303

porcine, purification of
actin, **85**, 371
prostaglandin-D 15-dehydrogenase (NADP$^+$), **86**, 149
rat
eicosanoid-metabolizing enzymes, inactivation by microwave irradiation, **86**, 635
preparation of
synaptic plasma membranes, **98**, 604
synaptosomes, **98**, 603
purification of
Ca^{2+}-activated phospholipid-dependent protein kinase, **99**, 292
glucose-6-phosphatase, **90**, 399
prostaglandin-H_2 D-isomerase, **86**, 74
stem subcellular fractions
preparation, **96**, 379
purification of myelin-specific proteins, **96**, 382
RNA extraction, **96**, 382
Broccoli
mitochondrial pyruvate dehydrogenase complex purification, **89**, 411
N-Bromosuccinimide
reactivity with rhodopsin tryptophan residues, assay, **81**, 285
Brown membranes
Halobacterium halobium
bacterio-opsin
conversion *in vivo* to bacteriorhodopsin, **88**, 394
pulse labeling with radioactive amino acids, **88**, 395
preparation, **88**, 268, 391
properties, **88**, 393
purple membrane formation, reconstitution, **88**, 394
retinal-free, preparation, **88**, 393
X-ray diffraction analysis, **88**, 268
Buffer systems
for amino acid analyzers with automated integrators, **91**, 3
Buformin
antibodies
production, **84**, 580
specificity, **84**, 582
radioimmunoassay, **84**, 577
characterization, **84**, 583
n-Butylbiguanide, *see* Buformin

C

Calcium
release from
binding sites in isolated rod outer segments after rhodopsin photoexcitation, **81**, 570
detection, **81**, 571
sensitivity and time resolution, **81**, 574
sonicated bovine disk vesicles
ionophore-induced, **81**, 581
light-induced, **81**, 577
stoichiometry, measurement, **81**, 580
uptake by pancreatic islet endoplasmic reticulum, **98**, 194
Caldariella acidophila
5′-methylthioadenosine phosphorylase purification, **94**, 357
Calmodulin
antiserum production and antibody purification, **84**, 141
purification from
bovine and ovine testes, **99**, 280
rat testis, **84**, 139
radioimmunoassay, **84**, 143
radioiodination, **84**, 140
Calorimetry
light-induced processes, **88**, 667
applications, **88**, 672
microtubules during assembly *in vitro*, **85**, 438
Candida boidinii
alcohol oxidase
production, **89**, 425
purification, **89**, 426
formaldehyde dehydrogenase
production, **89**, 528
purification, **89**, 528
Carbamoyl-phosphate synthase (ammonia)
mitochondrial
intracellular transport, **97**, 402
precursor posttranslational uptake and processing *in vitro*, **97**, 405
synthesis, **97**, 399

N-Carbamoylputrescine
large-scale preparation and purification, **94**, 429
radiolabeled, chemical synthesis and purification, **94**, 430
Carbocyclic thromboxane A_2
biological properties, **86**, 403
synthesis, **86**, 403
Carbohydrates
analytical methods, **89**, 3
binding proteins
glycolipid ligand detection by protein binding to thin-layer chromatograms, **83**, 235
in isolation of specific sugar sequences, **83**, 241
chemical and enzymatic synthesis, **89**, 59
complex, *see* Complex carbohydrates
epimerases, isomerases, and mutases, **89**, 547
oxidation–reduction enzymes, **89**, 129
polymeric, quantification by partition affinity ligand assay, **92**, 506
rhodopsin
acetolysis, **81**, 222
composition, **81**, 217
hydrolysis by endo-β-*N*-acetylglucosaminidase, **81**, 217
mass spectrometry, **81**, 219
oligosaccharide structure, **81**, 218, 223
Carbonate dehydratase
catalyzed CO_2 hydration, ^{18}O exchange analysis, **87**, 746
kinetic expressions, **87**, 739
Carbon dioxide
and water, ^{18}O exchange
carbonate dehydratase-catalyzed, **87**, 746
kinetic expressions, **87**, 739
uncatalyzed, **87**, 743
kinetic expressions, **87**, 735
Carbonic anhydrase, *see* Carbonate dehydratase
Carbon isotopes
enriched aldose phosphates, chemical synthesis, **89**, 73
enriched monosaccharides, chemical synthesis, **89**, 64
Carbonyl reductase (NADPH)
porcine kidney
assay, **86**, 143
properties, **86**, 145
purification, **86**, 144
γ-Carboxyglutamic acid
alkaline hydrolysis, **91**, 14
ion-exchange chromatography, **91**, 15
Carboxylesterase
avian salt gland, assay, **96**, 638
Carboxyl groups
protein, crosslinking reagents, **91**, 601
Carboxylic ionophores
inhibitory effects on lysosomes, **96**, 755
monovalent, Golgi complex perturbation effect on
intracellular protein transport, **98**, 51
morphology, **98**, 49
mechanism, **98**, 57
reversibility, **98**, 56
transport arrest site, **98**, 52
Carrier ampholytes
characterization, **91**, 291
noncommercial, for chromatofocusing, **91**, 299
synthesis for isoelectric focusing
by hydrolysis and catalytic hydrogenation, **91**, 288
manual, **91**, 288
semiautomatic, **91**, 284
Casein kinase
type I, rabbit reticulocyte
analysis with mixed casein and purified variants, **99**, 314
assay, **99**, 309
identification in cell lysates and homogenates, **99**, 316
inhibitors, **99**, 314
purification, **99**, 310
type II, rabbit reticulocyte
activation by polyamines, **99**, 328
assay, **99**, 319
properties, **99**, 320
autophosphorylation, **99**, 325
identification, **99**, 330
in cell lysates and homogenates, **99**, 328
inhibitors, **99**, 326
purification, **99**, 324

Cathepsin D
lysosomal
antibody preparation, **96**, 770
assay, **96**, 767
immunoprecipitation, **96**, 773
purification, **96**, 767
synthesis *in vitro*, **96**, 770
Cauliflower
mitochondrial pyruvate dehydrogenase complex purification, **89**, 409
Cell–cell channels, *see* Gap junctions
Cell cultures
biosynthetic matrices from
characterization, **82**, 532
preparation, **82**, 530
single-cell adhesion, **82**, 533
human endothelial cell
preparation, **92**, 40
supernatant
as hybridoma growth factor, **92**, 42
production and assay, **92**, 41
properties, **92**, 47
insoluble elastin biosynthesis, **82**, 617
protein detection, image intensification techniques, **98**, 266
smooth muscle, preparation, **82**, 617
soluble elastin biosynthesis, **82**, 716
Cell-fusion experiments
monoclonal hybridomas in, number estimation, **92**, 36
Cellobiose oxidase
Sporotrichum pulverulentum
assay, **89**, 129
properties, **89**, 134
purification, **89**, 130
Centrifugation
columns
construction, **91**, 278
salt removal from protein solutions, **91**, 279
gradient
coated vesicles from various tissues, **98**, 321, 341
Golgi complex from rat liver with
D_2O–sucrose, **98**, 62
H_2O–sucrose, **98**, 61
Cerebroside-sulfatase
human placental
effect on
slow-reacting substance, **86**, 20
spasmogenic mediators, **86**, 20
purification, **86**, 18
Cerebroside transfer protein
bovine spleen
assay, **98**, 614
properties, **98**, 620
purification, **98**, 617
Chaos carolinensis
nonmuscle myosin aggregate formation and assay, **85**, 38
Chemotherapeutic agents
antibody conjugates
assay, **93**, 330
linkage
amide, diazo, and disulfide, modification for, **93**, 323
strategy, **93**, 281
techniques, **93**, 288
comparison, **93**, 331
via intermediaries and spacers, **93**, 320
purification and characterization, **93**, 328
structural analysis, **93**, 282, 285
Chinese cabbage
putrescine aminopropyltransferase purification, **94**, 283
Chlordiazepoxide
antiserum preparation, **84**, 498
radioimmunoassay, **84**, 499
radiolabeling, **84**, 499
Chlorella pyrenoidosa
GDP-L-galactose epimerase preparation, **83**, 524
Chloroplasts
cytoplasmically synthesized proteins, *in vitro* reconstitution of
assembly, **97**, 507
synthesis, **97**, 502
uptake, **97**, 503
isolated pea, thylakoid membrane proteins
analysis and assembly assay, **97**, 495
synthesis, **97**, 487
optimum conditions, **97**, 492
maize, isolation of
DNA, **97**, 526
RNA, **97**, 530
proton ATPase
purification, **97**, 510

synthesis assay, **97**, 518
spinach leaf, fructose-bisphosphatase purification, **90**, 374
wheat, sedoheptulose-bisphosphatase purification, **90**, 394
CHO cells (Chinese hamster ovary)
cAMP-resistant mutants
characterization, **99**, 201
isolation, **99**, 199
in protein kinase analysis, **99**, 204
cell-free translation system
characteristics, **101**, 634
preparation, **101**, 631
translation of natural and synthetic mRNA templates, **101**, 633
growth and handling, **99**, 198
lectin-resistant mutants
nomenclature, **96**, 161
selection, **96**, 157
ornithine decarboxylase-deficient mutants, selection, **94**, 108
Cholera toxin
antibodies
purification, **84**, 239, 241
radioimmunoassay, **84**, 251
radioiodination, **84**, 241
solid phase-bound, preparation, **84**, 241
antiserum, preparation, **84**, 238
radioimmunoassay, **84**, 243
radiolabeled
binding to thin-layer chromatograms, glycolipid ligand detection, **83**, 238
preparation, **84**, 240
purification, **84**, 241
vaccine preparation with glutaraldehyde, **93**, 36
Cholesterol
content in synaptic plasma membranes, modification with lipid-transfer protein, **98**, 599
Cholesteryl esters
formation in cultured cells, assay, **98**, 257
Chondrus crispus
hexose oxidase purification, **89**, 147
Chromatography, *see also specific chromatographic techniques*
collagenous proteins on SDS gel, **82**, 420
gangliosides on DEAE-Sephadex column, **83**, 164
native collagens, **82**, 44, 50
nonmuscle myosins on
agarose column, **85**, 341
hydroxyapatite column, **85**, 339
nucleic acids on RPC-5 ANALOG column, **100**, 369
polyamines and histamine on phosphorylated cellulose column, **94**, 43
application to ornithine and histidine decarboxylase simultaneous assay, **94**, 45
proteins on reversed-phase column
applications, **91**, 188
columns
length effect, **91**, 186
testing, **91**, 184
mobile phases, **91**, 181
pore diameter effects, **91**, 183
preparative, **91**, 187
supports, **91**, 180
retinal pigment epithelium on glass bead column, **81**, 90
rhodopsin on
agarose column, **81**, 141
DEAE-cellulose column, **81**, 168
hydroxyapatite column, **81**, 145
rod outer segment intact disks on concanavalin A column, **81**, 61
rod outer segment phospholipids on silicic acid column, **81**, 812
SDS from proteins with ion-retardation resin, **91**, 259
in separation of free and bound fractions in radioimmunoassay, **92**, 366
α- and β-tropomyosins on hydroxyapatite column, **85**, 237
Chromium
–nucleotide complexes
applications, **87**, 180
in determination of enzyme specificity, **87**, 175
interactions with enzymes, EPR and NMR, **87**, 192
preparation, **87**, 163, 170
properties, **87**, 171, 180
purification, **87**, 166
stereochemistry, **87**, 159

radioisotopic, in tumor cell cytotoxicity assay, **93**, 233
Chymostatin
inhibitory effects on lysosomes, **96**, 757
Chymotrypsin
in analysis of
membrane insertion, **96**, 136
protein interaction with organelle membranes, **96**, 143
pulse-labeled spleen cells, **96**, 264
properties, **96**, 137
Cilia
lamellibranch gill, dynein ATPase preparation, **85**, 470
Strongylocentrotus, tubulin isolation, **85**, 388
Tetrahymena
dynein ATPase preparation, **85**, 466
tubulin isolation, **85**, 388
Circular dichroism
low-temperature, rhodopsin intermediates
at −85° or above, **81**, 639
instrumentation, **81**, 634
at liquid nitrogen temperature or below, **81**, 635
muscle protein systems for elucidating structure–function relationships, **85**, 679
problem areas, **85**, 696
units and calculations, **85**, 678
uv, rhodopsin in disk membranes and detergent solution, **81**, 629
Citrate synthase
porcine heart, acyl CoA-binding site photoaffinity labeling, **91**, 639
Clathrin
bovine brain
antibodies
affinity purification, **98**, 362
preparation, **98**, 359
basket assembly conditions, **98**, 357
biophysical properties, **98**, 353
cytochemical localization, **98**, 364
purification, **98**, 351
Clonazepam
antiserum preparation, **84**, 501
radioimmunoassay, **84**, 503
radiolabeled, preparation, **84**, 502
Cloning
adenosinetriphosphate complex structural genes from *Escherichia coli*, **97**, 176
bacterial glycosyltransferase genes
hybrid plasmid selection, **83**, 327
application, **83**, 331
hybrid strain characterization, **83**, 329
with bacteriophage λ vectors with positive selection of inserts, **101**, 3
with bacteriophage M13 vectors, **101**, 20
eukaryotic genes, vector construction from bovine papillomavirus DNA, **101**, 387
genes for polyamine biosynthetic enzymes from *Escherichia coli*, **94**, 117
human–human monoclonal antibody-producing hybridomas, **92**, 14
maize plastid genes, **97**, 535
mammalian mitochondrial genome, **97**, 426
mRNA 5′-terminal sequences: ds-cDNA synthesis with addition of dCMP tails, **100**, 285
Neurospora crassa middle-abundant mRNAs, **97**, 254
with pKC30 and derivatives for controlled gene expression, **101**, 123
into plants with *Agrobacterium* Ti plasmids as vectors, **101**, 527
into plasmids and single-stranded phages, integrated and simplified approach, **101**, 78
proton translocating ATPase from *Escherichia coli*, **97**, 188
with *trp* promoters from *Escherichia coli* and *Serratia marcescens* for regulating gene expression, **101**, 155
yeast
with *lac*Z fusions and yeast promoters for analysis of gene expression, **101**, 181
mutant genes and large segments of contiguous chromosomal sequences by integrative transformation, **101**, 290
STE genes in 2-μm vectors, **101**, 325

Clostridium
- *C. botulinum*, toxins A, B, and E, ELISA, **84**, 223
  - sensitivity, cross-reactivity, and practical application, **84**, 229
- *C. pasteurianum*, purification of
  - gluconate dehydratase, **90**, 284
  - phosphofructokinase, **90**, 84

Clostridium histolyticum collagenase
- assays, **82**, 456
- commercial preparation, purity tests, **82**, 465
- contamination assay, **82**, 463
- heterogeneity with regard to substrate, **82**, 454
- properties, **82**, 453
- protease-free, applications, **82**, 468
- purification, **82**, 461

CMP-*N*-acetylneuraminate:galactoseβ1-4-*N*-acetyl-D-glucosamine α2-3-sialyltransferase
- assay, **98**, 129

CMP-*N*-acetylneuraminate:galactoseβ1-4-*N*-acetyl-D-glucosamine α2-6-sialyltransferase
- assay, **98**, 127
- properties, **98**, 129
- purification from bovine colostrum, **98**, 128

Coat proteins
- bacteriophage f1, insertion in bacterial membranes, **97**, 139, 143
- bacteriophage M13
  - assembly in
    - *Escherichia coli* inner membrane, **97**, 130
    - intact cells, pulse labeling, **97**, 57
  - precursor isolation, **97**, 137
  - secretion in intact cells, pulse labeling, **97**, 57
  - transmembrane insertion criteria, **97**, 131

Cobalt
- –nucleotide complexes
  - applications, **87**, 180
  - in determination of enzyme specificity, **87**, 175
  - interactions with enzymes, EPR and NMR, **87**, 184
  - preparation, **87**, 163, 170
  - properties, **87**, 171, 180
  - purification, **87**, 166
  - stereochemistry, **87**, 159
- –phenanthroline complexes, in myosin active-site trapping of metal nucleotides, **85**, 109

Colchicine
- effect on secretory processes in
  - lacrimal and salivary glands
    - evaluation, **98**, 180
    - localization, **98**, 178
    - specificity, **98**, 180
  - liver
    - evaluation, **98**, 169
    - localization, **98**, 171
    - specificity, **98**, 173
- inhibitory effects on lysosomes, **96**, 762

Collagen
- antibodies
  - applications, **82**, 487
  - assays, **82**, 479
  - preparation, **82**, 472
  - purification, **82**, 476
  - specificity, **82**, 484
- chemistry, special features, **82**, 339
- CM-cellulose chromatography, **82**, 45, 55
- covalent crosslinks
  - mature
    - characterization, **82**, 370
    - isolation, **82**, 368, 370
  - reduced
    - identification, **82**, 364, 369
    - isolation, **82**, 361
    - preparation, **82**, 361, 369
    - quantitation, **82**, 368
- cuticle, *see* Cuticle collagen
- DEAE-cellulose chromatography, **82**, 49
- fibrous forms
  - ^{13}C- and ^{2}H-labeled, preparation, **82**, 177
  - characterization, **82**, 127
  - electron microscopy, **82**, 131
  - light microscopy with labeled antibodies, **82**, 137
  - molecular motion, solid-state NMR
    - backbone, **82**, 178
    - side chains, **82**, 183
  - preparation, **82**, 131

structure in
developing connective tissues, **82**, 167
diseases of connective tissues, **82**, 169
type I-containing tissues
electron microscopy, **82**, 146
X-ray diffraction, **82**, 147
type II-containing tissues
electron microscopy, **82**, 160
X-ray diffraction, **82**, 161
type III-containing tissues
electron microscopy, **82**, 163
X-ray diffraction, **82**, 164
X-ray diffraction, **82**, 138
hydroxylysine glycosides, quantitation, **82**, 343
identification with *Clostridium histolyticum* collagenase, **82**, 468
immunofluorescent localization in tissues, **82**, 487
interactions with cells during growth and differentiation, **82**, 499
invertebrate, *R* values, **82**, 95
calculation, **82**, 94
ion-exchange chromatography, **82**, 51
molecular sieve chromatography, **82**, 50, 58
nomenclature, **82**, 5
phosphocellulose chromatography, **82**, 58
–platelet adhesion
affinity chromatography, **82**, 512
column filtration assay, **82**, 511
membrane filtration assay, **82**, 509
turbidimetric assay, **82**, 512
polyacrylamide gel electrophoresis, **82**, 59
posttranslational modifications
extracellular enzymes in, **82**, 305
intracellular enzymes in, **82**, 245
procollagen processing, **82**, 27
purification, **82**, 9, 33, 425
quantitation by
hydroxyproline measurement, **82**, 393
immunoassay, **82**, 497
reconstituted, preparation, **82**, 505
size estimation by gel chromatography and electrophoresis, **82**, 410
soluble
hydrodynamic properties, **82**, 190
relaxation processes, **82**, 196
thermodynamic properties, **82**, 187
structural characterization, **82**, 18
type IV
degradation, **82**, 452
preparation, **82**, 451
urinary metabolites, **82**, 398
vertebrate, *R* values, **82**, 95
calculation, **82**, 94
Collagenase
bacterial, radiolabeled collagen determination, **82**, 394
Clostridium histolyticum, see Clostridium histolyticum collagenase
vertebrate
activation, **82**, 443
assays, **82**, 428
isolation, **82**, 439
purification, **82**, 444
synthetic substrates, **82**, 427
type IV, assay, **82**, 452
Collagen substrata
cell adhesion, **82**, 500, 535
adhesion protein requirement, **82**, 503
effect of collagen type and organization, **82**, 502
SEM, **82**, 537
TEM, **82**, 538
cell motility, **82**, 500
fibroblast adhesion, **82**, 503
assay, **82**, 505
characteristics, **82**, 506
requirements, testing, **82**, 507
hepatocyte adhesion, **82**, 513
antibody effects, **82**, 525
attachment determination, **82**, 521
intercellular attachment determination, **82**, 526
spreading determination, **82**, 524
platelet adhesion, **82**, 513
preparation, **82**, 504, 516, 536
SEM, **82**, 540
as substrate for cell growth and differentiation, **82**, 544
TEM, **82**, 540

Colostrum
bovine, purification of
CMP-*N*-acetylneuraminate:galactose-β1-4-*N*-acetyl-D-glucosamine α2-6-sialyltransferase, **98**, 128
UDP-*N*-acetyl-D-glucosamine:α-D-mannoside β1-2-*N*-acetyl-D-glucosamine transferases, **98**, 112, 116
Compactin, *see* Mevastin
Complement receptors
C3b, detection on
human lymphoblastoid cell surfaces by solid-phase radioimmunoassay, **93**, 155
lymphocytes with alkaline-labeled reagents, application to population enumeration in whole peripheral blood, **93**, 164
Complements
activation, components in
alternative pathway, purification and quantitation, **93**, 375
antibody preparation and isolation, **93**, 409
classical pathway, molecular titration, **93**, 339
C1q
assay, **82**, 321
properties, **82**, 323
purification, **82**, 321
serum concentration, **82**, 320
Complex carbohydrates
analytical methods, **83**, 3
biosynthesis, **83**, 399
chemical ionization mass spectrum, **83**, 18
^{13}C NMR, **83**, 58
degradation, **83**, 559
gas liquid chromatography-mass spectrometry, **83**, 21
glycosyl residues
acid-labile, attachment point determination, **83**, 34
anomeric configuration, **83**, 30
in determination of carbohydrate structure, **83**, 35
ring form, determination, **83**, 33
^{1}H NMR, **83**, 29
HPLC
–mass spectrometry, **83**, 13
with refractive index detection, **83**, 8
preparation, **83**, 5, 263
resolution by lectin affinity chromatography, **83**, 127
Computer analysis
bacteriorhodopsin, molecular modeling techniques
AIMS approach, **88**, 775
classification, **88**, 773
first-guess, selection, **88**, 776
refinement method
graphic demonstration, **88**, 783
mathematical formulation, **88**, 778
ELISA data on programmable desk-top calculator, **92**, 577
programs I and II, detailed description, **92**, 583, 585
ligand binding data with LIGAND
bumpy displacement curves, **92**, 558, 565
CORRECTION FACTOR for combination of results from several experiments, **92**, 561
curved Scatchard plots, **92**, 550
displacement curves, **92**, 555
linear Scatchard plots, **92**, 547
mathematical model, **92**, 575
statistical considerations, **92**, 570
protein sequence homologies
ALIGN program, **91**, 537
database searching, **91**, 530
RELATE program, **91**, 540
Concanavalin A
inhibitory effects on lysosomes, **96**, 763
quantification using partition affinity ligand assay, **92**, 508
Coomassie Blue, *see* Anazolene sodium
Corpora lutea
bovine
prostaglandin $F_{2\alpha}$ receptor determination, **86**, 202
subcellular organelles
isolation, **86**, 193
purity evaluation with marker enzymes, **86**, 197

specific binding of [^{3}H]prostaglandins
assessment, **86**, 200
determination, **86**, 199
ovine, prostaglandin $F_{2\alpha}$ receptor determination, **86**, 202

Corrinoids
binding assays, design, **84**, 379
binding proteins, properties, **84**, 385
biochemistry and ecology, **84**, 373
extraction, **84**, 380
nomenclature, **84**, 369
properties, **84**, 371
separation, **84**, 383

Cotinine
antibody production, **84**, 633, 651
carboxylic acid derivatives, preparation, **84**, 630
nicotinamide nucleotide analogs
preparation, **84**, 650
radioimmunoassay, **84**, 652
radioimmunoassay, **84**, 636
applications, **84**, 639
sensitivity and specificity, **84**, 637
radiolabeled, preparation, **84**, 634, 652

Cotton embryos
DNA extraction and purification, **101**, 433
exogenous DNA injection, **101**, 436

Countercurrent chromatography
apparatus, **91**, 337
peptide separation and purification, **91**, 341
principle, **91**, 335

Counterimmunoelectrophoresis
DNA detection and quantitation, **84**, 187

Creatine kinase
canine isozymes
assay, **90**, 186
purification, **90**, 195
chicken skeletal muscle
isolation, **90**, 491
M-line isoprotein
immunoassays, **85**, 142
properties, **85**, 142
purification, **85**, 140
spectrophotometric assay, **90**, 490
human isozymes
assay, **90**, 186
purification, **90**, 187

Crosslinking reagents
amino-specific, **91**, 591
carboxyl, **91**, 601
guanidino-specific, **91**, 600
indole-specific, **91**, 601
nonspecific, **91**, 601
photoactivable, **91**, 602
photolysis conditions, **91**, 604
selection, **91**, 580
sulfhydryl, **91**, 599

Cuticle collagen
Ascaris
extraction and purification, **82**, 89
properties, **82**, 92
subunits
properties, **82**, 92
purification, **82**, 91
oligochaete
electron microscopy, **82**, 80
extraction and purification, **82**, 83
properties, **82**, 86
polychaete
electron microscopy, **82**, 80
extraction, **82**, 66
properties, **82**, 75
subunits, purification by
gel filtration, **82**, 69
ion-exchange chromatography, **82**, 71
SDS-polyacrylamide gel electrophoresis, **82**, 71

Cyanoborohydride
rhodopsin reduction, **81**, 288

N-2-Cyanoethyldodecane-1,12-diamine
synthesis, **94**, 413

N-2-Cyanoethylpropane-1,3-diamine
synthesis, **94**, 412

Cyanogen bromide
and dimethyl sulfoxide–hydrochloric acid, protein cleavage at tryptophan residue, **91**, 318
comparison with other methods, **91**, 323
mechanism, **91**, 324
quantitation, **91**, 322
temperature effect, **91**, 321

3′,5′-Cyclic-GMP phosphodiesterase
retinal rod outer segment
bovine
activation by
transducin α-subunit, **96**, 623

trypsin, **96**, 626
and controller enzymes, real-time assay, **81**, 532
properties, **81**, 537; **96**, 621
purification, **96**, 625
by light-regulated protein binding, **81**, 564
γ-subunit, purification, **96**, 626
Rana catesbiana
properties, **81**, 530
purification, **81**, 527
3′,5′-Cyclic-GMP phosphodiesterase inhibitors
retinal rod outer segment
assay, **81**, 545
isolation, **81**, 543
preparation from purified enzyme, **81**, 543
role in visual transduction, **81**, 546
Cyclic nucleotides
binding sites in protein kinases, cooperativity and selectivity analysis with analogs, **99**, 168
retinal
acetylation, **81**, 555
concentration and distribution, **81**, 89
determination with quick-freeze techniques, **81**, 547
preparation, **81**, 87, 552
radioimmunoassay, **81**, 89, 553
Cycloheximide
inhibitory effects on
glycoprotein synthesis, **98**, 153
lysosomes, **96**, 763
Cysteine hydrochloride
preparation, **91**, 57
Cysteine residues
protein
crosslinking reagents, **91**, 601
hydrolysis, **91**, 26
Cytarabine
antiserum production, **84**, 474
radioimmunoassay, **84**, 471
applications, **84**, 477
influencing factors, **84**, 479
Cytochalasins
B or D, effect on secretory processes in lacrimal and salivary glands
evaluation, **98**, 180
localization, **98**, 178
specificity, **98**, 180
Cytochrome *b*
nuclear and mitochondrial mutants
complementation, **97**, 384
double, construction, **97**, 391
isolation, **97**, 377
mapping, **97**, 384
revertants, **97**, 393
screening, **97**, 382
types, **97**, 375
Cytochrome *c*
amino-terminal acetylamino acid, identification by HPLC, **91**, 88
biogenesis in *Neurospora crassa*, **97**, 261
cleavage at
aspartic acid residue with hydrochloric acid and formic acid, **91**, 324
tryptophan with *o*-iodosobenzoic acid, **91**, 316
eluted from polyacrylamide gel, amino acid composition determination, **91**, 40
posttranslational transfer into mitochondria, **97**, 418
purification from *Neurospora crassa*, **97**, 262
synthesis *in vitro*, **97**, 408
Cytochrome-*c* oxidase
subunit III-deficient mutants, as source for mitochondrial ATPase subunit 6, **97**, 304
Cytochrome *P*-450
biosynthesis in hepatocyte endoplasmic reticulum membrane, **96**, 538
Cytotoxicity assays
antibody–complement-mediated, cell-surface antigen quantitation, **93**, 253
antibody-dependent cell-mediated, in basic and clinical immunology, **93**, 245
single-cell, killer cell frequency determination, **93**, 270
applications, **93**, 279
tumor cells killed by
humoral factors, assessment with
^{51}Cr, **93**, 238
[^{125}I]iododeoxyuridine, **93**, 235

methyl[^{3}H]thymidine, **93**, 239
trypan blue, **93**, 234
immunocompetent cells, assessment with
^{51}Cr, **93**, 240
methyl[^{3}H]thymidine, **93**, 242

D

Dactinomycin
inhibition of glycoprotein synthesis, **98**, 153
Dactylium dendroides
galactose oxidase purification, **89**, 168
growth in culture, **89**, 164
Dansyl chloride
polyamine prechromatographic derivatization, **94**, 12
purification, **94**, 12
Datura stramonium
lectin preparation, **83**, 368
2-Dehydro-3-deoxy-L-pentonate aldolase
pseudomonad MSU-1
assay, **90**, 269
properties, **90**, 271
purification, **90**, 270
Dehydrogluconate dehydrogenase
Gluconobacter melanogenus membrane-bound
assay, **89**, 194
properties, **89**, 197
purification, **89**, 195
membrane-bound, 2-keto-D-gluconate microdetermination, **89**, 26
3-Dehydroretinal, *see* Vitamin A_2
N-Demethyldiazepam
antiserum production, **84**, 512
radioimmunoassay, **84**, 510
5′-Deoxy-5′-(γ-amino)propylthioadenosine
preparation, **94**, 288
5′-Deoxy-5′-chloroadenosine
preparation, **94**, 287
radiolabeled, preparation, **94**, 292
5′-Deoxy-5′-chlorotubercidin
preparation, **94**, 287
Deoxycytidine 5′-monophosphate
addition to ds-cDNA during synthesis for cloning of 5′-terminal mRNA sequences, **100**, 285
Deoxyhypusine[N^1-(4-aminobutyl)lysine]
chromatographic identification, **94**, 458
5′-Deoxy-5′-methylthioadenosine
preparation, **94**, 288
radiolabeled, preparation, **94**, 292
5′-Deoxy-5′-methylthiotubercidin
preparation, **94**, 288
radiolabeled, preparation, **94**, 293
3-Deoxy-*manno*-octulosonate cytidylyltransferase
Escherichia coli
assay, **83**, 535
properties, **83**, 539
purification, **83**, 537
3-Deoxy-*manno*-octulosonate-8-phosphate phosphatase
Escherichia coli
assay, **83**, 531
properties, **83**, 534
purification, **83**, 532
3-Deoxy-*manno*-octulosonate-8-phosphate synthase
Escherichia coli
assay, **83**, 526
properties, **83**, 530
purification, **83**, 528
Deoxyribonuclease I
in actin polymerization assay, **85**, 204
Deoxyribonucleases
functional, recovery in SDS gels, **91**, 268
Deoxyribonucleic acid
–2-acetylaminofluorene adducts, radioimmunoassay, **84**, 607
antibodies, passive hemagglutination and hemolysis tests, **84**, 201
applications, **84**, 208
characteristics, **84**, 207
bacteriophage M13mp7, cutting and excision of cloned single-stranded sequences by restriction endonucleases, **101**, 90
on bacteriophage M13 vectors
kilosequencing: creation of ordered nest of asymmetric deletions across large target sequence, **101**, 98
oligonucleotide-directed mutagenesis, **100**, 468
screening, **100**, 483

bacteriophage ϕW-14
digestion to mononucleotides, **94**, 426
isolation of
α-putrescinyldeoxythymidine, **94**, 426
α-putrescinylthymine, **94**, 423
binding proteins, recovery in SDS gels, **91**, 273
bovine papillomavirus, construction of eukaryotic cloning vectors, **101**, 387
cloning into
bacteriophage λ vectors with positive selection for inserts, **101**, 3
assays, **101**, 17
bacteriophage M13 vectors, **101**, 20
computer software for sequence analysis, **101**, 72
strategies, **101**, 40
plasmids and single-stranded phages, integrated and simplified approach, **101**, 78
pWR plasmids, **100**, 72
complementary double-stranded, synthesis with addition of dCMP tails for cloning of 5′-terminal mRNA sequences, **100**, 285
cotton seed
extraction and purification, **101**, 433
recombination with histone, **101**, 436
covalently coupled to macroporous supports, in hybrid selection of specific RNAs, **100**, 400
denatured and native
antibodies
alkaline phosphatase-conjugated, preparation, **84**, 194
ELISA, **84**, 197
preparation, **84**, 196
detection and quantitation by counterimmunoelectrophoresis, **84**, 187
directed mutagenesis with sodium bisulfite, **100**, 457
electrophoretic transfer to diazobenzyloxymethyl paper, **100**, 326
encapsulation in liposomes, **101**, 514
fragments
preparation, **101**, 11
radiolabeled, preparation, **100**, 312
transfer from polyacrylamide gels to
diazobenzyloxymethyl paper and hybridization with specific probe, **100**, 312
nitrocellulose paper and hybridization with specific probe, **100**, 325
Halobacterium halobium
bacterio-opsin gene identification by hybridization, **97**, 233
digestion with restriction endonucleases, **97**, 229
isolation, **97**, 230
hybridization with specific probe after transfer to diazobenzyloxymethyl paper, **100**, 317
efficiency, **100**, 319, 331
introduction into
cotton embryos by microinjection, **101**, 437
eukaryotic cells by protoplast fusion, **101**, 402
tissue culture cells by microinjection, **101**, 482
various cell types with liposomes as carriers, **101**, 521
viable mammalian cells with fusogenic reconstituted Sendai virus envelopes as vehicle, **101**, 492
maize chloroplast
cloning, **97**, 535
isolation, **97**, 526
linked transcription–translation, **97**, 551
nick translation, **97**, 545
restriction endonuclease cleavage sites, mapping, **97**, 541
mammalian mitochondrial
cloning, **97**, 431
genetic and transcription maps, **97**, 465
isolation, **97**, 428, 443
labeling, **97**, 437
restriction fragments, isolation, **97**, 445
–RNA hybridization, **97**, 455
strand separation, **97**, 443
transcription products, **97**, 465

multigene families, isolation and homology determination by filter hybridization methods, **100**, 266
mutagenesis *in vitro* with terminal deoxyribonucleotidyltransferase, **100**, 103
nucleotide-specific deletions with exonuclease III, **100**, 93
plasmid
characterization by agarose gel electrophoresis, **100**, 249
isolation by rapid alkaline extraction method, **100**, 243
screening at high colony density, **100**, 333
in polyacrylamide gels, silver staining, **96**, 235
–protein interaction, specific binding sequences, immunoassay, **92**, 138
quantitation in submicrogram amounts by rocket electrophoresis, **84**, 181
reactions with
T4 RNA ligase, **100**, 44
type II restriction endonucleases, **100**, 3
troubleshooting guide, **100**, 33
recombinant
analysis with
Escherichia coli minicells, **101**, 347
transposon $\gamma\delta$, **101**, 362
gene products, analytical methods, **100**, 411
identification with cell-free protein-synthesizing systems, **101**, 650
locus-reacting proteins with specialized functions, **100**, 183
monitoring systems
cell-free
transcription, **101**, 540
translation, **101**, 598
intact-cell, **101**, 347
vectors for cloning genes, **101**, 3
repair genes, amplification with plasmid pKC30, **101**, 138
in vitro replication systems, analysis of bacteriophage ϕX174 gene *A* protein, **100**, 217
–RNA hybrids of naturally occurring sequence
antibody radioimmunoassay, **84**, 216
preparation, **84**, 211
Semliki Forest virus membrane proteins
expression analysis in eukaryotic cell nuclei, **96**, 510
ligation and transformation, **96**, 499
in small-scale plasmids
microinjection into eukaryotic cell nuclei, **96**, 504
preparation, **96**, 503
sequence analysis with
exonuclease III, **100**, 60
mercury–thiol affinity system for generation of overlapping labeled fragments, **100**, 116
supercoiled, purification by RPC-5 ANALOG chromatography, **100**, 378
tailing with terminal deoxyribonucleotidyltransferase, **100**, 96
transcription in whole-cell extracts, **101**, 568
yeast
centromere sequences, selection procedures for isolation, **101**, 300
chromosomal rearrangements, construction, **101**, 269
high-copy cloning vector construction from plasmid 2-μm circle sequences, **101**, 307
library construction, **97**, 348
mutant genes and contiguous chromosomal sequences, cloning by integrative transformation, **101**, 290
pet mutants, transformation, **97**, 349, 355
plasmid, small-scale preparation, **97**, 351, 359
promoter sequences
application in gene expression regulation, **101**, 192
design for analysis of cloned gene expression, **101**, 181
STE genes, cloning in plasmid 2-μm vectors, **101**, 325

transformation with linear and gapped plasmids, genetic applications, **101**, 228

Depolymerization
- endwise linear, quantitative kinetic analysis with microtubule disassembly as model, **85**, 439

Desmin
- antiserum
  - production, **85**, 495
  - specificity analysis, **85**, 497
- identification and properties in different cell types, **85**, 490
- immunofluorescence, **85**, 503
- polyacrylamide gel electrophoresis, **85**, 494
- purification from smooth muscle, **85**, 492

Desmosine
- antibody preparation, **82**, 759

Detergents
- deuterated and hydrogenated, in rhodopsin analysis by neutron small-angle scattering, **81**, 742
- opsin and rhodopsin in, thermal stability, **81**, 256

Deuterium
- enriched aldoses, chemical synthesis, **89**, 79
- –hydrogen exchange in membrane-bound rhodopsin
  - comparison with tritium exchange, **88**, 681
  - infrared measurements, **88**, 676
- kinetic effects
  - in deduction of enzyme reaction mechanisms, **87**, 711
  - on initial rates of enzyme reactions, **87**, 607

Dextrans
- neutral, as marker for endocytosis and membrane recycling, **98**, 212

Diacetyl-1,6-diaminohexane
- synthesis, **94**, 422

Diacetyl reductase, *see* Acetoin dehydrogenase

Diacylglycerol lipase, *see* Lipoprotein lipase

Diamine oxidase, *see* Amine oxidase (copper-containing)

Diamines
- aliphatic, naturally occurring, quantitative determination with automated liquid chromatography, **94**, 29

3,5-Diaminobenzoic acid
- in fluorometric assay for gangliosides, **83**, 179

Diazepam
- radioimmunoassay, **84**, 491

1,8-Diazido-3-[*S*-acetyl]thiooctane
- synthesis, **94**, 290

Diazobenzyloxymethyl paper
- transfer of
  - DNA fragments from polyacrylamide gels, **100**, 309
  - DNA, RNA, and proteins by electrophoresis, **100**, 326

Diazomethanes
- inhibitory effects on lysosomes, **96**, 759

Dibenzo-*p*-dioxins
- chlorinated, radioimmunoassay, **84**, 619

Dichroism
- circular, *see* Circular dichroism
- linear, oriented rod suspensions in visible, uv, and ir, **81**, 605
- photo-, in measurement of rhodopsin and bacteriorhodopsin rotational motions, **88**, 741
- transient, bacteriorhodopsin
  - electric field-induced, **88**, 253
  - measurement, **88**, 249
  - in purple membranes, **88**, 251
  - in reconstituted membranes, **88**, 252

Dictyostelium discoideum
- adhesion assay, **98**, 426
- aggregation assay, **98**, 425
- mutants defective in
  - phagocytosis
    - adhesive revertants, isolation, **98**, 428
    - characterization, **98**, 427
    - enrichment, **98**, 426
  - secretion
    - characterization, **96**, 818, 821
    - isoelectric focusing, **96**, 823
    - isolation, **96**, 816
- myosin aggregate formation and assay, **85**, 39
- phagocytosis assay, **98**, 425
- pinocytosis assay, **98**, 422

plasma membrane glycoconjugates, labeling
with galactosyltransferase, **98**, 416
release with β-galactosidase, **98**, 419
Dicyclohexylcarbodiimide
reaction with D-fructose 1,6-bisphosphate, **89**, 103
Di(2-ethylhexyl) phthalate, *see* Bis(2-ethylhexyl) phthalate
Diffusion
protein elution from SDS gels, **91**, 454
1,5-Difluoro-2,4-dinitrobenzene
in myosin active-site trapping of metal nucleotides, **85**, 93
4,4′-Difluoro-3,3′-dinitrophenyl sulfone
in myosin active-site trapping of metal nucleotides, **85**, 93
α-Difluoromethylarginine
synthesis, **94**, 206
α-Difluoromethylornithine
^{14}C-labeled, ornithine decarboxylase labeling and quantitation, **94**, 206
experimental interruption of pregnancy, **94**, 213
synthesis, **94**, 200
therapeutic effects on
Eimeria tenella infection in chicken, **94**, 212
Trypanosoma b. brucei infection in mice, **94**, 210
Digitonin
extraction of visual pigments, **81**, 133
Diglyceride lipase, *see* Lipoprotein lipase
Diglyceryl tetraethers
Thermoplasma
cultivation, **88**, 398
degradation to C_{40} hydrocarbon chain derivatives, **88**, 403
isolation, **88**, 399
quantification, **88**, 405
Digoxin
antibodies
affinity assessment, **84**, 573
detection, **84**, 567
preparation, **84**, 564
specificity determination, **84**, 571
–protein conjugates
characterization, **84**, 562
preparation, **84**, 559
quantification with partition affinity ligand assay, **92**, 500
radioimmunoassay, **84**, 573
Dihydrolipoamide acetyltransferase
Bacillus
assay, **89**, 401
properties, **89**, 405
purification, **89**, 402
bovine kidney and heart
assay, **89**, 376
properties, **89**, 382, 386
purification, **89**, 377, 383
broccoli mitochondrial
assay, **89**, 408
properties, **89**, 412
purification, **89**, 411
cauliflower mitochondrial
assay, **89**, 408
properties, **89**, 410
purification, **89**, 409
Escherichia coli
assay, **89**, 392
behavior on ethanol–Sepharose 2B, **89**, 398
purification, **89**, 393
Hansenula miso
assay, **89**, 420
properties, **89**, 423
purification, **89**, 421
Neurospora crassa
assay, **89**, 387
properties, **89**, 390
purification, **89**, 389
pigeon breast muscle
assay, **89**, 414
properties, **89**, 418
purification, **89**, 416
Dihydrolipoamide dehydrogenase
Bacillus
assay, **89**, 401
properties, **89**, 405
purification, **89**, 402
bovine kidney and heart
assay, **89**, 376
properties, **89**, 382, 386
purification, **89**, 377, 383
broccoli mitochondrial
assay, **89**, 408
properties, **89**, 412
purification, **89**, 411

cauliflower mitochondrial
assay, **89**, 408
properties, **89**, 410
purification, **89**, 409
Escherichia coli
assay, **89**, 392
behavior on ethanol–Sepharose 2B, **89**, 398
purification, **89**, 393
Hansenula miso
assay, **89**, 420
properties, **89**, 423
purification, **89**, 421
Neurospora crassa
assay, **89**, 387
properties, **89**, 390
purification, **89**, 389
pigeon breast muscle
assay, **89**, 414
properties, **89**, 418
purification, **89**, 416
Dihydrolipoyl dehydrogenase, *see* Dihydrolipoamide dehydrogenase
Dihydrolipoyl transacetylase, *see* Dihydrolipoamide acetyltransferase
5α,7α-Dihydroxy-11-ketotetranorprostane-1,16-dioic acid
antibodies
production, **86**, 343
properties, **86**, 344
isolation from urine, **86**, 342
as prostaglandin production parameter in blood and urine, **86**, 351
quantitation in urine by gas chromatography–mass spectrometry, **86**, 579
radioimmunoassay, **86**, 339
3,3′-Diiodothyronine
antibody preparation, **84**, 280
radioimmunoassay, **84**, 272
minimization of serum iodothyronine-binding protein effects, **84**, 285
radiolabeled, preparation, **84**, 275
serum concentrations in health and disease, **84**, 298
6,15-Diketo-13,14-dihydro-2,3-dinor-prostaglandin $F_{1\alpha}$
quantitation in urine by gas chromatography–mass spectrometry, **86**, 585
Dimethylaminoazobenzene isothiocyanate
peptide amino-terminal analysis, **91**, 79
in polypeptide manual sequencing, **91**, 455
Dimethylaminoazobenzenesulfonyl chloride
amino acid precolumn derivatization, **91**, 43
5-Dimethylaminonapthalene-1-sulfonyl chloride, *see* Dansyl chloride
Dimethyl sulfoxide
–hydrochloric acid and cyanogen bromide, protein cleavage at tryptophan residue, **91**, 318
comparison with other methods, **91**, 323
mechanism, **91**, 324
quantitation, **91**, 322
temperature effect, **91**, 321
Dinitrophenol
bioluminescent immunoassay, **92**, 426
2,3-Dinor-6-keto-prostaglandin $F_{1\alpha}$
quantitation in urine by gas chromatography–mass spectrometry, **86**, 585
2,3-Dinor-thromboxane B_2
quantitation in urine by gas chromatography–mass spectrometry, **86**, 592
1,2-Dipalmitoyl-3-benzyl-*sn*-glycerol
synthesis, **98**, 625
Dipeptidyl aminopeptidase I, *see* Dipeptidyl peptidase I
Dipeptidyl aminopeptidase IV, *see* Dipeptidyl peptidase IV
Dipeptidyl carboxypeptidase I
polypeptide sequencing, **91**, 516
rabbit lung, preparation, **91**, 513
Dipeptidyl carboxypeptidase II
Escherichia coli, preparation, **91**, 514
polypeptide sequencing, **91**, 516
Dipeptidyl peptidase I
bovine spleen, preparation, **91**, 512
polypeptide sequencing, **91**, 516
Dipeptidyl peptidase IV
polypeptide sequencing, **91**, 516
porcine kidney, preparation, **91**, 513
Diphenylhexatriene
as membrane fluorescent probe, analysis, **81**, 678

2,2′-Dipyridyl disulfide
and related compounds, as two-protonic-state electrophile probes of enzyme mechanisms, characteristics, **87**, 441
5,5′-Dithiobis(2-nitrobenzoic acid)
chemistry in alkaline solutions, **91**, 51
equations for thiol titrations, **91**, 53
factors relevant to use, **91**, 54
in myosin active-site trapping of metal nucleotides, **85**, 93
preparation, **91**, 58
protein thiol concentration determination, **91**, 58
Dithiothreitol
reduction of
free methionine sulfoxide, **91**, 550
methionine sulfoxide residues in
peptides, **91**, 554
proteins, **91**, 556
ribulose-bisphosphate carboxylase reactivation, **90**, 520
DNA, *see* Deoxyribonucleic acid
DNA gyrase, *see* DNA topoisomerase (ATP-hydrolyzing)
DNA nucleotidylexotransferase
in tailing of DNA and for *in vitro* mutagenesis, **100**, 96
DNA polymerase
functional, recovery in SDS gels, **91**, 272
DNA terminase
bacteriophage λ
assay, **100**, 183
host factor, **100**, 189
purification, **100**, 190
properties, **100**, 189
purification, **100**, 185
DNA topoisomerase I
HeLa cell
assay, **100**, 133
properties, **100**, 136
purification, **100**, 134
rat liver, multiple forms
assays, **100**, 137
characterization, **100**, 141
DNA topoisomerase II
Drosophila melanogaster embryo
assay, **100**, 162
properties, **100**, 167
purification, **100**, 164
Escherichia coli phage T4
assay, **100**, 150
comparison with other type II topoisomerases, **100**, 159
properties, **100**, 154
purification, **100**, 145
DNA topoisomerase (ATP-hydrolyzing)
Escherichia coli
assays, **100**, 173
purification, **100**, 175
reactions and uses, **100**, 178
Dolichyl phosphate
synthesis inhibitors, effect on glycoprotein biosynthesis, **98**, 149
Dolichylpyrophosphoryl oligosaccharides, *see* Oligosaccharides, dolichylpyrophosphoryl
Dopamine β-monooxygenase
kinetic mechanisms, deduction from primary hydrogen isotope effects, **87**, 711
Drosophila melanogaster
cultivation, **89**, 297
embryos
preparation of purified nuclei, **96**, 591
purification of DNA topoisomerase II, **100**, 164
subnuclear fraction
comparison with vertebrate fractions, **96**, 596
morphological characteristics, **96**, 595
preparation, **96**, 594
purification of
alcohol dehydrogenase, **89**, 446
glycerol-3-phosphate dehydrogenase (NAD^+), **89**, 297
hexokinase, **90**, 17
phosphoglucose isomerase, **89**, 560
Duimycin
inhibition of glycoprotein lipid-linked saccharide formation, **98**, 148
Dynein ATPase
latent form
preparation from sea urchin sperm flagella, **85**, 453
properties, **85**, 463
subunits, properties, **85**, 465

preparation from
Chlamydomonas flagella, **85**, 467
lamellibranch gill cilia, **85**, 470
sea urchin sperm, **85**, 472
starfish sperm, **85**, 471
Tetrahymena cilia, **85**, 466
trout sperm, **85**, 473
properties, **85**, 468

E

E-64, *see* L-*trans*-Epoxysuccinylleucylagmatine
Ecdysones
antibodies
incubation and separation, **84**, 356
preparation, **84**, 354
specificity, **84**, 355
extraction from biological samples, **84**, 357
iodination, **84**, 353
radioimmunoassay, **84**, 350
succinylation, **84**, 350
Edeines
A and B
biological assay, **94**, 448
identification, **94**, 447
preparation and isolation, **94**, 443
separation, **94**, 445
Egasyn
microsomal membrane
antiserum preparation, **96**, 566
assay, **96**, 568
iodination, **96**, 567
isolation, **96**, 564
Eggs
sea urchin, ^{31}P NMR profiles, **85**, 674
Eicosanoids
metabolizing enzymes in brain, inactivation by microwave irradiation, **86**, 635
^{18}O-labeled, preparation for gas chromatography–mass spectrometry, **86**, 547
radioimmunoassay with iodinated derivatives as tracers, **86**, 297
radioiodinated derivatives
binding parameters, **86**, 301
preparation, **86**, 298
separation by open tubular glass capillary gas chromatography, **86**, 624
Eimeria tenella
infection in chickens, treatment with ornithine decarboxylase inhibitors, **94**, 212
Elastase
assays, **82**, 593
modulation by
elastin ligands, **82**, 601, 603
physical–chemical environment, **82**, 600
Elastase inhibitors
functional, measurement, **82**, 604
Elastin
amino acid composition, **82**, 560, 576
biosynthesis, **82**, 566
lysyl oxidase role, **82**, 637
crosslinks
effect of
[^{3}H]$NaBH_4$ reduction, **82**, 580
[^{14}C]NaCN and NH_3 treatment, **82**, 585
formation, **82**, 561
photolysis in ultraviolet light, **82**, 585
electron microscopy, **82**, 578
enzymatic digestion, **82**, 747
histochemical analysis, **82**, 578
immunoassays, **82**, 751
insoluble, *see* Insoluble elastin
isolation from
cartilage, **82**, 575
lung and other tissues, **82**, 571
modified, preparation, **82**, 593
mRNA
cell-free translation, **82**, 731
isolation, **82**, 732
soluble, *see* Soluble elastin
structure, **82**, 559
valylprolyl content, **82**, 577
Electric fields
high-voltage
analysis of membrane turnover by exocytosis and endocytosis, **98**, 33
cellular exposure methodology, **98**, 31
effect on cell suspension, **98**, 29

orientation of purple membranes, **88**, 287
measurement, **88**, 289
Electrodialysis
protein elution from SDS gels, **91**, 450
Electron microscopy
actin polymerization, **85**, 200
clathrin, **98**, 365
contractile proteins, micrograph optical diffraction and filtering, **85**, 511
microtubule assembly *in vitro*, **85**, 434
parotid plasma membranes, **98**, 78
planar membrane monolayers, **88**, 230
proteins in polarized epithelia, **98**, 386
secretion granule membranes from parotid gland, **98**, 83
virus-infected monolayers, **98**, 493
yeast secretory mutants, **96**, 808
Electron paramagnetic resonance
cAMP-dependent protein kinase catalytic subunit, **99**, 94
contractile proteins, **85**, 595
saturation-transfer technique, **85**, 599
spectrum modification after labeling, **85**, 598
spin probe-spin label methods, **85**, 621
Cr(III)– and Co(III)–nucleotide complexes and interactions with enzymes, **87**, 180
electron transfer components in methanogens, **88**, 413
spin-labeled rhodopsin, **81**, 703
light effect on rotational correlation time, **81**, 709
Electron transfer components
in methanogens, EPR, **88**, 413
Electrophiles
two-protonic-state
characteristics, **87**, 441
as enzyme mechanism probes, **87**, 427
synthesis and preparation, **87**, 447
Electrophoresis, *see also specific electrophoretic techniques*
diagonal, sulfhydryl peptides alkylated with *N*-ethylmaleimide, **91**, 392
ELISA, *see* Enzyme-linked immunosorbent assay
Ellman's reagent, *see* 5,5′-Dithiobis(2-nitrobenzoic acid)
Embryos
cotton, DNA
extraction and purification, **101**, 433
recombination with histone, **101**, 436
reintroduction, **101**, 437
Drosophila melanogaster
preparation of purified nuclei, **96**, 591
purification of DNA topoisomerase II, **100**, 164
subnuclear fraction
comparison with vertebrate fractions, **96**, 596
morphological characteristics, **96**, 595
preparation, **96**, 594
murine, gene transfer by pronuclear fusion, **101**, 411
Emission spectroscopy
rhodopsin and bacteriorhodopsin, **88**, 217
Emulphogene BC-720
extraction of visual pigments, **81**, 133
efficiency, **81**, 137
Endo-β-*N*-acetylglucosaminidase, *see* Mannosyl-glycoprotein endo-β-*N*-acetylglucosaminidase
Endocytosis, *see also* Protein transport, intracellular
in adrenal medullary cells, high-voltage analytical techniques, **98**, 33
adsorptive, plasma membrane components, **98**, 203
with coated vesicles, **98**, 9
fluid-phase markers, **98**, 209
in inside-out thyroid follicles, **98**, 453
membrane markers, **98**, 215
phagocytic, in *Dictyostelium discoideum*
assays, **98**, 425
defective mutants
characterization, **98**, 427
enrichment, **98**, 426
revertant isolation, **98**, 428
pinocytic
assay in *Dictyostelium discoideum*, **98**, 422

β-hexosaminidase B in Sandhoff disease fibroblasts, **98**, 292
membrane flow during, 5′-nucleotidase antibody as marker, **98**, 395
in polarized epithelia, immunochemical analysis, **98**, 379
receptor-mediated
epidermal growth factor, **98**, 283
in isolated perfused liver, **98**, 235
low-density lipoproteins in cultured cells, **98**, 251
role of actin microfilaments, **98**, 368
Semliki Forest virus in BHK-21 cells, **98**, 262
sorting sites and mechanisms, **98**, 11
Endo-β-galactosidase (keratan sulfate-degrading), *see* Keratan-sulfate endo-1,4-β-galactosidase
β-*N*-Endoglycosidase H
in assay for transport-coupled oligosaccharide processing *in vitro*, **98**, 42
Endoplasmic reticulum
hepatic, membrane proteins
antibody preparation, **96**, 525
characterization, **96**, 536
cotranslationally inserted polypeptides, **96**, 535
degradation after drug induction, **96**, 547
drug induction, **96**, 544
influencing factors, **96**, 543
isolation, **96**, 521
posttranslationally inserted polypeptides, **96**, 533
pancreatic
biochemical and morphological characterization, **98**, 188
Ca^{2+} uptake assay, **98**, 194
docking protein
assay, **96**, 693
purification, **96**, 696
transport pathways to Golgi complex, **98**, 8
Enolase
chicken muscle
assay, **90**, 490
isolation, **90**, 492
Entamoeba histolytica
purification of
galactose-1-phosphate uridylyltransferase, **90**, 553
6-phosphofructokinase (pyrophosphate), **90**, 99
Enterobacter aerogenes
purification of 5-methylthioribose kinase, **94**, 363
Enzyme function
analysis by affinity labeling, **87**, 469
Enzyme immunoassays
α_1-fetoprotein by
double antibody method, **84**, 20
sandwich method, **84**, 26
immunoglobulins and haptens with antigen-coated plastic tubes and labeled protein A, **92**, 411
prostaglandin $F_{2\alpha}$, **86**, 269
proteoglycan aggregate components, **83**, 229
separation of free and bound fractions
with activated thiol-Sepharose, **92**, 345
by affinity exclusion, **92**, 359
Enzyme intermediates
covalent, evaluation by
chemical trapping, **87**, 11
exchange kinetics, **87**, 3
initial rate kinetics, **87**, 9
reaction with substrate analogs, **87**, 14
stereochemical characterization, **87**, 17
thermal trapping, **87**, 13
enol
coupled elimination, **87**, 88
destruction analysis, **87**, 91
direct testing of enols as substrates, **87**, 92
with enols as normal products, **87**, 90
oxidation, **87**, 88
proton isotope exchange, **87**, 87
reaction-based inactivators, **87**, 90
role in reaction mechanisms, **87**, 84
as sources of enols, **87**, 94
spectroscopy, **87**, 88
structural homology, **87**, 85
transition-state inhibitors, **87**, 89

Enzyme-linked immunoelectrotransfer blot technique
specificity analysis of antigens and antibodies separated by gel electrophoresis, **92**, 377
Enzyme-linked immunosorbent assay
antibodies to native and denatured DNA, **84**, 194
Clostridium botulinum toxins A, B, and E, **84**, 223
sensitivity, cross-reactivity, and practical application, **84**, 229
with enzyme-labeled second antibody, monoclonal antibody production screening, **92**, 168
fibrinogen, **92**, 577
data analysis on programmable desk-top calculator, **92**, 578
plasma ferritin, **84**, 168
quantitative single-tube kinetic-dependent, **92**, 392
applications, **92**, 398
sources of error, **92**, 401
Enzyme reactions
chemical mechanisms, determination by pH studies, **87**, 390
equilibrium isotope exchange, fingerprint kinetic patterns for modifier action, **87**, 647
calculation programs, **87**, 671
Haldane relationships, analysis, **87**, 366
involving branched chain pathways
class members, **87**, 321
distinguishing kinetic characteristics, **87**, 319
rate equations
experimental determination, **87**, 340
inhibitor effects, **87**, 346
interrelationships, **87**, 335, 344
metabolic implications, **87**, 340
pH effects, **87**, 345
kinetic isotope effects
determination by
equilibrium perturbation method, **87**, 641
Northrop's method, **87**, 633
exchange at protein sites, kinetic analysis, **87**, 588
on hydrogen bonding, **87**, 577
on initial rates, **87**, 607
calculation, **87**, 616, 624
in isotopic waters
acid-base reactions, **87**, 566
deuterium content, determination, **87**, 592
enzyme kinetics, **87**, 583
free energies of transfer, **87**, 555
isotope fractionation factors, **87**, 559
pH adjustment, **87**, 594
protein structure and stability, **87**, 578
reaction rates and equilibria, **87**, 552
mechanistic interpretation, **87**, 599
in metal-ion environments, **87**, 575
in mixed isotopic waters, **87**, 564, 591
ionic stabilities and solvation, **87**, 571
product and rate isotope effects, **87**, 589
proton inventories, **87**, 581
primary and secondary, **87**, 626
kinetic mechanisms
deduction from primary hydrogen isotope effects, **87**, 711
error analysis, **87**, 730
two-protonic-state electrophiles as probes, **87**, 427
applications, **87**, 460
pH dependence, **87**, 431
pH-dependent processes, analysis with buffers of constant ionic strength, **87**, 405
rival kinetic models, discrimination
error analysis, **87**, 385
regression analysis, **87**, 380
statistical criteria, **87**, 371
terreactant mechanisms, initial velocity analysis, **87**, 353
transition-state structure, determination by isotope effects, **87**, 625
Enzyme structure
active site-directed reagents
design considerations, **87**, 481
evaluation, **87**, 486
reaction facilitation, **87**, 495
analysis by affinity labeling, **87**, 469

Ep-475, *see* L-*trans*-Epoxysuccinylleucylamidoisopentane
Epidermal growth factor
ferritin-conjugated, preparation, **98**, 287
fluorescent, preparation, **98**, 284
radiolabeled, preparation, **98**, 288
receptor-mediated endocytosis, **98**, 283
Epidermal growth factor receptors
function as tyrosyl kinase, **99**, 379
purification from A-431 cells, **99**, 379
spontaneous transfer into receptor-deficient mutants, **98**, 555
Epitopes
distinction with monoclonal antibodies, **92**, 77, 242
Epoxide hydrolase
biosynthesis and processing in endoplasmic reticulum membrane, **96**, 537
L-*trans*-Epoxysuccinylleucylagmatine
inhibitory effects on lysosomes, **96**, 759
L-*trans*-Epoxysuccinylleucylamidoisopentane
inhibitory effects on lysosomes, **96**, 759
EPR, *see* Electron paramagnetic resonance
Equilibrium perturbation method
determination of kinetic isotope effects, **87**, 641
applications, **87**, 645
Erythrocytes
chicken, isolation of nuclear pore complex–lamina structures, **96**, 604
human
biogenesis and processing of glycophorin A *in vitro*, **96**, 281
detection of spectrin oligomers, **96**, 298
orientation of glycophorins, **96**, 280
preparation of
ghosts, **96**, 314
inside-out vesicles, **96**, 315
purification of
ankyrin, **96**, 318
ankyrin fragment containing spectrin-binding site, **96**, 320
band-3 cytoplasmic domain, **96**, 321
erythroglycan, **83**, 311
fructose-bisphosphate aldolase, **90**, 251
glucose-6-phosphate isomerase, **89**, 552
glycophorins, **96**, 269
glyoxalase I, **90**, 536
phosphoglycerate kinase, **90**, 103
pyruvate kinase, **90**, 132
spectrin, **85**, 475; **96**, 306, 316
transketolase, **90**, 225
murine, anion transport protein *in vitro* synthesis and processing, **96**, 257
ovine
antibody-coated, detection of Fc receptors, **93**, 115
hapten and protein attachment
facilitation by conjugate succinylation, **92**, 265
with methyl-*p*-hydroxybenzimidate, **92**, 269
with *N*-succinimidyl 3-(2-pyridyldithio)propionate, **92**, 259
with synthetic lipopolysaccharide, **92**, 274
with 1,3,5-trichlorotriazine, **92**, 271
in rosette-forming assay for antibody-synthesizing hybridomas, **92**, 175
rabbit
hexokinase purification, **90**, 4
^{31}P NMR profiles, **85**, 674
rat, pyruvate kinase purification, **90**, 161
Xenopus laevis, isolation of nuclear pore complex–lamina structures, **96**, 604
Erythroglycan
preparation from human erythrocytes, **83**, 311
D-Erythrose
deuterated, preparation, **89**, 81
Erythrulose reductase
bovine liver
assay, **89**, 232
properties, **89**, 236
purification, **89**, 233

Escherichia coli
adenosinetriphosphatase complex
structural genes
cloning, **97**, 176, 188
restriction map, **97**, 193
chromosomal mutagenesis with bacteriophage Mu, **94**, 91
coupled transcription-translation system
applications, **101**, 678
preparation, **101**, 675
gene expression, analysis with
β-galactosidase gene fusions, **100**, 293
Mud(Ap, *lac*)-generated fusions, **100**, 501
inner membrane, analysis of bacteriophage M13 coat protein assembly, **97**, 130
lactose permease content estimation, **97**, 164
leader peptidase gene
isolation, **97**, 47
product role in cell growth, **97**, 54
structure and expression, **97**, 51
membrane biogenesis, analytical techniques
semi-*in vitro* systems, **96**, 79
in vitro systems, **96**, 81
in vivo systems, **96**, 75
minicells
preparation for recombinant DNA analysis, **101**, 348
purification, **101**, 355
outer membrane
lipoprotein mutants
characterization, **97**, 127
isolation, **97**, 124
separation, **97**, 128
site-specific mutagenesis, **97**, 126
OmpA and OmpF proteins
gel electrophoresis, **97**, 118
immunoprecipitation, **97**, 116
immunoreplicate electrophoresis, **97**, 121
synthesis and assembly, kinetic analysis, **97**, 112
polyamine biosynthetic enzymes
gene cloning, **94**, 117
deficient mutants, mass screening, **94**, 83
preparation of free and membrane-bound polysomes, **97**, 70
protein export, genetic analysis, **97**, 3, 11
protein-synthesizing systems
DNA extraction, **101**, 602
preparation, **101**, 599
protein synthesis *in vitro*, **101**, 602
RNA extraction, **101**, 601
purification of
S-adenosylmethionine decarboxylase, **94**, 229
altronate dehydratase, **90**, 289
D-altronate-NAD dehydrogenase, **89**, 212
L-arabinose-binding protein, **90**, 465
arginine decarboxylase, **94**, 127
3-deoxy-D-*manno*-octulosonate cytidylyltransferase, **83**, 537
3-deoxy-D-*manno*-octulosonate-8-phosphate phosphatase, **83**, 532
3-deoxy-D-*manno*-octulosonate-8-phosphate synthase, **83**, 528
diacetyl reductase, **89**, 519
DNA gyrase, **100**, 175
formate dehydrogenase, **89**, 540
D-galactose-binding protein, **90**, 465
galactose-1-phosphate uridylyltransferase, **87**, 30
glutathionylspermidine, **94**, 436
lactose permease, **97**, 168
lysine decarboxylase, **94**, 180
maltose-binding protein, **90**, 460
mannonate dehydratase, **90**, 289
D-mannonate-NAD dehydrogenase, **89**, 212
mannose-specific lectins, **83**, 387
methionine adenosyltransferase, **94**, 220
ornithine decarboxylase, **94**, 127
ornithine decarboxylase antizymes, **94**, 194
periplasmic D-xylose-binding protein, **90**, 474
phage T4 DNA topoisomerase, **100**, 145
phosphofructokinase, **90**, 63, 65

putrescine aminopropyltransferase, **94**, 267
pyruvate dehydrogenase complex, **89**, 393
pyruvate kinase, **90**, 173
*rec*A protein, **100**, 203
ribose-binding protein, **90**, 469
RNA polymerase, **101**, 541
structural genes coding for glycogen biosynthetic enzymes, **83**, 549
UDP-*N*-acetyl-D-glucosamine 2′-epimerase, **83**, 516
UDP-*N*-acetyl-D-mannosamine dehydrogenase, **83**, 520
synthesis of capsular polymers containing polysialic acid, **83**, 540
trp promoter-containing plasmids, as vector for expressing cloned genes, **101**, 155
Escherichia freundii
isolation of endo-β-galactosidase, **83**, 612
Esters
active, in polyacrylamide gels
applications, **83**, 310
preparation, **83**, 306
Estriol
homogeneous immunoassay with differential pulse polarography, **92**, 433
1,N_6-Etheno-cAMP
as probe for cAMP-dependent protein kinases
activation kinetic assay, **99**, 163
cyclic nucleotide exchange kinetic assay, **99**, 163
deactivation assays
equilibrium, **99**, 164
kinetic, **99**, 163
N-Ethylmaleimide
-alkylated sulfhydryl peptides, isolation, **91**, 392
α-Ethynylornithine
synthesis, **94**, 203
α-Ethynylputrescine
synthesis, **94**, 204
Exocytosis, *see also* Protein transport, intracellular
in adrenal medullary cells, high-voltage analytical techniques, **98**, 33
intracellular pathway, **98**, 1
in pancreatic zymogen granules, **98**, 74
Exodeoxyribonuclease III
for DNA sequence analysis and specific deletion of nucleotides, **100**, 60
Exonuclease III, *see* Exodeoxyribonuclease III
Extraction
arachidonic acid metabolites, **86**, 467, 477
Ascaris cuticle collagen, **82**, 89
corrinoids, **84**, 380
DNA from
cotton embryos, **101**, 433
Escherichia coli, **101**, 602
plasmids by alkaline technique, **100**, 243
ecdysones from biological fluids, **84**, 357
hydroxyprostaglandins, **86**, 170
glucose-6-phosphatase from rat brain, **90**, 400
lipids from
invertebrate photoreceptor membrane, **81**, 316
purple membranes, **88**, 74, 100, 107
oligochaete cuticle collagen, **82**, 83
pigment chromophores, **88**, 552
polychaete cuticle collagen, **82**, 66
prostaglandin endoperoxides, **86**, 378
proteoglycans, **82**, 779
retinal preparations, **81**, 792
RNA from
brain stem subcellular fraction, **96**, 382
Escherichia coli, **101**, 601
whole cells or tissues, **96**, 25
rod outer segment, **81**, 802
SDS from proteins by ion-pair method, **91**, 254
slow-reacting substance of anaphylaxis, **86**, 427
soluble elastin, **82**, 565, 723
structural glycoproteins, **82**, 843
visual pigments
with detergents, **81**, 133
from insects, **81**, 183

F

Factor VIII
coagulant protein, *see* Antihemophilic factor
related protein, *see* Von Willebrand factor
Factor X
conformation-specific antibodies
interactions with protein antigens, **84**, 76
preparation, **84**, 64
radioimmunoassay, **84**, 75
Factor B
human, in alternative complement pathway
hemolytic assay, **93**, 381
immunoquantitation, **93**, 381
purification, **93**, 379
Factor D
human, in alternative complement pathway
assay, **93**, 393
purification, **93**, 390
Factor H
human, in alternative complement pathway
hemolytic assay, **93**, 396
purification, **93**, 395
radial immunodiffusion assay, **93**, 396
Factor I
human, in alternative complement pathway
assay, **93**, 401
purification, **93**, 399
FAD, *see* Flavin adenine dinucleotide
Falling-ball assay
actin filament networks, **85**, 216
Fast Green FCF
procurement, **91**, 241
–protein complexes in gels
absorption spectra, **91**, 245
metachromasia, **91**, 245
photography, **91**, 246
scanning, **91**, 246
protein staining, **91**, 239
Fatty acids
-acylated proteins in eukaryotic cells
detection, **96**, 797
distribution, **96**, 796
hydrolysis, **96**, 799
site and acylation mechanism, **96**, 799
^{14}C-labeled, synthesis, **86**, 357
composition and pairing in rod outer segment phospholipids, analytical techniques, **81**, 806
monoacylglycerophosphocholine, positional distribution, **86**, 373
monoacylglycerophosphoethanolamine, positional distribution, **86**, 373
polyunsaturated, HPLC, **86**, 523
tritiated, synthesis, **86**, 365
Fc receptors
IgE
assay, **93**, 186
purification, **93**, 182, 188
radiolabeling, **93**, 179
solubilization, **93**, 181
IgG
assay, **93**, 186
detection
with anti-receptor antibodies, **93**, 141
in isolation and purification processes, **93**, 140
on lymphoblastoid cell surfaces by solid-phase radioimmunoassay, **93**, 155
on lymphocyte surface with alkaline-labeled reagents, **93**, 164
in material secreted from lymphoreticular cells, **93**, 139
by molecular immune complexes and aggregated Ig binding, **93**, 131
by rosette formation with antibody-coated erythrocytes, **93**, 115
on fetal rabbit yolk sac membrane
binding assays *in vitro*, **93**, 198
properties, **93**, 210
specificity, **93**, 203
on human B-lymphocytes, characterization, **93**, 226
purification, **93**, 182, 188, 213, 220
quantitative fluorometric assay, **93**, 147

application, **93**, 154
sensitivity, **93**, 153
radiolabeling, **93**, 179
solubilization, **93**, 181, 195
Fentanyl
radioimmunoassay, **84**, 526
Ferricyanide
reduction, transketolase determination, **89**, 43
Ferritin
antiserum
antibody purification, **84**, 155
preparation, **84**, 153
cationized, as membrane marker for endocytosis and membrane recycling, **98**, 215
conjugation with epidermal growth factor, **98**, 287
iodination, **84**, 157
native, as fluid-phase marker for endocytosis and membrane recycling, **98**, 214
plasma, immunoassays, **84**, 148
isoferritin composition effect, **84**, 170
purification from tissues, **84**, 151
purity criteria, **84**, 153
storage, **84**, 153
α_1-Fetoprotein
antibody production, **84**, 4, 9
concanavalin A variants, assay, **84**, 18
double antibody enzyme immunoassay, **84**, 20
purification, **84**, 4
radioimmunoassay, **84**, 8
radioiodination, **84**, 12
sandwich enzyme immunoassay, **84**, 26
Fibrinogen
ELISA, **92**, 577
Fibrinopeptides
antibodies
detection, **84**, 109
specificity determination, **84**, 111
antiserum
choice for radioimmunoassay, **84**, 113
preparation, **84**, 106
in clinical blood samples, processing for radioimmunoassay, **84**, 114
radioimmunoassay, **84**, 116
applications, **84**, 117
radiolabeled analogs, preparation, **84**, 107
Fibroblasts
adhesion to collagen substrates
assay, **82**, 505
characteristics, **82**, 506, 542
requirement determination, **82**, 507
cultured monolayers
cholesteryl ester formation assay, **98**, 257
hydroxymethylglutaryl-CoA reductase assay, **98**, 255
low-density lipoproteins
proteolytic degradation, assay, **98**, 254
surface binding and internalization, assays, **98**, 251
from I-cell disease patient, pinocytosis of
[2-^{3}H]mannose-labeled glycopeptides, **98**, 295
[2-^{3}H]mannose-labeled oligosaccharides, **98**, 296, 297
from Sandhoff disease patients, β-hexosaminidase B pinocytosis measurements, **98**, 292
from Tay–Sachs disease patients
preparation of [2-^{3}H]mannose-labeled glycopeptides, **98**, 293
purification of β-hexosaminidase B, **98**, 291
Fibronectin
attachment assay, **82**, 526, 828
chemical characteristics, **82**, 803
immunochemical properties, **82**, 816
isolation, **82**, 808
localization by immunofluorescence, **82**, 826
quantitation, **82**, 821
Filamin
avian, purification and properties, **85**, 323
mammalian, purification and properties, **85**, 326
Films
air–water interface
assembly of bacteriorhodopsin-containing planar membranes, **88**, 50
formation, **88**, 47

Filtration
in actin polymerization assay, **85**, 207
gel, *see* Gel filtration
microtubules during assembly *in vitro*, **85**, 438
Flagella
Chlamydomonas, dynein ATPase preparation, **85**, 467
sea urchin sperm
dynein ATPase preparation, **85**, 453, 472
tubulin isolation, **85**, 388
starfish sperm, dynein ATPase preparation, **85**, 471
trout sperm, dynein ATPase preparation, **85**, 473
Flavin adenine dinucleotide
as label in colorimetric immunoassays, **92**, 413
Flavobacterium keratolyticus
endo-β-galactosidase isolation, **83**, 619
Flotation
Ficoll, preparation of osmotically intact retinal rod outer segment disks, **81**, 57
Flow birefringence
actin polymerization assay, **85**, 192
microtubule assembly assay, **85**, 436
N-2-Fluorenylacetamide
–DNA adducts, radioimmunoassay, **84**, 607
induction of endoplasmic reticulum membrane components, **96**, 544
Fluorescamine
prechromatographic derivatization of polyamines, **94**, 22
Fluorescence
in actin polymerization assay, **85**, 195
depolarization with diphenylhexatriene, **81**, 678
data analysis, **81**, 679
methodological problems, **81**, 681
energy transfer measurements, for determination of distances in rhodopsin and purple membrane protein, **81**, 668
in quantitative binding assay for Fc receptors, **93**, 147
time-resolved, bacteriorhodopsin photocycle intermediates, **88**, 306, 617
Fluorescence microscopy
intensifier systems, **98**, 272
addendum, **98**, 635
in localization of contractile proteins in cultured cells, **85**, 514
for protein detection in cultured cells, **98**, 280
Fluorescence spectroscopy
as probe of contractile protein
accessibility, **85**, 580
conformation, **85**, 575
distance, **85**, 590
molecular rotation, **85**, 586
orientation, **85**, 581
Fluorography
for isotope detection in
agarose gels, **96**, 220
polyacrylamide gels, **96**, 215
solid porous supports, **96**, 221
membrane proteins in polyacrylamide gels, **96**, 189
Flurazepam
radioimmunoassay, **84**, 507
Follicle cells
thyroid
closed structures
with access to both sides of layer, formation in culture, **98**, 485
formation in culture, **98**, 480
inside-out
in analysis of transcellular transport, **98**, 453
preparation, **98**, 447
structural and functional polarity, **98**, 451
isolation, **98**, 480
membrane markers of polarity, **98**, 479
open structures, formation in culture, **98**, 484
organizational consequences, **98**, 479
polarization, **98**, 477
Foot-and-mouth disease virus
vaccine preparation with glutaraldehyde, **93**, 44
Formaldehyde
fixation of proteins in polyacrylamide gels, **91**, 247

labeling of proteins by reductive alkylation, **91**, 570
Formaldehyde dehydrogenase (glutathione)
Candida boidinii
assay, **89**, 527
production, **89**, 528
properties, **89**, 530
purification, **89**, 528
Formate dehydrogenase
Pseudomonas oxalaticus
assay, **89**, 531
flavin removal and reactivation, **89**, 536
properties, **89**, 535
purification, **89**, 532
Formate dehydrogenase (cytochrome)
Escherichia coli
assay, **89**, 538
properties, **89**, 542
purification, **89**, 540
Formic acid
protein cleavage at aspartic acid, **91**, 324
Formylglycinamide ribonucleotide amidotransferase, *see* Phosphoribosylformylglycinamidine synthase
N-Formylmethionyl-tRNA
^{35}S-labeled
preparation, **96**, 152
protein NH_2 terminus labeling, **96**, 154
N'-Formylnornicotine
radioimmunoassay, **84**, 641
sensitivity and specificity, **84**, 646
synthesis, **84**, 644, 645
N^5-Formyltetrahydrofolate
ligand-binding radioassay, **84**, 462
Fowl plague virus
interaction with MDCK cells, **98**, 264
Freeze-fracture techniques
freeze-substitution and freeze-etching, for photoreceptor membrane analysis, **88**, 118
for frog retinal photoreceptor analysis, **81**, 585
β-Fructofuranosidase
accumulated forms in yeast secretory mutant, electrophoretic analysis, **96**, 812
D-Fructose
^{13}C-enriched, enzymatic synthesis, **89**, 84
microdetermination with membrane-bound dehydrogenase, **89**, 24
Fructose-bisphosphatase
Bacillus licheniformis
assay, **90**, 385
properties, **90**, 388
purification, **90**, 386
bovine liver
assay, **90**, 330
properties, **90**, 333
purification, **90**, 331
bumblebee flight muscle
assay, **90**, 366
properties, **90**, 370
purification, **90**, 367
chicken muscle
assay, **90**, 340
properties, **90**, 344
purification, **90**, 341
murine intestinal mucosa
assay, **90**, 358
properties, **90**, 365
purification, **90**, 360
rabbit
intestinal mucosa
assay, **90**, 358
properties, **90**, 365
purification, **90**, 362
liver
assay, **90**, 327, 345
properties, **90**, 347
purification, **90**, 327, 346
zinc-free, preparation, **90**, 328
muscle
assay, **90**, 340
properties, **90**, 344
purification, **90**, 343
rat liver
assay, **90**, 352
properties, **90**, 355
purification, **90**, 353
Rhodopseudomonas palustris
assay, **90**, 379
properties, **90**, 383
purification, **90**, 379
snake muscle
assay, **90**, 349

properties, **90**, 351
purification, **90**, 350
spinach leaf
assay, **90**, 372
properties, **90**, 377
purification from
chloroplast, **90**, 374
cytoplasm, **90**, 375
turkey liver
assay, **90**, 334
properties, **90**, 337
purification, **90**, 335
D-Fructose 1,6-bisphosphate
^{13}C-enriched, enzymatic synthesis, **89**, 89
reaction with dicyclohexylcarbodiimide, **89**, 103
Fructose 2,6-bisphosphate
properties, **89**, 104
synthesis, **89**, 101
Fructose-bisphosphate aldolase
Ascaris suum
assay, **90**, 255
properties, **90**, 257
purification, **90**, 256
Bacillus subtilis
assay, **90**, 235
properties, **90**, 240
purification, **90**, 238
bovine muscle, simultaneous purification with glyceraldehyde-3-phosphate dehydrogenase, phosphoglyerate kinase, and phosphoglycerate mutase, **90**, 509
chicken muscle
assay, **90**, 490
purification, **90**, 492
class I
Mycobacterium smegmatis
assay, **90**, 242
properties, **90**, 245
purification, **90**, 243
Mycobacterium tuberculosis
assay, **90**, 242
properties, **90**, 250
purification, **90**, 249
class II, *Mycobacterium tuberculosis*
assay, **90**, 242
properties, **90**, 248
purification, **90**, 247
Helix pomatia
assay, **90**, 259
properties, **90**, 262
purification, **90**, 259
human erythrocyte
assay, **90**, 251
properties, **90**, 253
purification, **90**, 251
Fructose 5-dehydrogenase
Gluconobacter industrius membrane-bound
assay, **89**, 154
microdetermination of D-fructose, **89**, 24
properties, **89**, 158
purification, **89**, 156
Fructose 6-phosphate
determination with transketolase, **89**, 19
Fructuronate reductase
Escherichia coli
assay, **89**, 210
properties, **89**, 213
purification, **89**, 212
Fucose
binding protein, rat liver
assay, **98**, 314
properties, **98**, 318
purification, **98**, 315
α-L-Fucosidase
almond emulsin
assay, **83**, 626
properties, **83**, 630
purification, **83**, 628
(Fucosylα1→2)galactoside α1→3-*N*-acetylgalactosaminyltransferase
porcine submaxillary gland
assay, **83**, 488
properties, **83**, 493
purification, **83**, 489

G

D-Galactitol-6-phosphate dehydrogenase
Klebsiella pneumoniae
assay, **89**, 275
properties, **89**, 277
purification, **89**, 276

Galactocerebrosides
  radioisotopic assay, **83**, 191
Galactokinase
  Saccharomyces cerevisiae
    assay, **90**, 30
    properties, **90**, 33
    purification, **90**, 31
D-Galactonate dehydratase
  pseudomonad MSU-1
    assay, **90**, 294
    properties, **90**, 297
    purification, **90**, 295
D-Galactonate/D-fuconate dehydratase
  Klebsiella pneumoniae
    assay, **90**, 299
    properties, **90**, 302
    purification, **90**, 300
β-D-Galactopyranosylhydroxylysine
  production, **82**, 342
  quantitation in peptides and proteins, **82**, 343
  sequence analysis, **82**, 344
D-Galactose
  binding protein, *Escherichia coli*
    assay, **90**, 464
    purification, **90**, 465
Galactose dehydrogenase
  Pseudomonas fluorescens
    assay, **89**, 177
    properties, **89**, 181
    purification, **89**, 178
Galactose oxidase
  from commercial samples
    assay, **89**, 174
    properties, **89**, 175
    separation from contaminating proteins, **89**, 172
  Dactylium dendroides
    assay, **89**, 167
    properties, **89**, 170
    purification, **89**, 168
α-D-Galactose 1-phosphate
  ^{13}C-enriched, enzymatic synthesis, **89**, 91
D-Galactose 6-phosphate
  synthesis, **89**, 93
D-Galactose-6-phosphate isomerase
  Staphylococcus aureus
    assay, **89**, 562
    properties, **89**, 565
    purification, **89**, 563
Galactose-1-phosphate uridylyltransferase, *see* UDPglucose-hexose-1-phosphate uridylyltransferase; UTP–hexose-1-phosphate uridylyltransferase
β-Galactosidase
  assay, in diagnosis of genetic mucopolysaccharide storage disorders, **83**, 565
  Escherichia coli
    assay in SDS gels, **91**, 274
    (2*R*)-glycerol-*o*-β-D-galactopyranoside synthesis, **89**, 59
  gene fusions
    for analyzing gene expression in *Escherichia coli* and yeast, **100**, 293
    construction and application in yeast, **100**, 167
  in label release from plasma membrane glycoconjugates, **98**, 419
  linked immunoassay for prostaglandin $F_{2\alpha}$, **86**, 269
β-Galactoside α1→2-fucosyltransferase
  porcine submaxillary gland
    assay, **83**, 460
    properties, **83**, 469
    purification, **83**, 462
β-Galactoside α2→3-sialyltransferase
  porcine submaxillary gland
    assay, **83**, 471
    properties, **83**, 477
    purification, **83**, 472
β-Galactoside α2→6-sialyltransferase
  porcine colostrum
    assay, **83**, 495
    properties, **83**, 499
    purification, **83**, 496
Galactosyltransferase, *see* Lactose synthase
Gangliosides
  analytical techniques, **83**, 175
  cholera toxin-binding, detection, **83**, 238
  hydrolytic protein activators
    assay, **83**, 591
    properties, **83**, 594
    purification, **83**, 588, 592
  monoclonal antibody-binding, detection, **83**, 239
  purification, **83**, 155

resolution, **83**, 164
structure, **83**, 140
determination, **83**, 183
Gap junctions
hepatic
characterization, **98**, 508
isolation, **98**, 503, 521
purity criteria, estimation
biochemical methods, **98**, 532
morphological methods, **98**, 530
uterine smooth muscle
biogenesis, **98**, 537
detection of protein components, **98**, 545
Gas chromatography–mass spectrometry
eicosanoids, ^{18}O derivative preparation, **86**, 547
hexosamine-containing oligosaccharide alditols, **83**, 46
5-hydroxyeicosatetraenoic acid in biological fluids, **86**, 607
peralkylated oligosaccharide alditols, **83**, 21, 34
polyamines and polyamine conjugates, **94**, 48
prostacyclin dinor metabolites, **86**, 585
prostaglandin D_2 urinary metabolite, **86**, 559
prostaglandin E_2 plasma metabolite, **86**, 571
prostaglandin $F_{2\alpha}$ urinary metabolites, **86**, 557, 579
prostaglandin ω-hydroxylase products, **86**, 174
thromboxane urinary metabolites, **86**, 592, 605
Gas–liquid chromatography
eicosanoids by open tubular gas capillary method, **86**, 612
gangliosides, **83**, 179
myo-inositol 1-phosphate, **89**, 3, 14
permethylated *N*-trifluoroacetyl derivatives of hexosamine-containing oligosaccharide alditols, **83**, 46
sugar enantiomers, **89**, 3
GDP-fucose
synthesis, **98**, 103
GDP-fucose:β-*N*-acetylglucosaminide α1-6-fucosyltransferase
assay, **98**, 131
properties, **98**, 131
GDP-fucose:β-galactoside α1-2-fucosyltransferase
properties, **98**, 132
GDP-fucose:β-galactoside α1-3-fucosyltransferase
properties, **98**, 132
GDP-fucose:β-galactoside α1-4-fucosyltransferase
properties, **98**, 132
GDP-D-mannose epimerase
Chlorella pyrenoidosa
assay, **83**, 522
properties, **83**, 525
purification, **83**, 524
Gel electrophoresis
acetylcholine receptor subunits synthesized *in vitro*, **96**, 371
alcohol dehydrogenase SS isozyme, **89**, 431
cAMP-dependent protein kinase regulatory subunit, **99**, 238
for collagenous protein size estimation, **82**, 411
denatured collagen, **82**, 63
desmin, **85**, 494
diacetyl reductase, **89**, 517
DNA in submicrogram amounts, **84**, 181
erythrocyte band III proteolytic fragments, **96**, 262
L-glycol dehydrogenase, **89**, 524
glycophorin A, **96**, 289
invertase accumulated in yeast secretory mutants, **96**, 812
lymphocyte hybridoma glucose-6-phosphate isomerase isozymes, **92**, 237
lysosomal enzymes, **96**, 734
membrane proteins, **96**, 184, 194
monoclonal antibodies and specific antigens, **92**, 196
for mRNA size fractionation, **96**, 33
myelin-specific proteins, **96**, 384
myosin, **85**, 127
nascent chains, for analysis of cotranslational proteolytic processing, **97**, 77
nascent secretory proteins, **96**, 105
oligopeptides below M_r 10,000 with slab gel apparatus, **96**, 239

outer membrane proteins from *Escherichia coli*, **97**, 118
phosphorylated proteins in intact cells, **99**, 28
in preparation of stable radioiodinated polypeptide hormones and proteins, **92**, 309
preparative polyacrylamide, **91**, 214
applications, **91**, 218
proteins
in cell extracts, **96**, 208
translated in rabbit reticulocyte lysates, **96**, 63
pyruvate kinase isozymes with thin-layer polyacrylamide gel system, **90**, 155
radiolabeled proteins from cultured cells, **100**, 411
-separated antibodies and antigens, specificity analysis with enzyme-linked immunoelectrotransfer blot technique, **92**, 377
Sindbis virus membrane glycoproteins, **96**, 515
spectrin intermediate-sized peptides, **96**, 306
spectrin oligomers, **96**, 298
two-dimensional, on giant gels
advantages, **91**, 193
for protein separation, **91**, 190
vimentin, **85**, 494
Gel filtration
myosins, **85**, 335
oligosaccharides, **83**, 105
Gelmeter
in actin filament network characterization, **85**, 221
Gene fusions
to β-galactosidase *lacZ*
for analyzing gene expression in
Escherichia coli, **100**, 293
yeast, **100**, 293; **101**, 176
for cloning regulated genes in yeast, **101**, 253
construction
for application in yeast, **101**, 167
in *Escherichia coli*, **97**, 12
design for analyzing cloned gene expression in yeast, **101**, 181
for selection of
export-defective mutants in *Escherichia coli*, **97**, 25
signal sequence mutations in *Escherichia coli*, **97**, 5
Mud(Ap, *lac*)-generated, in analysis of gene expression, **100**, 501
Genes
bacterial glycosyltransferase, cloning and selection, **83**, 326
bacteriophage λ *int*, protein purification and properties, **100**, 210
bacteriophage φX174 *A*, protein analysis with *in vitro* DNA replication system, **100**, 217
bacteriorhodopsin, isolation by colony hybridization, **97**, 226
cloned
analysis with transposon γδ, **101**, 362
expression
regulation with plasmids containing *trp* promoters of *Escherichia coli* and *Serratia marcescens*, **101**, 155
in *Xenopus* oocytes, **101**, 370
cloning in pKC30 and derivatives for controlled expression, **101**, 123
DNA repair, amplification with pKC30, **101**, 138
Escherichia coli, expression analysis with
β-galactosidase fusions, **100**, 293
Mud(Ap, *lac*)-generated fusions, **100**, 501
eukaryotic
cloning vector construction from bovine papillomavirus DNA, **101**, 387
transcription with purified components, **101**, 582
with expression based on di- and tripeptide formation, analysis with coupled DNA-directed *in vitro* system, **101**, 690
leader peptidase
isolation from *Escherichia coli*, **97**, 46
structure and expression, **97**, 51
maize chloroplast
cloning, **97**, 535

physical mapping, **97**, 546
mitochondrial, in yeast
characterization and selection, **97**, 344
isolation, **97**, 363
physical characterization, **97**, 369
multigene families, isolation and homology determination by filter hybridization, **100**, 266
affecting parameters, **100**, 268
mutant, in yeast
cloning by integrative transformation, **101**, 290
eviction and transplacement, **101**, 211
pet, in yeast
mutant transformation, **97**, 349, 355
confirmation, **97**, 360
translation *in vitro*, **97**, 352
*rec*A, protein analysis in *Escherichia coli*, **100**, 197
regulated, in yeast, cloning with *lacZ* fusions, **101**, 253
STE, from yeast, cloning in 2-μm vectors, **101**, 325
structural, coding for
adenosinetriphosphatase complex in *Escherichia coli*, cloning, **97**, 176
glycogen biosynthetic enzymes, isolation from *Escherichia coli*, **83**, 549
transfer into murine embryos by pronuclear fusion, **101**, 411
transposon γδ resolvase protein
cloning, **100**, 192
nucleotide sequence, **100**, 195
yeast
expression analysis with β-galactosidase fusions, **100**, 293
one-step disruption, **101**, 202
Gentamicin
fluorescence fluctuation immunoassay, **92**, 458
Globomycin
in isolation of *Escherichia coli* outer membrane lipoprotein mutants, **97**, 124
Glucokinase
Streptococcus mutans
assay, **90**, 25
properties, **90**, 28
purification, **90**, 27
D-Gluconate
microdetermination with membrane-bound dehydrogenase, **89**, 25
Gluconate dehydratase
Clostridium pasteurianum
assay, **90**, 283
properties, **90**, 286
purification, **90**, 284
Gluconate 2-dehydrogenase
from acetic acid bacteria
assay, **89**, 203
properties, **89**, 208
purification, **89**, 204
bacterial membrane-bound
assay, **89**, 187
microdetermination of D-gluconate, **89**, 25
properties, **89**, 191
purification, **89**, 188
Gluconate 5-dehydrogenase
Gluconobacter suboxydans
assay, **89**, 199
properties, **89**, 201
purification, **89**, 200
Gluconobacter
G. industrius
cultivation, **89**, 155
D-fructose dehydrogenase purification, **89**, 156
G. liquefaciens
cultivation, **89**, 204
2-keto-D-gluconate reductase purification, **89**, 204
G. melanogenus
cultivation, **89**, 195
membrane-bound 2-keto-D-gluconate dehydrogenase purification, **89**, 195
G. suboxydans
cultivation, **89**, 142, 199, 292, 451
purification of
glucose dehydrogenase, **89**, 160
5-keto-D-gluconate reductase, **89**, 200
membrane-bound alcohol dehydrogenase, **89**, 452
membrane-bound aldehyde dehydrogenase, **89**, 493

membrane-bound D-sorbitol dehydrogenase, **89**, 142
6-phospho-D-gluconate dehydrogenase, **89**, 292
2-*O*-β-D-Glucopyranosyl-*O*-β-D-galactopyranosylhydroxylysine
conversion to β-D-galactopyranosylhydroxylysine, **82**, 342
isolation from sponge, **82**, 339
quantitation in proteins and peptides, **82**, 343
sequence analysis, **82**, 344
Glucosamine
inhibition of glycoprotein lipid-linked saccharide formation, **98**, 143
Glucosamine acetyltransferase
in diagnosis of genetic mucopolysaccharide storage disorders, **83**, 561
radioactive substrates, preparation, **83**, 566
D-Glucose
^{13}C-enriched, enzymatic synthesis, **89**, 84, 90
microdetermination with membrane-bound dehydrogenase, **89**, 23
transport assay in reconstituted system, **98**, 441
L-Glucose
transport assay in reconstituted system, **98**, 441
Glucose analogs
inhibition of glycoprotein lipid-linked saccharide formation, **98**, 142
Glucose dehydrogenase
Gluconobacter suboxydans
assay, **89**, 159
properties, **89**, 162
purification, **89**, 160
Glucose dehydrogenase (acceptor)
Gluconobacter suboxydans membrane-bound, microdetermination of D-glucose, **89**, 23
Pseudomonas fluorescens membrane-bound
assay, **89**, 149
microdetermination of D-glucose, **89**, 23
properties, **89**, 152
purification, **89**, 150
Glucose oxidase
purification, **89**, 46
Glucose-6-phosphatase
rat brain
assays, **90**, 398
comparison with hepatic form, **90**, 401
extraction and purification, **90**, 400
rat liver
differential effects of copper, **87**, 350
inhibitor effects, **87**, 346
kinetic mechanism, **87**, 327
pH kinetics, **87**, 345
rate constants
applicable equations, **87**, 334
experimental determination, **87**, 340
interrelationships, **87**, 335, 344
metabolic implications, **87**, 340
representative values, **87**, 343
spectrophotometric assay, **89**, 44
Glucose 6-phosphate
in 6-phosphogluconate preparation, **89**, 116
preparation, **89**, 112
Glucose-6-phosphate dehydrogenase
Bacillus subtilis vegetative and spore
assay, **89**, 258
properties, **89**, 261
purification, **89**, 259
Methylomonas M15
assay, **89**, 271
production, **89**, 272
properties, **89**, 274
purification, **89**, 272
murine
assay, **89**, 252
properties, **89**, 257
purification, **89**, 255
Pseudomonas fluorescens
activity measurement, **89**, 262
assay, **89**, 263
in Entner–Doudoroff pathway
properties, **89**, 266
purification, **89**, 264
Zwischenferment
properties, **89**, 269
purification, **89**, 267

Glucose-6-phosphate isomerase
catfish liver and muscle
assay, **89**, 550
purification, **89**, 555
Drosophila melanogaster
assay, **89**, 559
properties, **89**, 562
purification, **89**, 560
isozyme electrophoresis for detection of lymphocyte hybridomas, **92**, 237
mammalian tissue
allozyme resolution, **89**, 555
assay, **89**, 550
isozymes and allozymes, **89**, 551
purification, **89**, 552
murine muscle
assay, **89**, 559
properties, **89**, 562
purification, **89**, 560
Glucose-6-phosphate phosphatase, *see* Glucose-6-phosphatase
Glucose transporter
insulin-sensitive, in rat adipocytes
assay, **98**, 441
solubilization and reconstitution, **98**, 438
Glucosidase I
assay, **83**, 419, 429
properties, **83**, 431
purification from hen oviduct, **83**, 430
Glucosidase II
assay, **83**, 419
properties, **83**, 421
purification from bovine liver, **83**, 420
β-Glucosides
hydrolytic protein activator
assay, **83**, 598
properties, **83**, 601
purification, **83**, 599
Glucuronate reductase
porcine kidney
assay, **89**, 501
properties, **89**, 505
purification, **89**, 502
β-Glucuronidase
lysosomal
antibody preparation, **96**, 770
assay, **96**, 559, 765
immunoprecipitation, **96**, 773
purification, **96**, 560, 765
synthesis *in vitro*, **96**, 770
microsomal
assay, **96**, 559
purification, **96**, 561
purification from rat preputial glands, **98**, 302
uptake and binding assays in macrophages, **98**, 303
Glutamate–ammonia ligase
isotope exchange study, fingerprint kinetic patterns for modifier action, **87**, 648
Glutamate dehydrogenase
isotope exchange study, fingerprint kinetic patterns for modifier action, **87**, 652
Glutamine synthetase, *see* Glutamate–ammonia ligase
γ-Glutamylamine cyclotransferase
rabbit kidney
assay, **94**, 348
in L-γ-glutamylamine identification, **94**, 351
properties, **94**, 350
purification, **94**, 349
substrates, **94**, 457
L-γ-Glutamylamines
identification with γ-glutamylamine cyclotransferase, **94**, 351
N^1-(γ-Glutamyl)spermidine
ion-exchange chromatography, **94**, 456
preparation, **94**, 452
N^8-(γ-Glutamyl)spermidine
ion-exchange chromatography, **94**, 456
preparation, **94**, 452
N^1-(γ-Glutamyl)spermine
ion-exchange chromatography, **94**, 456
preparation, **94**, 452
γ-Glutamyltransferase
assay, **86**, 42
association with parotid plasmalemma and secretion granule membranes, **98**, 85
purification from porcine kidney, **86**, 43
γ-Glutamyl transpeptidase, *see* γ-Glutamyltransferase
Glutaraldehyde
in vaccine preparation, action on allergens, **93**, 50

bacterial cells, **93**, 38
toxins, **93**, 25
tumor cells, **93**, 56
venoms, **93**, 44
viruses, **93**, 41
Glutathionylspermidine
Escherichia coli
assay, **94**, 434
purification, **94**, 436
D-Glyceraldehyde 3-phosphate
preparation, **89**, 75
DL-Glyceraldehyde 3-phosphate
preparation, **89**, 76
Glyceraldehyde-3-phosphate dehydrogenase
bovine muscle, simultaneous purification with aldolase, 3-phosphoglycerate kinase, and phosphoglycerate mutase, **90**, 509
human tissue
assay, **89**, 302
properties, **89**, 304
purification, **89**, 303
pea seed
assay, **89**, 320
properties, **89**, 324
purification, **89**, 321
porcine liver and muscle
assay, **89**, 310; **90**, 499
properties, **89**, 314; **90**, 508
purification, **89**, 312
simultaneous purification with 3-phosphoglycerate kinase and phosphoglycerate mutase, **90**, 499
rabbit muscle
assay, **89**, 306
properties, **89**, 308
purification, **89**, 306
spinach leaf
assay, **89**, 316
properties, **89**, 319
purification, **89**, 317
Thermus thermophilus
assay, **89**, 335
properties, **89**, 340
purification, **89**, 336
yeast
assay, **89**, 326
properties, **89**, 333
purification, **89**, 330
Glyceraldehyde-3-phosphate dehydrogenase ($NADP^+$) (phosphorylating)
chicken muscle
assay, **90**, 490
purification, **90**, 492
Glycerol dehydrogenase ($NADP^+$)
rabbit muscle
assay, **89**, 238
properties, **89**, 241
purification, **89**, 239
(2*R*)-Glycerol-*o*-β-D-galactopyranoside
radiolabeled, preparation, **89**, 63
synthesis with β-galactosidase, **89**, 59
Glycerol oxidase
Aspergillus japonicus
assay, **89**, 243
properties, **89**, 245
purification, **89**, 244
Glycerol-3-phosphate dehydrogenase (NAD^+)
Drosophila melanogaster
assay, **89**, 297
properties, **89**, 301
purification, **89**, 297
Glycoconjugates
cell surface, ultrastructural visualization, **83**, 195
plasma membrane
labeling with galactosyltransferase, **98**, 416
label release with β-galactosidase, **98**, 419
prebound, internalization by macrophages, **98**, 304
preparation, **98**, 302
uptake by macrophages, **98**, 301
Glycogen biosynthetic enzymes
Escherichia coli, structural gene isolation, **83**, 549
Glycogen phosphorylase
isotope exchange study, fingerprint kinetic patterns for modifier action, **87**, 654
Glycogen synthase
peptide substrates, in analysis of phosphorylase kinase specificity, **99**, 275
Glycogen synthase kinase
calmodulin-dependent, purification from rabbit liver, **99**, 299

type 3, rabbit skeletal muscle
assay, **99**, 337
properties, **99**, 343
purification, **99**, 340
Glycolaldehyde phosphate
preparation, **89**, 74
L-Glycol dehydrogenase
hen muscle
assay, **89**, 523
properties, **89**, 525
purification, **89**, 524
Glycolytic enzymes
rabbit muscle
assays, **90**, 479
purification, **90**, 480
Glycopeptides
asparagine-linked sugar chains, hydrazinolysis to produce free oligosaccharides, **83**, 263
fractionation, **83**, 271
high-molecular-weight, preparation from human erythrocytes, **83**, 311
[2-^{3}H]mannose-labeled
pinocytosis by I-cell disease fibroblasts, **98**, 295
preparation from Tay–Sachs disease fibroblast secretions, **98**, 293
iodination, **98**, 156
preparation, **83**, 269
as probe for receptor studies, **98**, 154
sugar-labeled, preparation from procollagen, **82**, 347
Glycophorins
major
biogenesis and processing *in vitro*, **96**, 281
chemical and physicochemical properties, **96**, 277
glycosylated domains, preparation, **96**, 274
isolation, **96**, 269
orientation, **96**, 280
minor
chemical and physicochemical properties, **96**, 277
glycosylated domains, preparation, **96**, 274
isolation, **96**, 269
orientation, **96**, 280
Glycoprotein 4-β-galactosyltransferase
avian salt gland, assay, **96**, 638
Glycoproteins
asparagine-linked sugar chains, hydrazinolysis to produce free oligosaccharides, **83**, 263
composition determination in retinal interphotoreceptor matrix, **81**, 85
envelope, avian myeloblastosis virus
isolation, **83**, 321
properties, **83**, 323
lectin-binding, visualization on SDS gels, **81**, 71
membrane
immunologic detection, **96**, 205
synthesis sites, **96**, xxxv
myxovirus
assays, **96**, 436
biosynthesis, **96**, 434
mutants with transport defects
preparation, **96**, 440
protease activation, **96**, 442
processing inhibitors, **98**, 151
Semliki Forest virus, transport analysis with
organelle-specific lectins, **96**, 461
temperature-sensitive mutants, **96**, 453
Sindbis virus membrane
affinity chromatography, **96**, 519
biosynthesis *in vitro*, **96**, 512
gel electrophoresis, **96**, 515
posttranslational assays, **96**, 517
somatic cell
mutagenesis, **83**, 445
mutant selection for
cytotoxic lectin resistance, **83**, 447
tritiated carbohydrate suicide resistance, **83**, 451
structural
extraction and purification, **82**, 842
properties, **82**, 840
synthesis inhibitors, **98**, 135
uptake assays in macrophages, **98**, 301
vesicular stomatitis virus G, assay of transport-coupled oligosaccharide processing *in vitro*, **98**, 37

Glycosaminoglycans
composition determination in retinal interphotoreceptor matrix, **81**, 85
definition, **82**, 769
Glycosides
ω-acrylamidoalkyl, *see* ω-Acrylamidoalkyl glycosides
asparagine-linked, in procollagen
characterization, **82**, 348
preparation, **82**, 347
cardiac, radioimmunoassay, **84**, 558
hydroxylysine, *see* Hydroxylysine glycosides
Glycosphingolipids
high-resolution proton NMR, **83**, 69
Glycosylation
protein
dolichol pathway inhibition, **83**, 432
role of intracellular membrane systems, **98**, 91
rhodopsin, **81**, 783
Glycosyltransferase
bacterial, gene cloning, **83**, 326
mammalian, purification, **83**, 458
Glyoxalase I, *see* Lactoylglutathione lyase
Glyoxalase II, *see* Hydroxyacylglutathione hydrolase
Glyoxysomes
component analysis, **96**, 700
cytosol
detection of enzyme monomeric and oligomeric forms, **96**, 705
kinetic studies *in vivo*, **96**, 708
protein synthesis and import *in vitro*, **96**, 710
Golgi complex
murine plasma cell, perturbation by monovalent ionophores, **98**, 47
effect on
intracellular protein transport, **98**, 51
morphology, **98**, 49
mechanism, **98**, 57
reversibility, **98**, 56
transport arrest site, **98**, 52
rat liver, isolation with
D_2O–sucrose gradients, **98**, 62
H_2O–sucrose gradients, **98**, 61
thyroid follicle cell, endocytic pathways to, **98**, 453
transport pathways
from endoplasmic reticulum, **98**, 8
to lysosomes, **98**, 8
from plasmalemma, **98**, 3
Growth hormone, *see* Somatotropin
GTPase
from bovine rod cell outer segments
assay, **81**, 564
purification, **81**, 561, 567
Guanidino groups
protein, specific crosslinking reagents, **91**, 600
Guanidospermidine
chemical synthesis, **94**, 449
identification, **94**, 451
purification, **94**, 450
Guanylate cyclase
retinal
assays, **81**, 519, 525
localization, **81**, 524

H

Haldanes
applications, **87**, 369
kinetic, **87**, 367
nomenclature, **87**, 366
for Ping Pong mechanisms, **87**, 368
for random mechanisms, **87**, 368
thermodynamic, **87**, 366
Halobacteria
family definition, **88**, 361
Halobacterium
classification key, **88**, 365
description, **88**, 366, 369
genus definition, **88**, 364
H. cutirubrum
plasmid pHC1
molecular weight, **88**, 376
restriction map, **88**, 379
purple membrane lipids
composition, **88**, 101
extraction, **88**, 100
separation, **88**, 100
H. halobium
anaerobic growth
arginine-dependent, **88**, 418
light-dependent, **88**, 418
assay for bacteriorhodopsin and halorhodopsin, **88**, 13

brown membrane
bacterio-opsin
conversion *in vivo* to bacteriorhodopsin, **88**, 394
pulse labeling with radioactive amino acids, **88**, 395
preparation, **88**, 391
properties, **88**, 393
envelope vesicles
assay for
bacteriorhodopsin, **88**, 13
halorhodopsin, **88**, 13, 430
light-induced pH gradients, measurement, **88**, 334
preparation, **88**, 11, 427
properties, **88**, 430
reconstitution of bacteriorhodopsin and halorhodopsin, **88**, 14
fully deuterated, isolation of deuterated bacteriorhodopsin, **88**, 3
isolation of bacterio-opsin gene, **97**, 226
measurement of light-regulated phosphoproteins, **88**, 213
mutant strains with pigment synthesis defects
discrimination, **88**, 388
isolation, **88**, 379
spectroscopic determination, **88**, 383
photobehavioral analysis, **88**, 420
photophosphorylation, **88**, 349
reconstitution, **88**, 355
plasmids
molecular weights, **88**, 376
purification, **88**, 374
restriction mapping, **88**, 375
purple membranes
apomembrane preparation, **88**, 12
biogenesis, **97**, 218
electron microscopy, **97**, 221
ferritin labeling, **97**, 221
isolation, **88**, 12, 98; **97**, 220
lipids
composition, **88**, 101
extraction, **88**, 100
separation, **88**, 100
optical diffraction, **97**, 222
protein isolation by HPLC, **88**, 202
reconstituted, X-ray diffraction, **88**, 268
recrystallization, **97**, 223
silver decoration and shadowing, **97**, 222
reconstitution of bacteriorhodopsin and halorhodopsin, **88**, 14
secondary ion movements
atomic absorption spectroscopy and flame photometry, **88**, 346
dielectric dispersion measurements, **88**, 347
ion-selective electrode measurements, **88**, 345
radioactive isotopic determination, **88**, 346
spectrophotometric determination of halorhodopsin in membrane, **88**, 439
strain R_1mW lacking carotenoid, isolation of crystalline bacterio-opsin white membrane, **88**, 407
white membranes
bacteriorhodopsin reconstitution, **88**, 16
isolation, **88**, 12
wild-type and mutant vesicles, simultaneous measurements of proton movement and membrane potential, **88**, 356
isolation and cultivation, **88**, 370
phenotypic and genotypic characterization, **88**, 373
Halococcus
genus definition, **88**, 364
Haloperidol
radioimmunoassay, **84**, 532
comparison with gas chromatographic determination, **84**, 538
Halorhodopsin
discrimination from bacteriorhodopsin in cell suspensions by measuring light-induced pH changes, **88**, 388

by flash spectroscopy techniques, **88**, 434
reconstitution in intact cells and cell vesicles, **88**, 14
spectrophotometric determination, **88**, 13, 434, 439
Hansenula miso
pyruvate dehydrogenase complex purification, **89**, 421
Haptens
attachment to erythrocyte surface
facilitation by succinylation, **92**, 265
with methyl-*p*-hydroxybenzimidate, **92**, 269
with *N*-succinimidyl 3-(2-pyridyldithio)propionate, **92**, 259
with synthetic lipopolysaccharide, **92**, 274
with 1,3,5-trichlorotriazine, **92**, 271
Hapten-sandwich labeling
applications, **92**, 481
methods, **92**, 472
Heart
bovine, purification of
phosphorylase kinase, **99**, 255
pyruvate dehydrogenase complex, **89**, 383
pyruvate dehydrogenase phosphatase, **90**, 404
type II cAMP-dependent protein kinase
catalytic subunit, **99**, 53, 176
regulatory subunit, **99**, 56, 176
canine, purification of creatine kinase isozymes, **90**, 195
human, purification of
creatine kinase isozymes, **90**, 187
glyceraldehyde-3-phosphate dehydrogenase, **89**, 303
Hemagglutination
passive, as test for DNA antibodies, **84**, 201
applications, **84**, 208
Hemagglutinin
Aleuria aurantia
applications, **83**, 377
assays, **83**, 374
properties, **83**, 376
purification, **83**, 374
Hemoglobin
tryptic peptide mapping by ion-exchange and thin layer chromatography, **91**, 384
Hemolysis
passive, as test for DNA antibodies, **84**, 201
applications, **84**, 208
transfer template localized, in gel assay for detection of antibodies, **92**, 184
Heparin
as inhibitor for casein kinase I and II, **99**, 314, 327
Hepatocytes, *see also* Liver
collagen adhesion, **82**, 513
antibody effects, **82**, 525
determination, **82**, 521
cytoskeleton, role in protein secretion, **98**, 169
endoplasmic reticulum membrane
cotranslationally inserted polypeptides, **96**, 535
posttranslationally inserted polypeptides, **96**, 533
fibronectin adhesion, **82**, 513
antibody effects, **82**, 525
determination, **82**, 521
fractionation for preparation of rough microsomes and membrane-bound polysomes, **96**, 3
growth and differentiation with collagen gel as substrate, **82**, 544
intercellular adhesion determination after collagen attachment, **82**, 526
membrane endocytic compartments, selective labeling, **98**, 232
morphology after colchicine exposure, **98**, 171
nonsecreted proteins, intracellular localization, **98**, 171
peroxisomes
drug induction, **96**, 549
isolation, **96**, 553
phosphomannosyl receptors
assay, **83**, 392
properties, **83**, 396
purification, **83**, 394
protein secretion
colchicine effect *in vivo*, **98**, 169

measurement, colchicine side effects, **98**, 173
Hexokinase
Ascaris suum muscle
assay, **90**, 21
properties, **90**, 24
purification, **90**, 22
Drosophila melanogaster
assay, **90**, 16
properties, **90**, 21
purification, **90**, 17
rabbit erythrocyte
assay, **90**, 3
properties, **90**, 7
purification, **90**, 4
rat brain
activity assay, **97**, 471
binding assay, **97**, 470
binding protein isolation from outer mitochondrial membrane, **97**, 472
type Ia and Ib, rabbit reticulocyte
assay, **90**, 3
properties, **90**, 10
purification, **90**, 8
type II, rat skeletal muscle
assay, **90**, 11
properties, **90**, 15
purification, **90**, 12
yeast, isotope exchange study, kinetic patterns for modifier action, **87**, 654
β-Hexosaminidase B, *see* β-*N*-Acetylhexosaminidase B
Hexose oxidase
Chondrus crispus
assays, **89**, 145
properties, **89**, 148
purification, **89**, 147
3-Hexulose-6-phosphate synthase
Methylococcus capsulatus
assay, **90**, 314
properties, **90**, 317
purification, **90**, 316
Methylomonas M15
assay, **90**, 319
production, **90**, 320
properties, **90**, 322
purification, **90**, 321
High-performance liquid chromatography
acetylated oligosaccharides by reversed-phase technique, **83**, 132
adenosyl-sulfur compounds related to polyamine biosynthesis, **94**, 57
amino acids in picomole range, **91**, 41
amino-terminal acetylamino acids, **91**, 84
anionic oligosaccharides, **98**, 157
arachidonic acid metabolites, **86**, 511
argentation technique, advantages and disadvantages, **86**, 542
cellular decarboxylated *S*-adenosylmethionine, **94**, 69
complex carbohydrates, **83**, 3
dihydroxyeicosatetraenoic acids, **86**, 35
dolichylpyrophosphoryl oligosaccharides, **83**, 137
gangliosides, **83**, 167
halobacterial membrane proteins, **88**, 202
5-hydroxyeicosatetraenoic acid by reversed-phase technique, **86**, 610
12-hydroxyeicosatetraenoic acid by
normal-phase technique, **86**, 249
reversed-phase technique, **86**, 251
leukotrienes C_4 and D_4, **86**, 35
lipoxygenase products, **86**, 518
monohydroxyeicosenoic acids by argentation technique, **86**, 530
neutral oligosaccharides, **98**, 158
peptides by reversed-phase technique, **91**, 352
phenylthiohydantoins, **91**, 490
phosphopeptides, **99**, 37
polyamines with prechromatographic derivatization, **94**, 20
polyunsaturated fatty acids, **86**, 523
prostaglandin H synthase products, **86**, 518
prostaglandin ω-hydroxylase products with radioactive substrates, **86**, 177
prostaglandin ω-oxidation products, **86**, 171
prostaglandins by argentation technique, **86**, 530
proteins, **91**, 137

radiolabeled peptides for polypeptide hormone radioimmunoassay, **92**, 292
retinals, **88**, 491
side chain-protected phenylthiohydantoins, **91**, 494
6-sulfido-peptide-leukotrienes by reversed-phase technique, **86**, 252
High-pressure liquid chromatography, *see* High-performance liquid chromatography
Histamine
arylsulfatase effects, **86**, 20
phosphorylated cellulose column chromatography, **94**, 42
Histidine decarboxylase
assay by phosphorylated cellulose column chromatography, **94**, 45
Histocompatibility
in mixed lymphocyte reaction, early determination with ornithine decarboxylase assay, **94**, 396
Histone acetyltransferase
bovine liver
assay, **94**, 325
properties, **94**, 328
purification, **94**, 326
Histones
core, effect on cyclic nucleotide-dependent protein kinases, **99**, 206
preparative polyacrylamide gel electrophoresis, **91**, 214
Homarus americanus
L-lactate dehydrogenase purification, **89**, 360
Horseradish peroxidase
as marker for endocytosis and membrane recycling, **98**, 209
quantification by partition affinity ligand assay, **92**, 503
HPLC, *see* High-performance liquid chromatography
Hybridization
in bacteriorhodopsin gene isolation, **97**, 226
denatured RNA transferred or dotted to nitrocellulose paper, **100**, 255
multigene families, filter methods for isolation and homology determination, **100**, 266
pyruvate kinase types L and M, **90**, 147
small DNA fragments after transfer from polyacrylamide gels to diazo or nitrocellulose paper, **100**, 309
specific RNAs with DNA covalently coupled to macroporous supports, **100**, 400
Hybridomas
antigen-specific
frequency-enhancing methods, **92**, 17
spleen cell analysis and optimal immunization for high-frequency production, **92**, 26
growth-promoting effects of human endothelial culture supernatant, **92**, 39
immunoglobulins
light-chain secretion quantitation and loss variant selection, **92**, 147
quantitation, **92**, 156
lymphocyte, detection by electrophoresis of glucose-6-phosphate isomerase isozymes, **92**, 237
monoclonal antibody-producing
human–human, production, **82**, 3
number estimation in cell-fusion experiment, **92**, 36
screening by
immunosorption assay, **92**, 195
rosette-forming cell assay, **92**, 175
transfer template localized hemolysis in gel assay, **92**, 184
transfer template radioimmunoassay, **92**, 194
Hydrazinolysis
asparagine-linked sugar chains to produce free oligosaccharides, **83**, 263
Hydrochloric acid
–dimethyl sulfoxide and cyanogen bromide, protein cleavage at tryptophan residue, **91**, 318
comparison with other methods, **91**, 323

mechanism, **91**, 324
quantitation, **91**, 322
temperature effect, **91**, 321
protein cleavage at aspartic acid, **91**, 324
mechanism, **91**, 330
quantitation, **91**, 326
Hydrogen
–deuterium exchange in membrane-bound rhodopsin
comparison with tritium exchange, **88**, 681
infrared measurement, **88**, 676
movement in *Halobacterium halobium* vesicles, simultaneous measurement with membrane potential, **88**, 356
–tritium exchange in
bacteriorhodopsin
analytical techniques, **88**, 676
exchange-in, **88**, 674
exchange-out, **88**, 675
membrane-bound rhodopsin, comparison with deuterium exchange, **88**, 681
rhodopsin
analytical techniques, **88**, 676
exchange-in, **88**, 674
exchange-out, **88**, 675
15-L-Hydroperoxy-5,8,11,13-eicosatetraenoic acid
preparation, **86**, 387
purification, **86**, 388
structural identification, **86**, 390
thin-layer chromatography, **86**, 388
Hydroxyacylglutathione hydrolase
murine liver
assay, **90**, 547
properties, **90**, 550
purification, **90**, 547
25-Hydroxycholesterol
as dolichol synthesis inhibitor, **98**, 149
9α-Hydroxy-11,15-dioxo-2,3,18,19-tetranorprost-5-ene-1,20-dioic acid
incorporation of deuterium and tritium labels, **86**, 562
purification from human urine, **86**, 561
quantification by stable isotope dilution mass spectrometry, **86**, 565
5-Hydroxyeicosatetraenoic acid
measurement in biological fluids by gas chromatography–mass spectrometry, **86**, 607
purification by reversed-phase HPLC, **86**, 610
12-L-Hydroxyeicosatetraenoic acid
HPLC
normal-phase, **86**, 249
reversed-phase, **86**, 251
radioimmunoassay, **86**, 248
4-Hydroxy-2-ketopimelate aldolase
bacterial
assay, **90**, 277
properties, **90**, 279
purification, **90**, 278
Hydroxylysine
urinary, determination by
amino acid analysis, **82**, 401
colorimetric assay, **82**, 402
Hydroxylysine glycosides
isolation from sponge, **82**, 339
in proteins and peptides, automatic analysis, **82**, 343
sequence analysis, **82**, 344
urinary levels, determination, **82**, 403
Hydroxylysyl galactosyltransferase
molecular properties, **82**, 301
purification, **82**, 299
Hydroxylysyl glycosyltransferase
assay, **82**, 287
catalytic properties, **82**, 301
molecular properties, **82**, 297
purification, **82**, 294
Hydroxymethylglutaryl-CoA reductase (NADPH)
assay in cultured cells, **98**, 255
4-Hydroxy-4-methyl-2-ketoglutarate aldolase
Pseudomonas putida
assay, **90**, 273
properties, **90**, 276
purification, **90**, 274
Hydroxyproline
colorimetric assay, **82**, 375
detection with *o*-phthalaldehyde, **91**, 17
high-voltage paper electrophoresis, **82**, 385
hydrolysis, **82**, 374
ion-exchange chromatography, **82**, 387

measurements, in quantification of
collagen in unlabeled cells and
tissues, **82**, 393
synthesized collagen degradation,
82, 396
radiochemical assay, **82**, 380
urinary, determination by
amino acid analysis, **82**, 399
colorimetric assay, **82**, 400
9-Hydroxyprostaglandin dehydrogenase
(NAD^+)
rat kidney
assay, **86**, 114
properties, **86**, 117
purification, **86**, 115
15-Hydroxyprostaglandin dehydrogenase
(NAD^+)
assay with (15*S*)-[15-^{3}H]prostaglandin
E_2, **86**, 131
human placental
assay, **86**, 126
properties, **86**, 129
purification, **86**, 126
microassay in tissue samples, **86**, 135
15-Hydroxyprostaglandin dehydrogenase
($NADP^+$)
porcine kidney
assay, **86**, 143
properties, **86**, 145
purification, **86**, 144
prostaglandin D_2-specific, *see* Prostaglandin-D_2 15-dehydrogenase
($NADP^+$)
prostaglandin I_2-specific, rabbit kidney
assay, **86**, 152
properties, **86**, 155
purification, **86**, 153
Hydroxy-prostaglandins
chromatography, **86**, 171
extraction, **86**, 170
gas chromatography–mass spectrometry with selected ion monitoring,
86, 174
labeled, synthesis, **86**, 169
Hydroxyputrescine
identification by
automated cation-exchange liquid
chromatography, **94**, 29
nuclear magnetic resonance, **94**,
433
silica gel thin-layer chromatography,
94, 433
occurrence and properties, **94**, 431
preparation, **94**, 431
Hydroxypyruvate reductase
Pseudomonas acidovorans
assay, **89**, 341
properties, **89**, 344
purification, **89**, 342
Hypermyoglobinemia
associated serum myoglobin levels,
radioimmunoassay, **84**, 176
Hypolipidemic drugs
induction of peroxisomes, **96**, 549
Hypusine[*N*^{1}-(4-amino-2-hydroxybutyl)lysine]
ion-exchange chromatography, **94**, 458

I

I-cell disease
fibroblasts, pinocytosis of
[2-^{3}H]mannose-labeled glycopeptides, **98**, 295
[2-^{3}H]mannose-labeled oligosaccharides, **98**, 296, 297
D-Iditol dehydrogenase
Gluconobacter suboxydans membrane-bound
assay, **89**, 141
properties, **89**, 144
purification, **89**, 142
L-Iditol dehydrogenase
rat liver
assay, **89**, 135
properties, **89**, 137
purification, **89**, 136
Idoxuridine
radiolabeled, in cytotoxicity assays, **93**,
233
Iduronate sulfatase
human plasma
assay, **83**, 573
properties, **83**, 576
purification, **83**, 574
α-L-Iduronidase
human kidney
assay, **83**, 578
properties, **83**, 581
purification, **83**, 579

isozymes, human urine
properties, **83**, 586
separation, **83**, 585
uptake and binding assays, **83**, 583
Imidoesters
disc membrane permeability, **81**, 279
protein and peptide amidination, **91**, 559
rhodopsin modification, **81**, 285
storage, **81**, 279
synthesis, **81**, 276
Imino acids
detection with *o*-phthalaldehyde, **91**, 17
2-Imino-2-methoxyethyl 1-thioglycosides
in neoglycoprotein preparation, **83**, 278
storage and stability, **83**, 287
synthesis, **83**, 286
Immunoadsorption
for hybridoma rapid screening, **92**, 195
Immunoassays
dinitrophenol by bioluminescent method, **92**, 426
estriol in homogeneous assay by differential pulse polarography, **92**, 433
gentamicin by fluorescence fluctuation method, **92**, 458
human IgG by colorimetric method with FAD as label, **92**, 413
human serum albumin in heterogeneous assay by differential pulse anodic stripping voltammetry, **92**, 437
methotrexate by bioluminescent method, **92**, 426
model systems by fluorescence fluctuation method, **92**, 458
noncentrifugation, separation systems, **92**, 336
sequence-specific DNA–protein interaction, **92**, 138
serum thyroxine with chemiluminescent label, **84**, 303
theophylline by colorimetric method with FAD as label, **92**, 413
trinitrotoluene by bioluminescent method, **92**, 426
Immunoautoradiography
in desmin and vimentin antibody specificity assay, **85**, 497
Immunochemical techniques, *see also specific techniques*
in recognition of modifications on functionally related proteins, **98**, 159
related articles previously published in Methods in Enzymology
carbohydrates, simple and complex, **84**, 662
general assays, principles and procedures, **84**, 664
lipids, prostaglandins, and steroids, **84**, 663
low-molecular-weight compounds, **84**, 664
nucleic acids and related compounds, **84**, 662
proteins and peptides, **84**, 657
receptors, membranes, and cell surface antigens, **84**, 663
viruses, bacteria, and toxins, **84**, 663
Immunodiffusion
proteoglycan aggregate components by Ouchterlony technique, **83**, 223
Immunoelectron microscopy
with frozen sections, application to analysis of Semliki Forest virus glycoprotein transport, **96**, 466
prostaglandin H synthase, **86**, 220
viral proteins in infected MDCK cells, **98**, 497
Immunoelectrophoresis
membrane antigens by two-dimensional technique, **96**, 244
outer membrane proteins from *Escherichia coli,* **97**, 121
proteoglycan aggregate components, **83**, 223
Immunofixation electrophoresis
monoclonal immunoglobulins on cellulose acetate, **92**, 220
Immunofluorescence
clathrin in cells and tissues, **98**, 364
desmin in muscle cells, **85**, 503
fibronectin in cultured cells and tissues, **82**, 826
prostaglandin H synthase in tissue sections, **86**, 218
protein distribution and internalization in polarized epithelia, **98**, 381

Semliki Forest virus membrane protein expression in eukaryotic cell nuclei, **96**, 510
solid-phase assay for monoclonal antibodies to soluble antigens, **92**, 227
vimentin in muscle cells, **85**, 503
viral proteins in infected MDCK cells, **98**, 495
Immunoglobulins
aggregated, in binding assay for Fc receptors, **93**, 131
hybridoma, light-chain secretion quantitation and loss variant selection, **92**, 147
IgA
binding of precursor to transmembrane secretory component, **98**, 462
enzyme immunoassay with activated thiol-Sepharose in separation method, **92**, 355
IgE, antigen-specific monoclonal antibodies
assay, **92**, 58
ε heavy chain-specific, production, **92**, 63
physical and chemical characterization, **92**, 62
production, **92**, 49
purification, **92**, 60, 64
IgG
from anti-prostaglandin H synthase serum
in enzyme immunoelectron microscopy, **86**, 220
preparation, **86**, 215
colorimetric immunoassay with FAD as label, **92**, 424
purification from
fetal calf serum and culture medium, **86**, 234
hybridoma culture medium, **86**, 235
radioiodination, **86**, 235
monoclonal
quantitation, **92**, 156
screening by immunofixation on cellulose acetate, **92**, 220
nonspecific, production in ascitic fluids, **93**, 12
radioimmunoassay with antigen-coated plastic tubes and radiolabeled protein A, **92**, 404
Immunoprecipitation
acetylcholine receptor subunits synthesized *in vitro*, **96**, 370
cathepsin D, **96**, 773
erythrocyte band III proteolytic fragments, **96**, 265
β-glucuronidase, **96**, 773
glycophorin A, **96**, 286
lens fiber membrane polypeptide, **98**, 516
lysosomal enzymes, **96**, 732
outer membrane proteins from *Escherichia coli*, **97**, 116
prostaglandin H synthase, **86**, 216
proteins translated in cell-free system, **96**, 111
secretory component synthesized *in vitro*, **98**, 461, 465
type-II cAMP-dependent protein kinase, **99**, 190, 191
Immunoradiometric assay
antihemophilic factor, **84**, 56
ferritin, **84**, 159
prostaglandin H synthase, **86**, 236
prostaglandin I_2 synthase, **86**, 243
von Willebrand factor, **84**, 51
Influenza virus
-infected MDCK cells
electron microscopy, **98**, 493
viral protein localization by
immunoelectron microscopy, **98**, 497
immunofluorescence, **98**, 495
mutants with defects in glycoprotein processing
infectivity assays, **96**, 441
isolation, **96**, 441
preparation, **96**, 440
stock preparation, **98**, 490
titration, **98**, 491
Infrared spectroscopy
bacteriorhodopsin reaction cycle, **88**, 729
peptide hydrogen exchange in rhodopsin, **88**, 676
photoreceptor membranes, **88**, 699

purple membranes, **88**, 706
rhodopsin reaction sequence, **88**, 729
Initiation factor kinases
eIF-2α dsRNA-dependent, from rabbit reticulocyte lysates
assays, **99**, 349
properties, **99**, 357
purification, **99**, 351
Inorganic pyrophosphatase
yeast
active site studies, **87**, 539
assay, **87**, 526
binding sites, **87**, 529
catalysis of P_i:PP_i equilibration, **87**, 531
catalytic activities and substrate specificity, **87**, 528
mechanism, **87**, 545
purification, **87**, 526
structure, **87**, 527
Inositol epimerase
bovine brain
assay, **89**, 594
properties, **89**, 598
purification, **89**, 597
Inositol hexasulfate
inhibition of
casein kinase I, **99**, 314
casein kinase II, **99**, 326
myo-Inositol phosphate
derivatization, **89**, 3, 11
detectors, sensitivity, **89**, 10
gas chromatography, **89**, 5, 14
myo-Inositol-1-phosphate synthase
bovine testis
assay, **90**, 309
properties, **90**, 312
purification, **90**, 310
Inosose reductase
bovine brain
assay, **89**, 594
properties, **89**, 598
purification, **89**, 597
Insoluble elastin
antiserum preparation, **82**, 747
in cell culture
biosynthesis, **82**, 617
crosslink formation, **82**, 624
formation assessment, **82**, 618
reagent effect on formation, **82**, 630
tropoelastin incorporation, **82**, 630
ultrastructural analysis, **82**, 618
crosslinked peptides, isolation, **82**, 608
definition, **82**, 636
digestion, **82**, 608
in organ culture
biosynthesis, **82**, 632
rapid method, **82**, 634
crosslinks and intermediates, formation, **82**, 635
description, **82**, 633
desmosine formation, **82**, 633
preparation, **82**, 589, 622
primary structure, **82**, 606
proteolysis, **82**, 588
single-chain peptides
amino acid composition, **82**, 613
purification, **82**, 614
release from crosslinked peptides, **82**, 606
solubilization, **82**, 608, 744
assays, **82**, 594
as substrate for elastase assay, **82**, 594
Insulin
antibodies, enzyme immunoassay with activated thiol-Sepharose in separation method, **92**, 353
stable radioiodinated, preparation by polyacrylamide gel electrophoresis, **92**, 325
Insulin receptors
heavy-isotope density-shift study, **96**, 423
Intestine
mouse, fructose-bisphosphatase purification, **90**, 360
rabbit, fructose-bisphosphatase purification, **90**, 362
Invertase, *see* β-Fructofuranosidase
Iodine monochloride
iodination techniques
considerations, **92**, 288
macro method, **92**, 281
micro method, **92**, 284
with radioactive compound produced by oxidation, **92**, 287
scope of usefulness, **92**, 277
in shipping containers, **92**, 296

Iododeoxyuridine, *see* Idoxuridine
Iodopsin
chicken retina
assay, **81**, 154
comparison with rhodopsin, **81**, 159
purification, **81**, 155
o-Iodosobenzoic acid
amino acid reactivity, **91**, 315
protein cleavage at tryptophan residue, **91**, 316
purity evaluation, **91**, 314
Ion-exchange chromatography
S-adenosyl-(5′)-3-methylthiopropylamine, **94**, 74
aliphatic diamines by automated procedure, **94**, 29
γ-carboxyglutamic acid hydrolysates, **91**, 15
coated vesicles from rat liver, **98**, 327
collagen, **82**, 51
deoxyhypusine[N^1-(4-aminobutyl)lysine], **94**, 458
γ-glutamylamines, **94**, 456
hydroxyprolines, **82**, 387
hypusine[N^1-(4-amino-2-hydroxybutyl)-lysine], **94**, 458
methionine-containing peptides, **91**, 367
microscale peptide separations, **91**, 359
nonmuscle myosins, **85**, 337
polyamines
by automated procedure, **94**, 29
and monoacetyl derivatives, **94**, 25
proteins, **91**, 170
signal recognition particle, **96**, 685
and TLC, in peptide mapping, **91**, 384
Ionophores
carboxylic, *see* Carboxylic ionophores
Ion-retardation resins
preparation, **91**, 260
properties, **91**, 259
Ion transport
secondary, in *Halobacterium halobium*
atomic absorption spectroscopy and flame photometry, **88**, 346
dielectric dispersion measurements, **88**, 347
ion-selective electrode measurements, **88**, 345
radioactive isotopic determination, **88**, 346
Irradiation
microwave, to inactivate eicosanoid-metabolizing enzymes in brain, **86**, 635
Islets of Langerhans
rat
biochemical and morphological characterization, **91**, 187
isolation, **98**, 183
subcellular fractionation, **98**, 185
Isoelectric focusing
carrier ampholyte preparation for, **91**, 281
lysosomal enzymes from *Dictyostelium discoideum* secretory mutants, **96**, 823
with noncommercial carrier ampholytes, **91**, 299
Isomaltase, *see* Oligo-1,6-glucosidase
Isomerization
light-induced
all-*trans*-retinal to 11-*cis*-retinal, **81**, 459
retinal analogs, kinetic measurements, **88**, 533
Isotope effects
equilibrium, determination by equilibrium perturbation method, **87**, 641
applications, **87**, 645
exchange studies of regulatory enzymes, fingerprint kinetic patterns for modifier action, **87**, 647
intrinsic
determination by Northrop's method, **87**, 633
interpretation in terms of transition-state structure, **87**, 625
solvent
on enzyme systems, **87**, 551
mechanistic interpretation, **87**, 598
theory, **87**, 581
Isotopes, *see also specific isotopes*
detection
by fluorography in
agarose gels, **96**, 220

polyacrylamide gels, **96**, 215
solid porous supports, **96**, 221
with intensifying X-ray films, **96**, 221
in spark chamber, **96**, 221
heavy, in insulin receptor analysis, **96**, 423

J

Job plot
determination of binding stoichiometry, **87**, 509

K

Keratan-sulfate endo-1,4-β-galactosidase
Escherichia freundii
application for oligosaccharide structural analysis, **83**, 615
assay, **83**, 611
properties, **83**, 614
purification, **83**, 612
Flavobacterium keratolyticus
properties, **83**, 622
purification, **83**, 620
2-Ketoaldehyde dehydrogenase, *see* 2-Oxoaldehyde dehydrogenase
2-Keto-3-deoxy-L-fuconate dehydrogenase
porcine liver
assays, **89**, 219
properties, **89**, 224
purification, **89**, 222
2-Keto-3-deoxy-galactonate-6-phosphate aldolase, *see* 6-Phospho-2-dehydro-3-deoxygalactonate aldolase
2-Keto-3-deoxygluconate 6-phosphate
synthesis with 6-phosphogluconate dehydratase, **89**, 98
2-Keto-3-deoxy-D-xylonate aldolase, *see* 2-Dehydro-3-deoxy-L-pentonate aldolase
15-Keto-13,14-dihydro-prostaglandin E_2
chemical instability, **86**, 307
gas chromatography–mass spectrometry, **86**, 571
as prostaglandin E_2 parameter in biological samples, **86**, 319
radioimmunoassay, **86**, 308
tetradeuterated, biosynthesis, purification, evaluation, and storage, **86**, 572
15-Keto-13,14-dihydro-prostaglandin $F_{2\alpha}$
as prostaglandin $F_{2\alpha}$ parameter in plasma, **86**, 337
radioimmunoassay, **86**, 320
2-Keto-D-gluconate
microdetermination with membrane-bound dehydrogenase, **89**, 26
2-Keto-D-gluconate dehydrogenase, *see* Dehydrogluconate dehydrogenase
2-Keto-D-gluconate reductase, *see* Gluconate 2-dehydrogenase
5-Keto-D-gluconate reductase, *see* Gluconate 5-dehydrogenase
6-Ketoprostaglandin $F_{1\alpha}$
radioimmunoassay, **86**, 273
15-Ketoprostaglandin Δ^{13}-reductase
bovine lung
assay, **86**, 157
properties, **86**, 160
purification, **86**, 159
human placental
assay, **86**, 164
properties, **86**, 166
purification, **86**, 164
Ketose phosphate esters
^{13}C-enriched, enzymatic synthesis, **89**, 83
Ketoses
^{13}C-enriched, enzymatic synthesis, **89**, 83
Kidney
bovine, purification of pyruvate dehydrogenase kinase, **90**, 197; **99**, 333
human, purification of glyceraldehyde-3-phosphate dehydrogenase, **89**, 303
microassay for NAD$^+$-dependent 15-hydroxyprostaglandin dehydrogenase, **86**, 135
murine, purification of glucose-6-phosphate dehydrogenase, **89**, 255

β-glucuronidase, **96**, 559
L-lactate dehydrogenase B_4, **89**, 357
microsomal egasyn, **96**, 564
ornithine decarboxylase, **94**, 159
porcine, purification of
aldehyde reductase, **89**, 502
cAMP-dependent protein kinase catalytic subunit from
cytosol, **90**, 203
particulate fraction, **90**, 205
15-hydroxyprostaglandin dehydrogenase ($NADP^+$), **86**, 144
phosphoprotein phosphatase, **90**, 410
prostaglandin 9-ketoreductase, **86**, 144
rabbit, purification of
γ-glutamylamine cyclotransferase, **94**, 349
prostaglandin I_2-specific 15-hydroxyprostaglandin dehydrogenase ($NADP^+$), **86**, 153
rat, purification of
γ-glutamyl transpeptidase, **86**, 43
9-hydroxyprostaglandin dehydrogenase, **86**, 115
Killer cells
frequency determination by single-cell cytotoxicity assay, **93**, 270
Klebsiella pneumoniae
purification of
class II D-tagatose-1,6-bisphosphate aldolase, **90**, 232
D-galactitol-6-phosphate dehydrogenase, **89**, 276
D-galactonate/D-fuconate dehydratase, **90**, 300
gluconate 2-dehydrogenase, **89**, 190
L-sorbose-1-phosphate reductase, **89**, 249

L

Lacrimal gland
cytoskeleton role in protein secretion, **98**, 175
β-Lactamase
bacterial, membrane-bound forms
processing analysis, *in vitro* methods, **97**, 157
synthesis and isotopic labeling *in vivo*, **97**, 154
D-Lactate, *see* D-Lactic acid
D-Lactate dehydrogenase
in D-lactic acid determination, **89**, 29
L-Lactate dehydrogenase
Homarus americanus
assay, **89**, 359
properties, **89**, 361
purification, **89**, 360
isozymes
murine
assay, **89**, 352
properties, **89**, 357
purification, **89**, 355
sweet potato root
assay, **89**, 345
properties, **89**, 349
purification, **89**, 346
in L-lactic acid determination, **89**, 29
Streptococcus cremoris
assay, **89**, 362
properties, **89**, 365
purification, **89**, 363
Lactate–malate transhydrogenase
Veillonella alcalescens
assays, **89**, 368
properties, **89**, 372
purification, **89**, 370
Lactate-oxaloacetate transhydrogenase, *see* Lactate–malate transhydrogenase
D-Lactic acid
assay in plasma, **89**, 35
automated enzymatic determination, **89**, 29
D(−)-Lactic acid, *see* D-Lactic acid
L-Lactic acid
automated enzymatic determination, **89**, 29
L(+)-Lactic acid, *see* L-Lactic acid
Lactoperoxidase
in intracellular iodination of lysosomal membrane, **98**, 404
application to membrane recycling, **98**, 414

uptake and determination in internalized plasma membranes, **98**, 546
Lactose permease
Escherichia coli
assays, **97**, 164
purification and reconstitution, **97**, 168
radiolabeling, **97**, 165
substrate specificity, **97**, 160
Lactose synthase
in plasma membrane glycoconjugate labeling, **98**, 416
Lactoylglutathione lyase
human erythrocyte
assay, **90**, 536
properties, **90**, 540
purification, **90**, 536
murine liver
assay, **90**, 542
properties, **90**, 545
purification, **90**, 543
Laminin
biological activity, **82**, 838
biosynthesis and tissue distribution, **82**, 837
immunological properties, **82**, 837
purification, **82**, 832
solubilization, **82**, 835
Lathyrus sativus seeds
putrescine synthase purification, **94**, 337
Leader peptidase
Escherichia coli
assay, **97**, 41
molecular genetics, **97**, 46
properties, **97**, 43
purification, **97**, 41
requirement in cell growth, **97**, 54
role in membrane assembly, **97**, 44
Lectins, *see also specific lectins*
–antibody conjugates for quantitation and titration of antigens and antibodies, **92**, 489
applications, **92**, 495
in binding assay for proteins in polyacrylamide gels, **96**, 190
Datura stramonium
assay, **83**, 369
properties, **83**, 372
purification, **83**, 369
in intracellular localization of Semliki Forest virus glycoproteins, **96**, 461
mannose-specific, *Escherichia coli*
assay, **83**, 386
properties, **83**, 391
purification, **83**, 387
Pseudomonas aeruginosa
applications, **83**, 384
assay, **83**, 379
properties, **83**, 383
purification, **83**, 381
resistant mutants, animal cell
nomenclature, **96**, 161
selection, **96**, 157
in retinal photoreceptor membrane analysis, **81**, 65
tomato
assay, **83**, 363
properties, **83**, 367
purification, **83**, 364
Lens
bovine
epithelial cell culturing, **98**, 518
fiber membrane polypeptide
cell-free translation, **98**, 515
immunoprecipitation, **98**, 516
RNA fractionation and extraction, **98**, 511
Leucovorin
radioimmunoassay, **84**, 409
Leukocytes
peritoneal polymorphonuclear, arachidonic acid 15-lipoxygenase purification, **86**, 46
Leukotriene B_4
characterization, **86**, 414
production in mastocytoma cells, **86**, 409
purification, **86**, 412
Leukotriene C_3
characterization, **86**, 40
preparation, **86**, 38
Leukotriene C_4
analogs
immunochromatography, **86**, 256
preparation, **86**, 253
radioimmunoassay, **86**, 254

bioassay, **86**, 34
characterization, **86**, 40, 414
HPLC, **86**, 35
immunochromatography, **86**, 256
preparation, **86**, 253
production in mastocytoma cells, **86**, 38, 409
purification, **86**, 39, 412
radioimmunoassay, **86**, 254
synthetic, arylsulfatase effects, **86**, 28
Leukotriene C_5
characterization, **86**, 40
preparation, **86**, 38
Leukotriene D_3
characterization, **86**, 40
Leukotriene D_4
analogs
immunochromatography, **86**, 256
preparation, **86**, 253
radioimmunoassay, **86**, 254
antiserum
preparation, **86**, 253
serological specificity, **86**, 254
bioassay, **86**, 34
characterization, **86**, 40, 414
HPLC, **86**, 35
immunochromatography, **86**, 256
preparation, **86**, 253
production in mastocytoma cells, **86**, 409
purification, **86**, 412
radioimmunoassay, **86**, 254
synthetic, arylsulfatase effects, **86**, 28
Leukotriene D_5
characterization, **86**, 40
Leukotriene E_4
analogs
immunochromatography, **86**, 256
preparation, **86**, 253
radioimmunoassay, **86**, 254
characterization, **86**, 414
immunochromatography, **86**, 256
preparation, **86**, 253
production in mastocytoma cells, **86**, 409
purification, **86**, 412
radioimmunoassay, **86**, 254
Leupeptin
inhibitory effects on lysosomes, **96**, 757
Ligand binding assays
data analysis with computer program LIGAND
bumpy displacement curves, **92**, 558, 565
CORRECTION FACTOR for analysis of combined results from several experiments, **92**, 561
curved Scatchard plots, **92**, 550
displacement curves, **92**, 555
linear Scatchard plots, **92**, 547
mathematical model, **92**, 575
statistical considerations, **92**, 570
fluorometric, for Fc receptor detection, **93**, 147
N^5-formyltetrahydrofolate, **84**, 462
haptens, macromolecules, and cells by partition affinity technique, **92**, 498
N^5-methyltetrahydrofolate, **84**, 458
Ligand–receptor interactions, *see* Receptor–ligand interactions
Light
induced isomerization of
all-*trans*-retinal to 11-*cis*-retinal, **81**, 461
retinal analogs, kinetic measurements, **88**, 533
induced lysis of rhodopsin, calorimetric measurements, **88**, 667
induced volume changes in bacteriorhodopsin, **88**, 272
interactions with bacteriorhodopsin, X-ray diffraction, **88**, 282
regulation of protein binding to photoreceptor membrane, application to protein purification, **81**, 556
rhodopsin excitation, associated calcium release, measurement in isolated rod outer segments, **81**, 569
Light microscopy
collagen fibrils with labeled antibodies, **82**, 137
microtubules during assembly *in vitro* by dark-field technique, **85**, 437
Light scattering
in actin polymerization assay, **85**, 198

contractile proteins
kinetic analysis, **85**, 706
sample preparation for, **85**, 562
high-molecular-weight proteins and large macromolecular assemblies by quasi-elastic techniques, **85**, 573
in isolated rod disk size and shape determination, **81**, 616
microtubules during assembly *in vitro* by laser techniques, **85**, 437
Lipids
interactions with bacteriorhodopsin and rhodopsin, analysis with diffusible spin labels, **88**, 762
photoreceptor membrane, invertebrate
extraction and analysis of major classes, **81**, 316
structural analysis, **81**, 318
polar, synthesis sites in cellular membranes, **96**, xxxiii
purple membrane
from extreme halophiles
composition, **88**, 101, 105
extraction, **88**, 74, 99
separation, **88**, 100
from methanogenic bacteria
extraction, **88**, 107, 108
separation, **88**, 107
retinal
cell-free synthesis, **81**, 791
extraction, **81**, 792
separation, **81**, 792
synthesis *in vivo*, **81**, 791
rod outer segment
extraction, **81**, 802
interconversions, **81**, 805
quantitation, **81**, 804
separation, **81**, 803
turnover rates, **81**, 804, 805
spin-labeled analogs, synthesis, **88**, 763
Thermoplasma acidophilum
degradation, **88**, 403
isolation, **88**, 399
quantitation, **88**, 405
Lipid transfer proteins
bovine liver
activity assay, **98**, 575, 602
in modification of synaptic membrane cholesterol content, **98**, 599
purification, **98**, 580, 600
function, **98**, 572
nomenclature, **98**, 565
rat liver
activity assay, **98**, 575, 592
properties, **98**, 595
purification, **98**, 578, 594
related proteins, **98**, 596
from various sources
activity assay, **98**, 568
applications, **98**, 572
properties, **98**, 566
Lipopolysaccharides
bacterial
derivatized
iodination, **84**, 269
preparation, **84**, 267
isolation, **84**, 266
solid-phase radioimmunoassay, **84**, 264
synthetic, in attachment of haptens and proteins to erythrocytes, **92**, 274
Lipoprotein lipase
human platelet
assay, **86**, 12, 653
properties, **86**, 15
Lipoproteins
Escherichia coli outer membrane
assembly mutants
characterization, **97**, 127
isolation, **97**, 124
site-specific mutagenesis, **97**, 126
separation on gels, **97**, 128
low-density, *see* Low-density lipoproteins
β-Lipoproteins
reduction with *N*-methylmercaptoacetamide, **91**, 557
Liposomes
as biodegradable and harmless adjuvants, **93**, 83
advantages, **93**, 93
as carriers for nucleic acid intracellular delivery, **101**, 512
Lipoxygenase
brain, inactivation by microwave irradiation, **86**, 635

reaction products
HPLC, **86**, 518
TLC, **86**, 32
Liver, *see also* Hepatocytes
bovine, purification of
aldehyde dehydrogenase, **89**, 498
diacetyl reductase, **89**, 517
D-erythrulose reductase, **89**, 233
fructose-bisphosphatase, **90**, 331
lipid transfer proteins, **98**, 580
membrane phosphomannosyl receptor, **98**, 307
phosphatidylcholine transfer protein, **98**, 583
3-phosphoglycerate kinase, **90**, 116
pyruvate kinase, **90**, 146
spermidine *N*-acetyltransferase, **94**, 326
catfish, purification of glucose-6-phosphate isomerase, **89**, 555
equine, purification of
alcohol dehydrogenase, **89**, 436
alcohol dehydrogenase steroid-active SS isozyme, **89**, 432, 439
aldehyde dehydrogenase isozymes, **89**, 475
human, purification of
aldehyde reductase, **89**, 508
glyceraldehyde-3-phosphate dehydrogenase, **89**, 303
pyruvate kinase, **90**, 134
murine, purification of
β-glucuronidase, **96**, 559
glyoxalase I, **90**, 543
glyoxalase II, **90**, 547
microsomal egasyn, **96**, 564
pigeon, purification of diacetyl reductase, **89**, 518
porcine, purification of
glyceraldehyde-3-phosphate dehydrogenase, **89**, 312
2-keto-3-deoxy-L-fuconate dehydrogenase, **89**, 222
phosphoglycerate kinase, glyceraldehyde-3-phosphate dehydrogenase, and phosphoglycerate mutase simultaneously, **90**, 499
transketolase, **90**, 220
UDP-*N*-acetyl-D-glucosamine:α-D-mannoside β1-2-*N*-acetyl-D-glucosamine transferase I, **98**, 113
rabbit, purification of
calmodulin-dependent glycogen synthase kinase, **99**, 301
fructose-bisphosphatase, **90**, 327, 346
prostaglandin-D_2 11-ketoreductase, **86**, 119
UDP-*N*-acetyl-D-glucosamine:α-D-mannoside β1-2-*N*-acetyl-D-glucosamine transferase I, **98**, 113
rat
excision and perfusion, **98**, 229
isolated, perfused
endocytic compartments, selective labeling, **98**, 232
function assessment, **98**, 234
purification of
acetylspermidine deacetylase, **94**, 329
S-adenosylmethionine decarboxylase, **94**, 235
coated vesicles, **98**, 327
DNA topoisomerase I, **100**, 140
fructose-bisphosphatase, **90**, 353
fucose-binding protein, **98**, 315
gap junctions, **98**, 503, 521
β-glucuronidase, **96**, 559
Golgi complex, **98**, 60
2-ketoaldehyde dehydrogenase, **89**, 514
methionine adenosyltransferase isozymes, **94**, 224
nonspecific lipid transfer proteins, **98**, 578, 594
nuclear membrane pore complex–lamina structures, **96**, 607
ornithine decarboxylase, **94**, 156
ornithine decarboxylase antizyme, **94**, 187
ornithine decarboxylase antizyme inhibitor, **94**, 190
phosphatidylcholine transfer protein, **98**, 587

pyruvate kinase type L, **90**, 160
sorbitol dehydrogenase, **89**, 136
transketolase, **90**, 217
turkey, purification of fructose-bisphosphatase, **90**, 335
Low-density lipoprotein receptors
human fibroblast
expression, influencing factors, **98**, 244
mutations and their detection, **98**, 243
synthesis pathway, **98**, 242
Low-density lipoproteins
deficient serum, preparation, **98**, 250
proteolytic digestion, assay in cultured cells, **98**, 254
radioiodination, **98**, 247
receptor-mediated endocytosis, assay in cultured cells, **98**, 251
Lung
bovine, purification of
cGMP-dependent protein kinase, **99**, 64
15-ketoprostaglandin Δ^{13}-reductase, **86**, 159
guinea pig, purification of slow-reacting substance of anaphylaxis, **86**, 426
human, purification of glyceraldehyde-3-phosphate dehydrogenase, **89**, 303
rabbit, macrophage preparation, **98**, 301
rat, macrophage preparation, **98**, 301
Lupinus luteus seeds
purification of 5′-methylthioadenosine nucleosidase, **94**, 366
Lymph nodes
inoculation technique for antibody production, **93**, 3
Lymphoblastoid cells
surface C3b and Fc receptor detection by solid-phase radioimmunoassay, **93**, 155
Lymphocytes
B-cell, biologically active Fc receptors
properties, **93**, 226
purification, **93**, 220
biosynthetic labeling in tissue culture, **91**, 414
cytotoxic T-cell
activation by multivalent antibodies with dual-specificity, **92**, 538
frequency determination by single-cell cytotoxicity assay, **93**, 270
tumor cell metabolic responses, analysis by immobilization to poly-L-lysine-coated plastic surfaces, **93**, 260
Fc receptor quantitation by fluorometric assay, **93**, 154
mitogen-activated, physiological response to polyamine biosynthetic inhibitors, assays, **94**, 373
in mixed reaction, histocompatibility determination by ornithine decarboxylase assay, **94**, 396
surface marker demonstration with alkaline-labeled reagents, **93**, 164
application to population enumeration in whole blood, **93**, 170
T-cell, proliferation inhibition by monoclonal anti-Ia antibodies, **92**, 82
Lysine decarboxylase
Escherichia coli
assay, **94**, 181
deficient mutants, mass screening, **94**, 89
properties, **94**, 183
purification, **94**, 182
Lysosomal enzymes, *see also specific lysosomal enzymes*
biosynthesis in cultured cells, **96**, 729
cell-free synthesis, **96**, 770
gel electrophoresis, **96**, 734
immunoprecipitation, **96**, 732, 773
phosphorylated mannose residues, detection, **96**, 736
uptake by cultured fibroblasts, determinants, **98**, 290
Lysosomes
function, inhibitory effects of
ionophores, **96**, 755
lectins and polymers, **96**, 762
3-methyladenine, **96**, 761
microtubule poisons, **96**, 762
peptidic proteinase inhibitors, **96**, 756
proteinase inhibitory peptides, **96**, 756

protein synthesis inhibitors, **96**, 763
vanadate, **96**, 760
weak-base amines, **96**, 737
membranes, intracellular iodination, **98**, 404
application to membrane recycling, **98**, 414
thyroid follicle cell, endocytic pathways to, **98**, 453
transport to
biosynthetic pathway from Golgi complex, **98**, 8
endocytic pathway from plasmalemma, **98**, 4
Lysyl hydroxylase, *see* Procollagen-lysine, 2-oxoglutarate 5-dioxygenase
Lysyl oxidase
bovine aortic
assay, **82**, 639
properties, **82**, 645
purification, **82**, 642
chicken embryo
assays, **82**, 315
purification, **82**, 317

M

Macromolecules
large assemblies
falling-ball viscometry, **85**, 572
non-Newtonian viscosity measurements, **85**, 572
quasi-elastic light scattering, **85**, 573
sedimentation analysis, **85**, 570
turbidimetric ultracentrifugation, **85**, 572
Macrophages
alveolar
internalization of prebound glycoconjugate, **98**, 304
preparation from rat and rabbit by pulmonary lavage, **98**, 301
purification of actin-binding protein, **85**, 483
uptake and binding assays, **98**, 303
bone marrow-derived
internalization of prebound glycoconjugate, **98**, 304
preparation from murine and rat femurs, **98**, 301
uptake and binding assays, **98**, 303
frequency determination by single-cell cytotoxicity assay, **93**, 270
Malate dehydrogenase
isotope exchange study, kinetic patterns for modifier action, **87**, 653
Maltose
binding protein, *Escherichia coli*
assay, **90**, 459
properties, **90**, 462
purification, **90**, 460
Mammary gland
polyamine metabolic and functional analyses, **94**, 389
Mannonate dehydratase
Escherichia coli K12
assay, **90**, 288
properties, **90**, 291
purification, **90**, 289
D-Mannonate-NAD dehydrogenase, *see* Fructuronate reductase
Mannosamine
inhibition of glycoprotein lipid-linked saccharide formation, **98**, 143
Mannose
analogs, inhibition of glycoprotein lipid-linked saccharide formation, **98**, 142
–bovine serum albumin, prebound, internalization by macrophages, **98**, 304
α-Mannosidase I
α1,2-specific, rat liver
assay, **83**, 423
properties, **83**, 426
purification, **83**, 424
α-Mannosidase II
rat liver
assay, **83**, 427
properties, **83**, 428
purification, **83**, 428
α-D-Mannoside β1→2-N-acetylglucosaminyltransferase, *see* α-1,3-Mannosyl-glycoprotein β-1,2-N-acetylglucosaminyltransferase
α-1,3-Mannosyl-glycoprotein β-1,2-N-acetylglucosaminyltransferase
rabbit liver
assay, **83**, 508
properties, **83**, 513
purification, **83**, 510

Mannosyl-glycoprotein endo-β-*N*-acetylglucosaminidase
type D, hydrolysis of rhodopsin glycoprotein T1, **81**, 217
type L, *Streptomyces plicatus*
assay, **83**, 604
properties, **83**, 607
purification, **83**, 606
Mass spectroscopy
phenylthiohydantoins by chemical ionization method, **91**, 502
purple membrane retinals, **88**, 542
Matrices
biosynthetic, from cells in culture
characterization, **82**, 532
preparation, **82**, 530
single cell adhesion, **82**, 533
MDCK cells (Madin-Darby canine kidney)
endocytosis and membrane recycling, immunochemical analysis, **98**, 379
interactions with
fowl plague virus, **98**, 264
vesicular stomatitis virus, **98**, 264
source and culture conditions, **98**, 488
virus-infected monolayers
electron microscopy, **98**, 493
viral protein localization by
immunoelectron microscopy, **98**, 497
immunofluorescence, **98**, 495
Membranes
adrenal medullary, exocytosis and endocytosis, high-voltage analytical techniques, **98**, 33
bacterial
biogenesis
semi-*in vitro* systems, **96**, 79
in vitro systems, **96**, 81
in vivo systems, **96**, 75
protein insertion and transport, role of membrane potential, **97**, 146
basolateral, avian salt gland, enzyme markers, **96**, 634
brown, *see* Brown membranes
cellular
antigens, two-dimensional immunoelectrophoresis, **96**, 244
assembly
modes, **96**, 1
procedures, multiplicity, **96**, xlii
topography, **96**, lii
biogenesis of proteins
characteristics, **96**, xxix
consequences of monopolies, **96**, xxxvi
immunochemical analysis in polarized epithelia, **98**, 379
chemistry, **96**, xxx
component synthesis sites, **96**, xxxiii
continuity principles, **96**, xxxii
endocytosis, markers, **98**, 215
exocytosis, high-voltage analytical techniques, **98**, 28
fluidity
role in receptor–ligand interactions, **93**, 95
significance, **96**, liii
function as own template, **96**, liv
general function, **96**, xxxi
glycosylation and trimming of proteins, targeting, **98**, 91
insertion of proteins
cotranslational
analysis with proteases, **96**, 136
overview, **96**, 122
mechanisms, **96**, 672
posttranslational
analysis with proteases, **96**, 136
overview, **96**, 129
phylogeny, **96**, 676
proteins
gel electrophoresis, **96**, 184, 194
immunochemical identification, **96**, 192
in polyacrylamide gels, analytical techniques, **96**, 189
with protein translocation systems, **96**, 665
recycling
fluid-phase markers, **98**, 209
high-voltage analytical techniques, **98**, 28
membrane markers, **98**, 215
transcellular transport pathways, **98**, 7, 447
transfer of
phospholipids, **98**, 565

proteins
mechanisms, **96**, 669
regulation, **96**, xxxvii
targeting, **96**, 663
vesicular, **96**, xlv
disk, tritium–hydrogen exchange kinetics, **88**, 673
endoplasmic reticulum
degradation after drug induction, **96**, 547
drug induction, **96**, 544
influencing factors, **96**, 543
protein biogenesis and assembly, **96**, 520
Golgi, avian salt gland
enzyme assays, **96**, 638
preparation, **96**, 635
halobacterial envelope
electrical gradients
measurement with spin probes, **88**, 755
production in dark, **88**, 758
pH gradients
measurement with spin probes, **88**, 754
production in dark, **88**, 758
proton movement and electrical potential changes, simultaneous measurements, **88**, 356
surface area determination, **88**, 759
volume measurement with spin probes, **88**, 752
hepatic
endocytic compartments, selective labeling, **98**, 232
phosphomannosyl receptor purification, **98**, 307
intestinal brush border, aminopeptidase integration mode, relation to protein
biosynthesis, **96**, 417
structure, **96**, 407
intracellular
role in protein glycosylation, **98**, 91
transport pathways
from endoplasmic reticulum to Golgi complex, **98**, 8
exocytic, **98**, 1
from Golgi complex to lysosomes, **98**, 8
from plasmalemma to
Golgi complex, **98**, 3
lysosomes, **98**, 4
role of coated vesicles, **98**, 9
sites and mechanisms of sorting, **98**, 11
lens fiber, major polypeptide
cell-free synthesis, **98**, 510
immunoprecipitation, **98**, 516
lysosomal, iodination, **98**, 404
application to membrane recycling, **98**, 414
microsomal
core glycosylation activity, **96**, 92
integration of *in vitro* synthesized acetylcholine receptor subunits, **96**, 374
preparation, **96**, 87, 94
protein insertion, verification, **96**, 146
protein translocation
machinery, **96**, 85
posttranslational proteolytic assay, **96**, 92, 101
purification of signal recognition particle, **96**, 684
in translation systems, **96**, 90, 100
mitochondrial, protein synthesis and assembly, **97**, 245
natural and artificial, preparation of oriented multilamellar arrays, **88**, 326
nuclear, protein biogenesis and assembly, **96**, 589
pancreatic zymogen granule
biogenesis, **98**, 74
characterization, **98**, 72
isolation, **98**, 68
role in exocytosis, **98**, 74
peroxisomal, preparation, **96**, 553
photoreceptor, *see* Photoreceptor membranes
planar bilayer, bacteriorhodopsin-containing, assembly by sequential transfer of air–water interface films, **88**, 45
planar monolayer, preparation for spectroscopy and electron microscopy, **88**, 230
plasma, *see* Plasma membranes

purple, *see* Purple membranes
retinal rod outer segment, purification, **81**, 124
rhodopsin-containing, electrical responses to light, **81**, 431
correlation with spectral intermediates, **81**, 438
sarcoplasmic reticulum, protein biosynthesis, **96**, 570
secretion granule, rat parotid gland
associated enzyme activities, **98**, 85
characterization, **98**, 83
preparation, **98**, 81
Sendai virus fusogenic reconstituted envelope, as vehicle for DNA transfer into mammalian cells, **101**, 492
skeletal muscle, acetylcholine receptor role in recycling, **96**, 348
small-intestinal brush border, sucrase–isomaltase complex
anchoring segment, orientation of N-terminal amino acid, **96**, 392
assembly, **96**, 399
biosynthesis
in cell-free system, **96**, 400
mechanism, **96**, 405
lipid bilayer-embedded subunit identification, **96**, 388
splenocyte, preparation, **96**, 263
thylakoid
grana and stroma lamellae, preparation, **97**, 538
herbicide binding, **97**, 559
identification of phosphorylated polypeptides, **97**, 560
preparation, **97**, 557
protein synthesis and assembly, **97**, 487
pulse-labeling, **97**, 559
viral, protein biogenesis and assembly, **96**, 434
visual pigment-containing, isolation, **81**, 29
white, with crystalline bacterio-opsin, isolation, **88**, 407
Mercaptoacetic acid, *see* Thioglycolic acid
Mercaptoethanol
reduction of
free methionine sulfoxide, **91**, 550
methionine sulfoxide residues in
peptides, **91**, 554
proteins, **91**, 556
Mercenaria mercenaria
paramyosin preparation, **85**, 160
Mercury
–thiol affinity system, in generation of overlapping labeled DNA fragments for sequencing, **100**, 116
Meromyosin
heavy
equilibrium binding to F-actin, **85**, 714
preparation, **85**, 59
properties, **85**, 68
single-headed
preparation, **85**, 65
properties, **85**, 68
light
aggregate formation and assay, **85**, 51
preparation, **85**, 66
properties, **85**, 68
Metal decoration
purple membranes, **88**, 235
Metalloimmunoassays
analytical methods, **92**, 454
principles, **92**, 445
techniques, **92**, 448
Metallothionein
purification from liver, **84**, 122
radioimmunoassay, **84**, 121
Metal nucleotides
myosin active-site, trapping with thiol crosslinking reagents, **85**, 93
applications, **85**, 114
Metals
binding proteins, immunoassays, **84**, 121
in proteins, X-ray fluorescence spectrometry, **91**, 119
Metarhodopsin
types I and II
bovine, bleaching intermediate kinetics, **81**, 374
insect, absorbance properties, **81**, 187
Methanobacterium bryantii
electron transfer components, EPR detection, **88**, 413

Methanospirillum hungatei GPI
cultivation, **88**, 105
total lipids
composition, **88**, 108
extraction, **88**, 107
separation, **88**, 107
Methionine
-containing peptides, purification, **91**, 367
in erythrocytes, radiolabeling, **96**, 259
Methionine adenosyltransferase
Escherichia coli
assay, **94**, 219
deficient mutants, screening, **94**, 91
gene cloning, **94**, 117
properties, **94**, 221
purification, **94**, 220
isozymes, rat liver
assay, **94**, 223
properties, **94**, 226
purification, **94**, 224
Methionine sulfoxide
free, reduction, **91**, 550
in peptides, reduction, **91**, 554
in proteins, reduction, **91**, 556
Methotrexate
bioluminescent immunoassay, **92**, 426
competitive protein binding assay, **84**, 447
clinical applications, **84**, 454
radioimmunoassays, **84**, 409
with antigen-coated plastic tubes and labeled protein A, **92**, 409
clinical applications, **84**, 420
with magnetizable solid-phase particles, **84**, 422
3-Methyladenine
inhibitory effects on lysosomes, **96**, 761
3-Methylcholanthrene
induction of endoplasmic reticulum membrane components, **96**, 544
1,1′-[(Methylethanediylidene)dinitrilo]bis(3-aminoguanidine)
S-adenosylmethionine decarboxylase inhibition, **94**, 244
specificity, **94**, 245
synthesis, **94**, 244
Methylglyoxal bis(guanylhydrazone), *see* Mitoguazone
Methyl groups
chiral
analytical techniques, **87**, 130
biochemical application, **87**, 146
stereochemical course, **87**, 126
synthesis, **87**, 139
Methyl-*p*-hydroxybenzimidate
in attachment of haptens to erythrocytes, **92**, 269
Methyl-α-mannopyranoside
quantification by partition affinity ligand assay, **92**, 506
N-Methylmercaptoacetamide
reduction of
free methionine sulfoxide, **91**, 550
methionine sulfoxide residues in
peptides, **91**, 554
proteins, **91**, 556
Methylococcus capsulatus
purification of
3-hexulose-6-phosphate synthase, **90**, 316
6-phospho-3-ketohexulose isomerase, **89**, 568
Methylomonas M15
production of
glucose-6-phosphate dehydrogenase, **89**, 272
3-hexulose-6-phosphate synthase, **90**, 320
purification of
glucose-6-phosphate dehydrogenase, **89**, 272
3-hexulose-6-phosphate synthase, **90**, 321
15-Methyl prostaglandins
handling, **86**, 458
physical chemistry, **86**, 456
5-Methyltetrahydrofolate
radioimmunoassays, **84**, 409
with antigen-coated plastic tubes and labeled protein A, **92**, 409
clinical applications, **84**, 420
N^5-Methyltetrahydrofolate
ligand-binding radioassay, **84**, 458
5′-Methylthioadenosine
aminopropyltransferase inhibition, **94**, 296

metabolism, **94**, 355
quantification in biological samples, **94**, 61
5′-Methylthioadenosine nucleosidase
Lupinus luteus seeds
assay, **94**, 365
properties, **94**, 367
purification, **94**, 366
5′-Methylthioadenosine phosphorylase
Caldariella acidophila
assay, **94**, 356
properties, **94**, 359
purification, **94**, 357
5-Methylthioribose
metabolism, **94**, 361
5-Methylthioribose kinase
Enterobacter aerogenes
assay, **94**, 362
properties, **94**, 364
purification, **94**, 363
Methyl thymidine
radiolabeled, as marker in cytotoxicity assays, **93**, 233
Metyrapone
fluorometric assay, **84**, 587
and reduced derivative
bioassay, **84**, 587
radioimmunoassay, **84**, 588
spectrophotometric assay, **84**, 588
Mevastin
as dolichol synthesis inhibitor, **98**, 149
Micrococcus
D-apiitol dehydrogenase purification, **89**, 229
M. rubens, putrescine oxidase purification, **94**, 302
Microfilaments
lacrimal and salivary gland, role in protein secretion, **98**, 175
β_2-Microglobulin
quantification by partition affinity ligand assay, **92**, 508
Micromagnetorheometer
for characterizing actin filament networks, **85**, 226
Microsomal aminopeptidase
rabbit intestinal brush border membrane
mode of integration, relation to biosynthesis, **96**, 417
structure, relation to mode of integration, **96**, 407
Microsomes
membranes
integration of *in vitro* synthesized acetylcholine receptor subunits, **96**, 374
proteins, **96**, 146
purification of signal recognition particle, **96**, 684
rough
core-glycosylation activity, **96**, 92
inactive, preparation, **96**, 698
inverted, ribophorin recovery, **96**, 523
preparation from canine pancreas, **96**, 88, 94
for protein synthesis *in vitro*
preparation, **96**, 3
properties, **96**, 12
ribosome-binding capacity, correlation with ribophorin content, **96**, 526
solubilized, ribophorin isolation, **96**, 521
in translation systems, **96**, 90
spleen cell, preparation, **96**, 264
Microspectrophotometer
field portable
design goals and specifications, **81**, 648, 654
detection/analysis module, **81**, 651
monochromator control module, **81**, 650
optics module, **81**, 649
Microspectrophotometry
insect visual pigments, **81**, 729
polarized, visual pigment orientation and concentration, **81**, 642
Microtubule-associated proteins
assay, **85**, 415
enzyme activities, **85**, 430
interaction with guanine nucleotides, **85**, 416
purification, **85**, 409
Microtubules
assembly *in vitro*, analytical techniques, **85**, 433
disassembly, quantitative kinetic analysis, **85**, 439

role in secretory processes of
lacrimal and salivary glands, **98**, 175
liver, **98**, 169
Microwaves
irradiation techniques to inactivate eicosanoid-metabolizing enzymes in brain, **86**, 635
Milk
bovine, purification of UDP-galactose:*N*-acetyl-D-glucosamine β1-4-galactosyltransferase, **98**, 123
Mitochondria
broccoli, pyruvate dehydrogenase complex purification, **89**, 411
cauliflower, pyruvate dehydrogenase complex purification, **89**, 409
hepatic
induction by
di(2-ethylhexyl) phthalate, **96**, 557
thyroid hormones, **96**, 555
isolation of hexokinase-binding protein from outer membrane, **97**, 469
membrane biogenesis, assembly and targeting, **97**, 245
pancreatic islet, biochemical characterization, **98**, 190, 192
plant
protein synthesis *in vitro*, **97**, 481
purification, **97**, 478
protein import
analytical techniques, **96**, 148
overview, **97**, 245
skeletal muscle, induction by thyroid hormones, **96**, 555
yeast, polypeptide precursor processing, **97**, 311
Mitoguazone
and congeners
inhibitory effects on *S*-adenosylmethionine decarboxylase, **94**, 240
quantitation, **94**, 244
specificity, **94**, 241
synthesis, **94**, 240
enzyme inhibition assays, **94**, 247
with adenosylmethionine decarboxylase, **94**, 247
clinical applications, **94**, 251
with diamine oxidase, **94**, 250
Modeccin
antibodies, production and quantitation, **83**, 362
assay, **83**, 357
purification, **83**, 359
Molybdate
aldose interconversion catalysis, **89**, 72
Monensin
applications, **98**, 57
Golgi complex perturbation
effect on
intracellular protein transport, **98**, 51
morphology, **98**, 49
mechanism, **98**, 57
reversibility, **98**, 56
transport arrest site, **98**, 52
inhibitory effects on lysosomes, **96**, 755
Monoacetyl-1,3-diaminopropane monohydrochloride
N[1]-Monoacetylspermidine
synthesis, **94**, 421
Monoacylglycerophosphocholine
preparation and positional distribution of fatty acids, **86**, 373
synthesis, **94**, 420
Monoacylglycerophosphoethanolamine
preparation and positional distribution of fatty acids, **86**, 373
Monoclonal antibodies
carbohydrate-specific, assay in gel by transfer template localized hemolysis, **92**, 184
cell surface-specific
immunoadsorption assay, **92**, 195
immunoglobulin isotype determination, **92**, 166
indirect assay with ^{125}I-labeled protein A, **92**, 160
transfer template radioimmunoassay, **92**, 194
in epitope distinction by
binding patterns on different molecules, **92**, 248
competition solid-phase antibody binding assay, **92**, 244
additive effect, **92**, 252
with labeled antigen, **92**, 251

double diffusion in agar, **92**, 252
iodination sensitivity, **92**, 250
sandwich assay, **92**, 251
α-fetoprotein, preparation, **84**, 9
glycolipid antigen detection by binding to thin-layer chromatograms, **83**, 239
to guinea pig Ia antigens
epitope binding assay, **92**, 77
fluorescence-activated cell sorter analysis, **92**, 81
immunoprecipitation studies, **92**, 72
production, **92**, 68
in T-lymphocyte proliferation inhibition, **92**, 82
high-resolution two-dimensional gel electrophoresis, **92**, 206
to HLA gene products
characterization, **92**, 125
in immunoaffinity chromatography, **92**, 128
affinity considerations, **92**, 137
production, **92**, 114, 123, 136
purification, **92**, 123
immunoglobulin screening by immunofixation on cellulose acetate, **92**, 220
murine antigen-specific, of IgE class
assay, **92**, 55
determination by passive cutaneous anaphylaxis, **92**, 58
ε heavy chain-specific, production, **92**, 63
physical and chemical characterization, **92**, 62
production, **92**, 49
purification, **92**, 60, 64
production screening by ELISA with enzyme-labeled second antibody, **92**, 168
to prostaglandin H synthase
in immunoradiometric assay, **86**, 236
preparation, **86**, 230
producing hybridomas
IgG purification, **86**, 235
selection, **86**, 232
to prostaglandin I_2 synthase
assay, **86**, 242
in immunoradiometric assay, **86**, 243
preparation, **86**, 241
with selected applications
detection and assessment, **92**, 147
production, **92**, 3
to soluble antigens, solid-phase immunofluorescence assay, **92**, 227
α-Monofluoromethylputrescine
synthesis, **94**, 201
Monohydroxyeicosenoic acids
argentation-HPLC, **86**, 530
extraction with octadecylsilyl silica, **86**, 467
Monosaccharides
^{13}C-enriched, chemical synthesis, **89**, 64
Morphine
radioimmunoassay, **84**, 551
Motilin
radioimmunoassay, **84**, 359
Mucopolysaccharide storage disorders
genetic, enzymatic diagnosis, **83**, 559
Muscles
abdominal, murine, phosphoglycerate kinase-1 purification, **90**, 123
adductor, oyster, phosphofructokinase purification, **90**, 40
Ascaris suum, purification of
fructose-bisphosphate aldolase, **90**, 256
hexokinase, **90**, 22
phosphofructokinase, **90**, 45
bovine, simultaneous purification of aldolase, glyceraldehyde-3-phosphate dehydrogenase, phosphoglycerate kinase, and phosphoglycerate mutase, **90**, 509
breast
chicken, purification of
creatine kinase, **85**, 140; **90**, 490
enolase, **90**, 490
fructose-bisphosphatase, **90**, 341
fructose-bisphosphate aldolase, **90**, 490
glyceraldehyde-3-phosphate dehydrogenase, **90**, 490
myomesin, **85**, 143
phosphoglyceromutase, **90**, 490
phosphorylase, **90**, 490
triose-phosphate isomerase, **89**, 580, **90**, 490
pigeon, pyruvate dehydrogenase complex purification, **89**, 416

cardiac
bovine, purification of
troponin, **85**, 249
troponin subunits, **85**, 253
rabbit, tropomyosin purification, **85**, 234
catfish, glucose-6-phosphate isomerase purification, **89**, 555
cultured, acetylcholinesterase
intracellular transport and fate, **96**, 363
molecular forms, localization and assembly, **96**, 365
subcellular distribution, **96**, 354
synthesis, **96**, 361
fast-twitch, chicken and rabbit, preparation of myosin light chains, **85**, 74
flight
bumblebee, fructose-bisphosphatase purification, **90**, 367
grasshopper, cyclic nucleotide-dependent protein kinase purification, **99**, 73
fresh, myofibril preparation, **85**, 11
Helix pomatia, fructose-bisphosphate aldolase purification, **90**, 259
human, purification of
glucose-6-phosphate isomerase, **89**, 552
glyceraldehyde-3-phosphate dehydrogenase, **89**, 303
intact and extracted, ^{31}P NMR, **85**, 624
invertebrate, paramyosin preparation, **85**, 150
leg, grasshopper, cyclic nucleotide-dependent protein kinase purification, **99**, 73
Mercenaria, paramyosin preparation, **85**, 160
murine, purification of
L-lactate dehydrogenase A_4, **89**, 357
phosphoglucose isomerase, **89**, 560
neck, bovine, pyruvate kinase purification, **90**, 143
porcine, purification of
glucose-6-phosphate isomerase, **89**, 552
glyceraldehyde-3-phosphate dehydrogenase, phosphoglycerate kinase, and phosphoglycerate mutase simultaneously, **90**, 499
psoas, rabbit, preparation of
myofibrils, **85**, 9
thick and thin filaments, **85**, 17
rabbit, purification of
glucose-6-phosphate isomerase, **89**, 552
triose-phosphate isomerase, **89**, 582
scallop, preparation of myosin light chains, **85**, 81
skeletal, *see* Skeletal muscle
smooth, *see* Smooth muscle
snake, fructose-bisphosphatase purification, **90**, 350
stored, myofibril preparation, **85**, 11
vertebrate, phosphoglycerate kinase purification, **90**, 103
Mutagenesis
bacteriophage M13 DNA, creation of ordered nest of asymmetric deletions across large target sequence, **101**, 98
DNA *in vitro* with
sodium bisulfite, **100**, 457
terminal deoxynucleotidyltransferase, **100**, 96
Escherichia coli chromosome with
bacteriophage Mu, **94**, 91
restriction enzymes, **97**, 30
synthetic oligonucleotides as site-specific mutagens, **97**, 6, 33, 126
oligonucleotide-directed, DNA fragments cloned in M13 vectors, **100**, 468
yeast genome by
construction of site-specific rearrangements, **101**, 269
integrative transformation
in cloning mutant genes and large segments of contiguous sequences, **101**, 290
with linear and gapped plasmids, genetic applications, **101**, 228
one-step disruption, **101**, 202
yeast mitochondrial genes with ethidium bromide, **97**, 364

Mutants
Chinese hamster ovary cell
cAMP-resistant
characterization, **99**, 201
isolation, **99**, 199
in protein kinase analysis, **99**, 204
ornithine decarboxylase-deficient, selection, **94**, 108
Dictyostelium discoideum
with phagocytic defects
adhesive revertants, isolation, **98**, 428
characterization, **98**, 427
enrichment, **98**, 426
secretory
characterization, **96**, 818, 821
isoelectric focusing, **96**, 823
isolation, **96**, 816
epidermal growth factor receptor-deficient, spontaneous transfer of exogenous receptors, **98**, 555
Escherichia coli, with defects in polyamine biosynthesis, mass screening, **94**, 83
Halobacterium halobium
defective in pigment synthesis, isolation, **88**, 379
R_1mW lacking carotinoid, isolation of crystalline bacterio-opsin white membrane, **88**, 407
myxovirus, with glycoprotein processing defects
preparation, **96**, 440
selection, **96**, 442
Neurospora crassa arginase-deficient, in analysis of polyamine metabolism, **94**, 112
Semliki Forest virus transport-defective
isolation, **96**, 453
screening, **96**, 458
somatic cell
with altered glycoproteins, selection, **83**, 444
lectin-resistant
nomenclature, **96**, 161
selection, **96**, 157
yeast
with defects in polyamine biosynthesis, mass screening, **94**, 104
gene eviction and transplacement, **101**, 211
nuclear respiratory deficient, transformation, **97**, 349, 355
confirmation, **97**, 360
secretory
classification, **96**, 806
cytologic and biochemical characterization, **96**, 808
electrophoretic analysis, **96**, 812
screening, **96**, 805
selection, **96**, 804
Mycobacterium
M. smegmatis, purification of
class I fructose-bisphosphate aldolase, **90**, 243
D-ribose isomerase, **89**, 547
M. turberculosis, purification of
class I fructose-bisphosphate aldolase, **90**, 249
class II fructose-bisphosphate aldolase, **90**, 247
Myelin-specific proteins
antibody preparation, **96**, 385
biosynthesis, **96**, 378
purification, **96**, 382
Myofibrils
preparation, **85**, 9
purification of nebulin and titin, **85**, 264
Myoglobin
methionine-containing peptides, purification, **91**, 371
radioimmunoassay, **84**, 172
serum levels associated with hypermyoglobinemia, **84**, 176
Myomesin
immunological detection, **85**, 146
properties, **85**, 145
purification, **85**, 143
Myosin
Acanthamoeba castellanii
assays, **85**, 357
properties, **85**, 360, 363
purification, **85**, 359, 361
active-site trapping with
thiol crosslinking reagents, **85**, 93
applications, **85**, 114
vanadate, **85**, 116
applications, **85**, 122

aggregation
assays, **85**, 29
by dialysis, **85**, 23
by dilution, **85**, 24
ionic strength effects, **85**, 26
pH effects, **85**, 25
protein concentration effects, **85**, 29
time effects, **85**, 27
ATP hydrolysis
steady-state rate measurements, **85**, 698
transient kinetic measurements, **85**, 701
crude
actin removal, **85**, 134
phosphofructokinase removal, **85**, 134
preparation, **85**, 133
separation from C, H, and X proteins, **85**, 135
divalent cation-binding sites, detection with Mn^{2+} probe, **85**, 619
EPR, **85**, 594
filament hybrids, copolymerization, **85**, 41
gel electrophoresis, **85**, 127
interaction with
actin, EPR, **85**, 613
substrates, optical activity measurements, **85**, 679
light chains
alkali, labeling, **85**, 92
essential
exchanges, **85**, 78
purification, **85**, 84
fractionation, **85**, 76
mixed, isolation, **85**, 74
regulatory, purification, **85**, 82
storage, **85**, 78
minifilaments, aggregation, **85**, 42
negative staining for electron microscopy, **85**, 129
nonfilamentous, aggregation, **85**, 45
nonmuscle
concentration, **85**, 352
phosphorylation effect on thick filament stabilization, **85**, 364
purification, **85**, 331, 364
storage, **85**, 354
properties, **85**, 68
purification from
platelets, **85**, 355
skeletal muscle, **85**, 56; **98**, 371
smooth muscle, **85**, 292
rod aggregates, formation and assay, **85**, 48
saturation transfer studies, **85**, 612
spin labeling, **85**, 595, 607
selectivity and stoichiometry, **85**, 596
subfragments
equilibrium binding to
F-actin, **85**, 710
regulated actin, **85**, 715
interactions with substrates, optical activity measurements, **85**, 679
properties, **85**, 68
purification from skeletal muscle, **85**, 59; **98**, 373
steady-state binding to
F-actin, **85**, 717
regulated actin, **85**, 722
sulfhydryl groups
intramolecular crosslinking, **85**, 95
modification and labeling, **85**, 84
Myosin ATPase
assays, **85**, 123
Myosin light-chain kinase
smooth muscle
assays, **85**, 307; **99**, 279
properties, **85**, 305; **99**, 286
purification, **85**, 298; **99**, 281
regulation, **99**, 287
substrate specificity, **99**, 286
Myxoviruses
glycoproteins
biological activities, assays, **96**, 436
biosynthesis, **96**, 434
mutants with defects in glycoprotein processing
preparation, **96**, 440
selection, **96**, 442
replication in cell culture, **96**, 437

N

NADPH-cytochrome *P*-450 reductase
hepatocyte endoplasmic reticulum, biosynthesis, **96**, 536

Na^+/K^+-transporting ATPase, *see* ATPase, Na^+,K^+-
Naphthalene-1,5-monophosphate
in active-site trapping of myosin metal nucleotides, **85**, 93
Natural killer cells
frequency determination by single-cell cytotoxicity assay, **93**, 270
Nebulin
molecular properties, **85**, 274
purification, **85**, 264
Neocarzinostatin, *see* Zinostatin
Neoglycoproteins
preparation with
ω-aldehydoalkyl 1-thioglycosides, **83**, 289
cyanomethyl 1-thioglycosides, **98**, 302
2-imino-2-methoxyethyl 1-thioglycosides, **83**, 278
uptake and binding by macrophages, **98**, 303
Nephritic factor
human, in alternative complement pathway
assay, **93**, 407
purification, **93**, 406
Neurospora crassa
arginase-deficient mutant, in analysis of polyamine metabolism, **94**, 112
biosynthesis and assembly of
cytochrome *c*, **97**, 261
nuclear-coded mitochondrial membrane proteins, **97**, 275
middle-abundant mRNAs, cloning, **97**, 254
polyamine pools, double-isotope derivative assay, **94**, 36
purification of
cytochrome *c*, **97**, 262
mitochondrial outer membrane porin, **97**, 287
pyruvate dehydrogenase complex, **89**, 389
Neutron diffraction
bacteriorhodopsin, **88**, 241
retinal rod outer segments, **81**, 593
rhodopsin in hydrogenated and deuterated detergents, **81**, 742
Nicotinamide-adenine dinucleotide
–alcohol dehydrogenase complexes
assays, **89**, 461
properties, **89**, 464
synthesis, **89**, 459
Nicotine
nicotinamide nucleotide analogs, radioimmunoassay, **84**, 650
radioimmunoassay, **84**, 628
Nigericin
inhibitory effects on lysosomes, **96**, 755
Nitrazepam
radioimmunoassay, **84**, 505
Nitrocellulose paper
transfer and hybridization of
denatured RNA, **100**, 255
small DNA fragments, **100**, 309
N′-Nitrosonornicotine
radioimmunoassay, **84**, 641
NMR, *see* Nuclear magnetic resonance
Normeperidine
radioimmunoassay, **84**, 516
applications, **84**, 523
Nuclear magnetic resonance
bacteriorhodopsin, **88**, 310
cAMP-dependent protein kinase, **99**, 93
collagen fibrils by solid-state method, **82**, 174
complex carbohydrates by ^{13}C method, **83**, 58
contractile systems by ^{31}P method, **85**, 624
Cr(III)– and Co(III)–nucleotide complexes and interactions with enzymes, **87**, 180
enzyme mechanisms involving phosphorus by $^{31}P(^{18}O)$, $^{31}P(^{17}O)$, and ^{17}O methods, **87**, 235
glycosphingolipids by high-resolution proton method, **83**, 69
in hydroxyputrescine identification, **94**, 433
retinal analog *cis/trans* isomers, **88**, 546
rhodopsin by ^{13}C method, **81**, 698
rhodopsin–lipid interactions in retinal rod outer segment membranes, **81**, 714
Nuclei
from *Drosophila melanogaster* embryos
characterization, **96**, 595

purification, **96**, 589
erythrocyte, isolation of pore complex–lamina structures, **96**, 604
eukaryotic cell, expression of Semliki Forest virus cDNA after microinjection, **96**, 496
hepatic, isolation of membrane pore complex–lamina structures, **96**, 607
oocyte, isolation of nuclear envelope, **96**, 598
Nucleocapsids
Semliki Forest virus, uncoating assay, **98**, 264
Nucleoside-diphosphate kinase
microtubule-associated, assay, **85**, 431
Nucleosides
alkylated and natural, determination of antibody affinity and specificity by competitive radioimmunoassay, **92**, 589
Nucleoside-triphosphatase
stereochemical course, analytical techniques, **87**, 301
5′-Nucleotidase
as marker for membrane flow during pinocytosis, **98**, 395
Nucleotides
binding proteins, photoaffinity labeling with 8-azidopurine analogs, **91**, 613
applications, **91**, 629
complexes with Cr(III) and Co(III)
applications, **87**, 180
interactions with enzymes, EPR and NMR analyses, **87**, 184, 192
properties, **87**, 180
sugar, isolation and synthesis, **83**, 332
Nucleotidyltransferases
stereochemistry, analytical techniques, **87**, 197, 213

O

Oat seedlings
purification of
arginine decarboxylase, **94**, 178
polyamine oxidase, **94**, 312
Octadecylsilyl silica
in arachidonic acid extraction from biological samples, **86**, 467
Octyl glucoside
in visual pigment extraction, **81**, 133
Oligo-1,6-glucosidase
–sucrase complex, small-intestinal brush border membrane
anchoring segment, orientation of N-terminal amino acid, **96**, 392
assembly, **96**, 399
biosynthesis
in cell-free system, **96**, 400
mechanism, **96**, 405
lipid bilayer-embedded subunit, identification, **96**, 388
Oligonucleotides
mutagenic
design, **100**, 472
in mutagenesis of DNA fragments cloned in M13 vectors, **100**, 468
sequence determination, **100**, 475
synthesis, **100**, 475
synthetic, purification by RPC-5 ANALOG chromatography, **100**, 388
Oligopeptides
below M_r 10,000, slab gel system for resolution, **96**, 239
synthetic
preparation, **99**, 121
in protein kinase assay, **99**, 134
Oligosaccharide alditols
hexosamine-containing, as permethylated *N*-trifluoroacetyl derivatives
gas-liquid chromatography–mass spectrometry, **83**, 52
linkage determination, **83**, 57
preparation, **83**, 46
sugar sequence determination, **83**, 52
Oligosaccharides
acetylated, resolution by reverse-phase HPLC, **83**, 132
anionic, ion-exchange HPLC, **98**, 157
N-asparagine-linked
biosynthesis
in vitro, **83**, 406
in vivo, **83**, 399
processing, **83**, 416

release from glycoproteins or glycopeptides by hydrazinolysis, **83**, 263
transfer to peptides and proteins *in vitro*, **83**, 408
dolichylpyrophosphoryl, resolution by HPLC, **83**, 137
erythrocyte high-molecular-weight, preparation, **83**, 311
gel filtration analysis, **83**, 105
application for monosaccharide sequencing, **83**, 125
N-glycosidically linked, elongation-associated glycosyltransferases, **98**, 98
[2-^{3}H]mannose-labeled
pinocytosis by I-cell disease fibroblasts, **98**, 296, 297
preparation, **98**, 296
species separation, **98**, 297
neutral, amine adsorption-HPLC, **98**, 158
as probes for receptor studies, preparation, **98**, 154
rhodopsin, structural analysis, **81**, 218, 223
sugar-labeled
characterization, **82**, 352
preparation from
lipid intermediates, **82**, 348
procollagen, **82**, 347
transport-coupled processing *in vitro*, assays, **98**, 37
Oocytes
Xenopus laevis
in cloned gene expression, **101**, 370
extraction of phosphorylated proteins, **99**, 226
isolation, **99**, 223
microinjection techniques to study protein kinases and protein phosphorylation, **99**, 219
nuclear envelope preparation, **96**, 598
properties, **101**, 370, 372
Opsin
biosynthesis in retina, immunocytochemical analysis, **96**, 495
elution profile on agarose, **81**, 141
thermal stability in detergents, **81**, 256
Optical absorption spectroscopy
bacteriorhodopsin photocycle kinetics, **88**, 133
Optical diffraction
contractile proteins, **85**, 511
Optical rotatory dispersion
structure–function relationships in muscle protein systems, **85**, 677
Organ cultures
biosynthesis of
insoluble elastin, **82**, 632
soluble elastin, **82**, 716
Ornithine carbamoyltransferase
hepatic mitochondrial
intracellular transport, **97**, 402
posttranslational uptake and processing *in vitro*, **97**, 405
synthesis, **97**, 396
Ornithine decarboxylase
deficient mutants
Chinese hamster ovary cell, selection, **94**, 108
Escherichia coli, mass screening, **94**, 88
Saccharomyces cerevisiae, mass screening, **94**, 104
in early determination of histocompatibility in mixed lymphocyte reaction, **94**, 397
Escherichia coli
biosynthetic and biodegradative forms
assay, **94**, 125
properties, **94**, 133
purification, **94**, 127
gene cloning, **94**, 120
from germinated barley seeds
assay, **94**, 162
properties, **94**, 165
purification, **94**, 163
modifying protein, *Physarum polycephalum*
assay, **94**, 145
characteristics, **94**, 146
preparation, **94**, 145
murine kidney
assay, **94**, 159

autoradiographic localization, **94**, 169
immunocytochemical localization, **94**, 166
labeling and quantitation with α-[5-^{14}C]-difluoromethylornithine, **94**, 206
properties, **94**, 161
purification, **94**, 159
Physarum polycephalum
assay, **94**, 141
phosphorylation by polyamine-dependent protein kinase, **99**, 366
properties, **94**, 144
purification, **94**, 141
rat liver
assay, **94**, 154
labeling and quantitation with α-[5-^{14}C]difluoromethylornithine, **94**, 206
properties, **94**, 158
purification, **94**, 156
rat ovary and placenta, immunocytochemical localization, **94**, 166
Saccharomyces cerevisiae
assay, **94**, 135
properties, **94**, 139
purification, **94**, 136
simultaneous assay with histidine decarboxylase, **94**, 45
Ornithine decarboxylase antizymes
Escherichia coli
assay, **94**, 193
properties, **94**, 197
purification, **94**, 194
protein inhibitor, rat liver
assay, **94**, 189
properties, **94**, 191
purification, **94**, 190
rat liver
assay, **94**, 185
properties, **94**, 187
purification, **94**, 187
Ovalbumin
amino-terminal acetylamino acids, HPLC, **91**, 88
Oxazepam
radioimmunoassay, **84**, 513
2-Oxoaldehyde dehydrogenase
rat liver
assay, **89**, 514
purification, **89**, 514
Oxygen isotopes
enriched aldoses, chemical synthesis, **89**, 79
exchange in carbonic anhydrase catalysis of CO_2 hydration, **87**, 732

P

1-Palmitoyl-2-acyl-*sn*-phospholipids
labeled, synthesis, **98**, 630
1-Palmitoyl-*sn*-glycero-3-phosphoric acid alkyl esters
synthesis, **98**, 627
Pancreas
canine, isolation of
microsomal signal peptidase, **96**, 786
mRNA, **96**, 97
rough microsomes, **96**, 87, 94
guinea pig
lobule preparation for acinar cell functional analysis *in vitro*, **98**, 17
zymogen granule isolation, **98**, 68
islets of Langerhans, *see* Islets of Langerhans
Papillomaviruses
bovine, construction of eukaryotic cloning vectors, **101**, 387
Parabactin
Paracoccus denitrificans
assay, **94**, 441
isolation, **94**, 440
properties, **94**, 440
Paracoccus denitrificans
parabactin isolation, **94**, 440
Paramyosin
assays, **85**, 155
isolation from invertebrate muscles, **85**, 150
problems unique to *Mercenaria*, **85**, 160
Parinaric acid
fluorescent probes, in analysis of disk membrane phospholipid organization, **81**, 685

Parotid gland
rat
cytoskeleton role in protein secretion, **98**, 175
preparation of
plasma membranes, **98**, 76
secretion granule membranes, **98**, 81
Pea seeds
glyceraldehyde-3-phosphate dehydrogenase purification, **89**, 321
Penicillinase, *see* β-Lactamase
Penicillium wortmanni
β-D-xylosidase purification, **83**, 641
Pepstatin
inhibitory effects on lysosomes, **96**, 756
Peptide mapping
hemoglobin by combination of ion-exchange and thin-layer chromatography, **91**, 384
membrane proteins in gels by limited proteolysis, **96**, 191
phosphopeptides by HPLC, **99**, 37
polypeptides in gels by limited proteolysis, **100**, 424
proteins
in gels by limited proteolysis, **100**, 424
by silica thin-layer method, **91**, 466
radioiodinated polypeptides by HPLC, **92**, 304
SDS-solubilized proteins by limited proteolysis, **96**, 222
Peptides
amidination, **91**, 559
reversal, **91**, 566
amino-terminal analysis with dimethylaminoazobenzene isothiocyanate, **91**, 79
collagen crosslinked
isolation and characterization, **82**, 370
preparation and identification, **82**, 369
elastin
aggregate structure
circular dichroism, **82**, 689
electron microscopy, **82**, 680
structural characterization, **82**, 691
temperature-dependent aggregation, characterization, **82**, 676
glycogen synthase, in analysis of phosphorylase kinase specificity, **99**, 275
imino acid analysis with *o*-phthalaldehyde, **91**, 17
insoluble elastin
crosslinked
isolation, **82**, 608
sequencing, **82**, 615
single-chain
purification, **82**, 614
release and separation from crosslinked peptides, **82**, 613
sequencing, **82**, 614
methionine-containing, purification, **91**, 367
methionine sulfoxide reduction, **91**, 554
phosphorylase, in analysis of phosphorylase kinase specificity, **99**, 272
in polyacrylamide gels, detection after formaldehyde fixation, **91**, 247
purification by countercurrent chromatography, **91**, 335
radiolabeled
HPLC for characterization of precursor–product relationships in pulse-chase studies, **92**, 306
molecular sieving by HPLC, **92**, 301
purification by HPLC, **92**, 292
coupling with radioimmunoassay and microsequencing, **92**, 299
for radioimmunoassay studies, **92**, 294
retinyl
characterization, **81**, 213
isolation, **81**, 211
rhodopsin
preparation, **81**, 251
purification, **81**, 254
separation by
countercurrent chromatography, **91**, 335
ion-exchange chromatography, **91**, 359

reversed-phase HPLC, **91**, 352
sequencing
with gas-phase sequenator, **91**, 399
at nanomole level, **91**, 470
with spinning-cup sequenators, efficiency improvements, **91**, 434, 443
spectrin intermediate-sized
electrophoretic elution, **96**, 309
generation, **96**, 306
identification, **96**, 307
sulfhydryl, alkylated with *N*-ethylmaleimide, isolation by diagonal electrophoresis, **91**, 392
transfer of *N*-asparagine-linked oligosaccharides *in vitro*, **83**, 408
tryptic, affinity chromatography with immobilized anhydrotrypsin as adsorbent, **91**, 378
Peroxisomes
drug inducers, **96**, 548
hepatic
drug induction, **96**, 549
isolation, **96**, 553
protein biogenesis
in vitro studies, **96**, 726
in vivo studies, **96**, 721
pH
dependent enzymatic processes, analysis with buffers of constant ionic strength, **87**, 405
in determination of chemical mechanisms of enzyme-catalyzed reactions, **87**, 390
effect on prostaglandin extraction by octadecylsilyl silica, **86**, 472
light-induced gradients in halobacterial envelope vesicles, measurement with amine and carboxylic acid probes, **88**, 334
Phagocytosis, *see* Endocytosis, phagocytic
o-Phenanthroline
–cobalt complexes, in active-site trapping of myosin metal nucleotides, **85**, 109
Phenobarbital
induction of endoplasmic reticulum membrane components, **96**, 544
N,*N*′-*o*-Phenylenedimaleimide
in active-site trapping of myosin metal nucleotides, **85**, 103
N,*N*′-*p*-Phenylenedimaleimide
in active-site trapping of myosin metal nucleotides, **85**, 103
Phenylpiperidine carboxylic acid esters
N-substituted, and dealkylated metabolites, radioimmunoassay, **84**, 516
applications, **84**, 523
Phenylthiohydantoin amino acids
chemical ionization–mass spectrometry, **91**, 502
side chain-protected
HPLC, **91**, 496
application to solid-phase peptide synthesis, **91**, 497
preparation, **91**, 495
ultrasensitive gradient HPLC, **91**, 486
Phosphate
demonstration in glycophorin A, **96**, 291
in proteins
content determination, **99**, 7
removal by reverse reaction with protein kinases, **99**, 14
Phosphatidylcholine
with labeled polyunsaturated fatty acids, preparation, **86**, 370
Phosphatidylcholine transfer protein
bovine liver
assay, **98**, 582
properties, **98**, 585
purification, **98**, 583
rat liver
applications, **98**, 598
properties, **98**, 588
purification, **98**, 587
radioimmunoassay, **98**, 589
in sphingomyelin synthesis, **98**, 596
Phosphatidylethanolamine
with labeled polyunsaturated fatty acids, preparation, **86**, 370
Phosphatidylinositol
selectively labeled
generation, **86**, 4
purification, **86**, 5

1-Phosphatidylinositol phosphodiesterase
human platelet, assay, **86**, 8
6-Phospho-2-dehydro-3-deoxygalactonate aldolase
Pseudomonas saccharophila
assay, **90**, 264
purification, **90**, 265
Phosphoenolpyruvate:glycose phosphotransferase system
Salmonella typhimurium
assays, **90**, 421, 423
description, **90**, 417
Enzyme I
properties, **90**, 433
purification, **90**, 431
Enzyme III^{Glc}
properties, **90**, 444
purification, **90**, 439
HPr
properties, **90**, 437
purification, **90**, 436
Staphylococcus aureus
assays, **90**, 421, 447
description, **90**, 417
Enzyme III^{Lac}
properties, **90**, 455
purification, **90**, 453
HPr
properties, **90**, 452
purification, **90**, 450
1-Phosphofructokinase
Clostridium pasteurianum
assay, **90**, 82
properties, **90**, 85
purification, **90**, 84
6-Phosphofructokinase
Ascaris suum muscle
assay, **90**, 44
properties, **90**, 47
purification, **90**, 45
Bacillus licheniformis
assay, **90**, 71
properties, **90**, 73
purification, **90**, 72
Ehrlich ascites tumor
assay, **90**, 35
properties, **90**, 37
purification, **90**, 36
Escherichia coli types 1 and 2
assay, **90**, 60
immunological relationship to other bacterial phosphofructokinases, **90**, 68
properties, **90**, 64, 67
purification, **90**, 63, 65
oyster adductor muscle
assay, **90**, 39
properties, **90**, 44
purification, **90**, 40
rabbit muscle, isolation from crude myosin preparation, **85**, 134
rat liver, substrate-directed regulation of cAMP-dependent phosphorylation, **99**, 215
Streptococcus lactis
assay, **90**, 77
properties, **90**, 80
purification, **90**, 78
yeast
assay, **90**, 49
properties, **90**, 54
purification, **90**, 50
6-Phosphofructo-2-kinase/fructose-2,6-bisphosphatase
rat liver, substrate-directed regulation of cAMP-dependent phosphorylation, **99**, 217
6-Phosphofructokinase (pyrophosphate), *see* Pyrophosphate–fructose-6-phosphate 1-phosphotransferase
Phosphoglucomutase
yeast
assay, **89**, 599
properties, **89**, 603
purification, **89**, 600
6-Phosphogluconate
preparation from glucose 6-phosphate, **89**, 116
in preparation of
ribose 5-phosphate, **89**, 117
ribulose 1,5-bisphosphate, **89**, 117
6-Phosphogluconate dehydratase
Pseudomonas putida
2-keto-3-deoxygluconate 6-phosphate synthesis, **89**, 98
purification, **89**, 99
Phosphogluconate dehydrogenase (decarboxylating)
Bacillus stearothermophilus
assay, **89**, 283

properties, **89**, 286
purification, **89**, 283
Gluconobacter suboxydans
assay, **89**, 291
properties, **89**, 294
purification, **89**, 292
Pseudomonas fluorescens
assay, **89**, 278
properties, **89**, 281
purification, **89**, 279
Phosphoglucose isomerase, *see* Glucose-6-phosphate isomerase
Phosphoglycerate kinase
Bacillus stearothermophilus
assay, **90**, 126
properties, **90**, 129
purification, **90**, 127
from blue-green algae
assay, **90**, 110
concentration determination, **90**, 110
properties, **90**, 112
purification, **90**, 111
bovine liver
assay, **90**, 115
properties, **90**, 117
purification, **90**, 116
bovine muscle, simultaneous purification with aldolase, glyceraldehyde-3-phosphate dehydrogenase, and phosphoglycerate mutase, **90**, 509
isozymes and genetic variants from murine muscle and testis
assay, **90**, 121
properties, **90**, 126
purification, **90**, 122
porcine liver and muscle
assay, **90**, 499
properties, **90**, 508
simultaneous purification with glyceraldehyde-3-phosphate dehydrogenase and phosphoglycerate mutase, **90**, 499
spinach
assay, **90**, 110
concentration determination, **90**, 110
properties, **90**, 112
purification, **90**, 110
vertebrate tissue
assay, **90**, 103
concentration determination, **90**, 103
properties, **90**, 107
purification, **90**, 103
yeast
assay, **90**, 110, 115
concentration determination, **90**, 110
properties, **90**, 112
purification, **90**, 111, 118
Phosphoglycerate mutase
bovine muscle, simultaneous purification with aldolase, glyceraldehyde-3-phosphate dehydrogenase, and phosphoglycerate kinase, **90**, 509
chicken muscle
assay, **90**, 490
phosphorylated intermediate
evidence for participation in enzyme reaction, **87**, 43
kinetic competence, **87**, 47
phosphoprotein bond, characterization, **87**, 46
preparation, **87**, 50
primary structure at active site, homology with bisphosphoglycerate synthase, **87**, 49
purification, **90**, 492
porcine liver and muscle
assay, **90**, 499
properties, **90**, 508
simultaneous purification with glyceraldehyde-3-phosphate dehydrogenase and phosphoglycerate kinase, **90**, 499
Phosphoglyceromutase, *see* Phosphoglycerate mutase
6-Phospho-3-ketohexulose isomerase
Methylococcus capsulatus
assay, **89**, 566
properties, **89**, 570
purification, **89**, 568
Phospholipase A_2
applications in photoreceptor membrane studies, **81**, 320
human platelet, assay, **86**, 653
Phospholipase C
applications in photoreceptor membrane studies, **81**, 320
human platelet
assay, **86**, 653
preparation, **86**, 8

phosphatidylinositol-specific, *see* 1-Phosphatidylinositol phosphodiesterase
Phospholipase D
applications in photoreceptor membrane studies, **81**, 320
Phospholipids
biosynthetic enzymes, cytochemical localization in avian salt gland, **96**, 648
labeled, synthesis in high yield, **98**, 623
rod outer segment
fatty acid composition and pairing, **81**, 806
molecular species determination, **81**, 297
organizational analysis with parinaric acid fluorescent probes, **81**, 685
transmembrane topology, analytical methods, **81**, 309
substitution in purple membranes, **88**, 26
transfer between membranes, **98**, 565
Phospholipid vesicles
incorporation of
bacterio-opsin, **88**, 81
bacteriorhodopsin, **88**, 17, 78
monomeric bacteriorhodopsin, **88**, 31
rhodopsin, **81**, 149
Phosphomannosyl receptors
bovine hepatic membrane
assays, **83**, 392; **98**, 305
properties, **83**, 396; **98**, 308
purification, **83**, 394; **98**, 307
Phosphopeptides
mapping by HPLC, **99**, 37
purification, **99**, 43
Phosphoprotein phosphatase
porcine kidney
assay, **90**, 408
properties, **90**, 412
purification, **90**, 410
turkey gizzard smooth muscle
assay, **85**, 314
purification, **85**, 308
Phosphoproteins
light-regulated, measurement in *Halobacterium halobium*, **88**, 213
Phosphoribosylformylglycinamidine synthase
covalent reaction with substrates and antimetabolites, **87**, 76
Phosphoric monoesters
chiral
stereochemical analysis, **87**, 288
as stereochemical probes of phosphotransferases, **87**, 279
synthesis, **87**, 282
Phosphorus
enzyme mechanistic analysis by $^{31}P(^{18}O)$, $^{31}P(^{17}O)$, and ^{17}O NMR methods, **87**, 235
applications, **87**, 246
Phosphorylase
chicken muscle
assay, **90**, 490
isolation, **90**, 492
peptide substrates, in analysis of phosphorylase kinase specificity, **99**, 272
Phosphorylase kinase
bovine heart
assay, **99**, 252
properties, **99**, 259
purification, **99**, 255
rabbit skeletal muscle
assays, **99**, 244
properties, **99**, 248
purification, **99**, 246, 269
specificity analysis with peptide substrates, **99**, 268
subunit dissociation, **99**, 262
assays, **99**, 260
physicochemical characterization, **99**, 261
preparation of
active $\alpha\gamma\delta$ complex, **99**, 263
active $\gamma\delta$ complex, **99**, 265
Phosphorylation
auto-, *see* Autophosphorylation
cAMP-dependent, substrate-directed regulation, **99**, 212
nonmuscle myosin, and thick filament stabilization, **85**, 364
photo-, *see* Photophosphorylation
proteins
in amphibian oocytes, analytical microinjection techniques, **99**, 219

hormone-stimulated, measurement in intact cells, **99**, 20
rhodopsin, assays
in vitro, **81**, 489
in vivo, **81**, 493
Phosphotransferases
stereochemistry, analytical techniques, **87**, 197, 213
with chiral [^{16}O, ^{17}O, ^{18}O]phosphoric monoesters as probes, **87**, 279
Phosphotyrosine
physical and chemical properties, **99**, 390
in proteins
base hydrolysis and amino acid analysis, **99**, 402
detection, **99**, 391
quantitation, **99**, 401
separation, **99**, 397
Photobehavior
Halobacterium halobium, analytical techniques, **88**, 420
Photocalorimetry, *see* Calorimetry
Photocycle
bacteriorhodopsin
conformational changes, low-temperature uv spectroscopy, **88**, 265
kinetic optical absorption spectroscopy, **88**, 133
low-temperature spectrophotometry, **88**, 163
molecular aspects, comparison with rhodopsin, **88**, 470
in native hexagonal and orthorhombic purple membranes, cooperativity kinetics, **88**, 193
protonation state, spectroscopic determination, **88**, 254
in rotating cell, kinetic resonance Raman spectroscopy, **88**, 643
in thin layers, kinetic and spectroscopic analysis, **88**, 180
time-resolved protein fluorescence measurements, **88**, 306, 624
time-resolved resonance Raman spectroscopy, **88**, 617
viscosity effect, **88**, 297
rhodopsin, molecular aspects, comparison with bacteriorhodopsin, **88**, 470
Photon correlation spectroscopy
contractile proteins, preparation methods, **85**, 562
Photophosphorylation
in *Halobacterium halobium*, **88**, 349
ATP assay, **88**, 354
reconstitution, **88**, 355
Photoreceptor membranes
analysis with lectins, **81**, 65
functional and structural analysis with phospholipases, **81**, 320
infrared absorption spectroscopy, **88**, 696
protein synthesis, immunochemical analysis, **96**, 609
renewal, analytical techniques, **81**, 763
ultrastructural analysis with freeze-substitution and freeze-etching methods, **88**, 118
Photoreceptors
insect, noninvasive optical probing techniques, **81**, 752
invertebrate
lipid composition determination, **81**, 315
morphology, **81**, 17
vertebrate
electrical signaling, **81**, 403
morphology, **81**, 3
sensitivity, relation to bleaching, **81**, 423
Photosystem II
assembly and function, role of thylakoid polypeptide phosphorylation and turnover, **97**, 554
o-Phthalaldehyde
in amino acid analysis, **91**, 46
fluorescent methods for protein quantitation, **91**, 110
in imino acid analysis, **91**, 17
Physarum polycephalum
myosin aggregate formation and assay, **85**, 38
purification of
ornithine decarboxylase, **94**, 141
ornithine decarboxylase-modifying protein, **94**, 145
polyamine-dependent protein kinase, **94**, 149; **99**, 369

Physical mapping
maize chloroplast genes, **97**, 546
plasmids in halobacteria, **88**, 376
Picornaviruses
attachment assays, **96**, 445, 448
eclipse and uncoating assay, **96**, 450
penetration assay, **96**, 450
Picosecond spectroscopy
rhodopsin bleaching intermediate kinetics, **81**, 368
visual pigments, **81**, 354
Pigment P_{588}, *see* Halorhodopsin
Pimozide
radioimmunoassay, **84**, 542
Pinane thromboxane A_2
biological properties, **86**, 403
synthesis, **86**, 401
Pinocytosis, *see* Endocytosis, pinocytic
Placenta
human, purification of
arylsulfatase A and B, **86**, 18
coated vesicles, **98**, 321
glucose-6-phosphate isomerase, **89**, 552
15-hydroxyprostaglandin dehydrogenase (NAD^+), **86**, 126
15-ketoprostaglandin Δ^{13}-reductase, **86**, 164
ribonuclease inhibitor, **96**, 98
Plants, *see also specific plants*
genetic engineering with *Agrobacterium* Ti plasmids as vectors, **101**, 527
vacuole preparation
from lysed protoplasts, **96**, 581
by tissue mechanical shearing, **96**, 583
Plasma
platelet-rich and -poor, preparation, **86**, 642
purification of
amine oxidase, **94**, 315
arachidonic acid and metabolites with octadecylsilyl silica, **86**, 467
quantitative determination of
5-hydroxyeicosatetraenoic acid by gas chromatography–mass spectrometry, **86**, 607
imino acids with *o*-phthalaldehyde, **91**, 17
D-lactate, **89**, 35
prostaglandin E_2 metabolite by gas chromatography–mass spectrometry, **86**, 571
Plasma membranes
component internalization during adsorptive endocytosis, **98**, 203
differentiation, **98**, 477
eukaryotic, protein biogenesis and assembly, **96**, 257
flow during pinocytosis, antibody to 5′-nucleotidase as marker, **98**, 395
halobacterial, protein isolation by HPLC, **88**, 202
internalized
analytical techniques, **98**, 553
iodinated polypeptides, analytical techniques, **98**, 552
iodination, **98**, 548
localization, **98**, 550
lactoperoxidase uptake and enzymatic determination, **98**, 546
labeling with galactosyltransferase, **98**, 416
release with β-galactosidase, **98**, 419
pancreatic islet
biochemical and morphological characterization, **98**, 187
Ca^{2+},Mg^{2+}-ATPase assay, **98**, 192
parotid gland
associated enzyme activities, **98**, 85
electron microscopy, **98**, 78
purification, **98**, 76
yield and organelle contaminant analysis, **98**, 80
picornavirus attachment and penetration, assays, **96**, 445, 450
protein recycling, **98**, 379
salt gland
adaptation, **96**, 628
deadaptation, **96**, 655
enzyme assays, **96**, 638
preparation, **96**, 635
synaptic
cholesterol content, modification with lipid transfer protein, **98**, 599
preparation from rat forebrain, **98**, 604

transport pathways to
Golgi complex, **98**, 3
lysosomes, **98**, 4
Plasmids
Agrobacterium Ti, as vectors for plant genetic engineering, **101**, 527
chloramphenicol amplification, **100**, 340
cloning into, integrated and simplified approach, **101**, 78
DNA
characterization by agarose gel electrophoresis, **100**, 249
isolation by rapid alkaline extraction method, **100**, 243
Escherichia coli, small-scale preparation, **96**, 503
freezing, **100**, 340
in halobacteria
molecular weights, **88**, 376
purification, **88**, 374
restriction mapping, **88**, 375
linear and gapped, in yeast transformation, genetic applications, **101**, 228
pKC30
in amplification of DNA repair genes, **101**, 138
and derivatives, for controlled expression of genes, **101**, 123
screening at high colony density, **100**, 333
with *trp* promoters of *Escherichia coli* and *Serratia marcescens*, in expressing cloned genes, **101**, 155
yeast
high copy, construction with 2-μm circle sequences, **101**, 307
linear, construction, **101**, 245
with negative selection, construction, **101**, 278
promoter sequence-containing
design for analysis of cloned gene expression, **101**, 181
in regulation of gene expression, **101**, 192
2-μm vectors, cloning of STE genes, **101**, 325
Platelet factor 4
antiserum preparation, **84**, 87
isolation, **84**, 85
radioimmunoassay, **84**, 87
radioiodination, **84**, 86
Platelets
bovine
myosin aggregate formation and assay, **85**, 40
purification of
arachidonic acid 12-lipoxygenase, **86**, 51
thromboxane A synthase, **86**, 108
human
aggregation, **86**, 646
as assay for
diglyceride lipase, **86**, 653
phospholipase A_2 and C, **86**, 653
prostaglandins D_2, E_1, and I_2, **86**, 652
prostaglandin endoperoxides, **86**, 651
prostaglandin endoperoxide synthase, **86**, 653
thromboxane A_2, **86**, 651
thromboxane synthase, **86**, 653
drug effects, **86**, 654
pattern analysis, **86**, 648
quantification, **86**, 650
–collagen adhesion, assays, **82**, 509
myosin aggregate formation and assay, **85**, 40
myosin rod aggregate formation and assay, **85**, 50
^{31}P NMR profiles, **85**, 674
purification of
diacylglycerol lipase, **86**, 15
myosin, **85**, 355, 364
phospholipase C, **86**, 8
receptors for prostaglandin D_2 and I_2, binding assays, **86**, 179
resuspending buffers, choice, **86**, 645
wash techniques, **86**, 643
porcine, myosin purification, **85**, 364
Polarography
differential pulse, in homogeneous immunoassay, **92**, 433
Poliovirus
vaccine preparation with glutaraldehyde, **93**, 42

Pollen
timothy grass, vaccine preparation with glutaraldehyde, **93**, 52
Polyacrylamide gels
with active esters
applications, **83**, 310
preparation, **83**, 306
contamination in eluted protein amino acid composition determination, **91**, 36
with copolymerized ω-acrylamidoalkyl glycosides, preparation, **83**, 294
destaining, **96**, 232
DNA in, silver staining, **96**, 235
isotope detection by fluorography, **96**, 215
peptides in
detection after formaldehyde fixation, **91**, 247
protease mapping, **100**, 424
spectrin intermediate-sized, elution, **96**, 309
proteins in
Amido Black staining, **91**, 237
autoradiography, **91**, 213; **96**, 190
Coomassie Blue staining, **91**, 211, 238; **96**, 189
detection after formaldehyde fixation, **91**, 247
detection systems, comparison, **91**, 243
electrophoretic transfer to paper, **96**, 205
elution, **91**, 222
in microgram quantities for amino acid sequencing, **91**, 227
Fast Green staining, **91**, 239
fixation and long-term storage, **91**, 244
fluorography, **96**, 189
immunodetection, **96**, 191, 212
lectin-binding assay, **96**, 190
peptide mapping, **96**, 191; **100**, 424
silver staining, **91**, 211; **96**, 189, 230
effects on
autoradiography, **96**, 234
fluorography, **96**, 234
sensitivity, **96**, 233
intensification, **96**, 232
small DNA fragments from, transfer to diazo or nitrocellulose paper and hybridization, **100**, 309
with thioglycoside, preparation, **83**, 299
Polyamine oxidase
oat seedling
assay, **94**, 312
properties, **94**, 313
purification, **94**, 312
rat liver
assay, **94**, 306
properties, **94**, 309
purification, **94**, 307
Polyamines
acetylated derivatives, synthesis, **94**, 420
aliphatic, naturally occurring, quantitative determination by automated liquid chromatography, **94**, 29
biosynthesis
in *Escherichia coli*
mutant mass screening, **94**, 83
related enzymes, gene cloning, **94**, 117
irreversible inhibitors, synthesis, **94**, 199
in mitogen-activated lymphocytes, physiological effects of inhibitors, **94**, 373
related adenosyl-sulfur compounds, HPLC, **94**, 57
in *Saccharomyces cerevisiae*, mutant mass screening, **94**, 104
–bovine plasma amine oxidase system, structure–activity relationships, **94**, 416
and conjugates, gas chromatography–mass spectrometry, **94**, 48
dependent protein kinase in *Physarum polycephalum*
assay, **94**, 148
purification, **94**, 149
and derivatives
staining with fluorescamine and ninhydrin, **94**, 7
thin-layer chromatography, **94**, 3
thin-layer electrophoresis, **94**, 5
2,3-dihydroxybenzoyl derivatives
assays, **94**, 439, 441

properties, **94**, 439, 440
purification, **94**, 438, 440
double-isotope derivative assay, **94**, 36
applications, **94**, 41
function in mammary gland tissue, analytical techniques, **94**, 389
inhibitory effects on aminopropyltransferases, **94**, 295
liquid chromatography with prechromatographic derivatization, **94**, 10
metabolism in
mammary gland tissue, analytical techniques, **94**, 389
Neurospora crassa, analysis with arginaseless cells, **94**, 112
and monoacetyl derivatives, ion-pair chromatographic separation, **94**, 25
phosphorylated cellulose column chromatography, **94**, 42
Thermus thermophilus
analytical techniques, **94**, 401
chemical synthesis, **94**, 405
^{1}H-NMR, ^{13}C-NMR, and mass spectrum, **94**, 406
isolation and identification, **94**, 403
Poly-L-arginine
effects on cyclic nucleotide-dependent protein kinases, **99**, 206
Poly-L-lysine
in tumor cell immobilization to plastic for analysis of metabolic responses to cytotoxic T-cell attack, **93**, 260
Polymerization
actin
assays, **85**, 185
mechanism, **85**, 182
Polymers
capsular, containing polysialic acid, synthesis in *Escherichia coli* 07-K1, **83**, 540
Polypeptide hormones
radiolabeled tracers
molecular sieving by HPLC, **92**, 301
preparation by polyacrylamide gel electrophoresis, **92**, 309
purification by HPLC, **92**, 294
Polypeptides
amino acid analysis in picomole range by precolumn derivatization and HPLC, **91**, 41
bacteriorhodopsin
helical segments, identification, **88**, 81
modified positions, determination by HPLC, **88**, 89
cotranslational proteolytic processing, analytical techniques, **97**, 77
iodinated, in internalized plasma membranes, analytical techniques, **98**, 552
lens fiber membrane
cell-free synthesis, **98**, 510
immunoprecipitation, **98**, 516
nascent secretory
chain separation, posttranslational analysis, **96**, 101
cotranslational membrane insertion
analysis with proteases, **96**, 136
overview, **96**, 122
posttranslational membrane insertion
analysis with proteases, **96**, 136
overview, **96**, 129
pancreatic, stable radioiodinated, preparation by polyacrylamide gel electrophoresis, **92**, 332
sequencing with
dimethylaminoazobenzene isothiocyanate, **91**, 455
dipeptidyl peptidases, **91**, 511
thylakoid, phosphorylation and turnover in photosystem II assembly and function, **97**, 554
yeast mitochondrial
in vitro import assay, **97**, 329
precursor processing, **97**, 311
Polyribonucleotide synthase (ATP)
bacteriophage T4
assay, **100**, 40
DNA joining reactions, **100**, 44
purification, **100**, 41
RNA joining reactions, **100**, 52
Polysaccharides
ester substituents
identification, **83**, 251
isolation, **83**, 250

localization, **83**, 255
quantitative analysis, **83**, 249
ether substituents
identification, **83**, 258
isolation, **83**, 257
localization, **83**, 259
Rhizobium meliloti strain 1021, glycosyl sequence analysis, **83**, 35, 44
Rhizobium trifolii NA30, glycosyl sequence analysis, **83**, 36, 41
structural analysis by X-ray fiber diffraction, **83**, 87
xyloglucan nonasaccharide, glycosyl sequence analysis, **83**, 37
Polysialic acid
-containing capsular polymers, synthesis in *Escherichia coli* 07-K1, **83**, 540
Polysomes
free
isolation from
Escherichia coli, **97**, 71
various bacteria, **97**, 63
protein synthesis *in vitro*, **97**, 66, 74
translational products
identification, **97**, 67
preferential synthesis, **97**, 68
proteolytic processing, **97**, 76
membrane-bound
isolation from
Escherichia coli, **97**, 71
various bacteria, **97**, 63
preparation for protein synthesis *in vitro*, **96**, 3
properties, **96**, 12
protein synthesis *in vitro*, **97**, 66, 74
translational products
identification, **97**, 67
preferential synthesis, **97**, 68
proteolytic processing, **97**, 76
Porin
outer membrane
bacterial
channel properties, **97**, 99
liposome reconstitution for evaluation of
exclusion limits, **97**, 91
penetration rates, **97**, 94
properties, **97**, 86
purification, **97**, 87, 90
reconstitution into planar films, **97**, 98
Neurospora crassa mitochondria
properties, **97**, 290
purification, **97**, 287
Phaseolus aureus mitochondria
channel properties, **97**, 99
liposome reconstitution for evaluation of
exclusion limits, **97**, 91
penetration rates, **97**, 94
properties, **97**, 99
purification, **97**, 90
reconstitution into planar films, **97**, 98
Pregnancy
experimental interruption by ornithine decarboxylase inhibitors, **94**, 213
Preputial gland
rat, purification of
β-glucuronidase, **96**, 765; **98**, 302
mRNA, **96**, 770
Procollagen chains
multiple types
characterization, **82**, 116
purification, **82**, 111
noninterstitial, cell-free synthesis, **82**, 225
Procollagen C-proteinase
assays, **82**, 305
purification, **82**, 314
Procollagen-lysine,2-oxoglutarate 5-dioxygenase
assays, **82**, 248
catalytic properties, **82**, 280
molecular properties, **82**, 276
purification, **82**, 274
Procollagen N-proteinase
assays, **82**, 305
purification, **82**, 312
Procollagen-proline,2-oxoglutarate 3-dioxygenase
assays, **82**, 248
catalytic properties, **82**, 280
molecular properties, **82**, 279
purification, **82**, 278
Procollagen-proline,2-oxoglutarate 4-dioxygenase
assays, **82**, 248, 262
catalytic properties, **82**, 280

molecular properties, **82**, 267
purification, **82**, 266
β subunit-related protein
molecular properties, **82**, 273
purification, **82**, 271
Procollagens
antibodies
applications, **82**, 487
assays, **82**, 479
preparation and characterization, **82**, 472
purification, **82**, 476
asparagine-linked glycosides
characterization, **82**, 352
preparation, **82**, 347
biosynthesis, **82**, 218
characterization, **82**, 22, 116
interstitial, mRNA
preparation, **82**, 218
translation, **82**, 222
isolation from tissues, **82**, 9, 106
processing to collagen, **82**, 27
purification, **82**, 111
quantitative determination, **82**, 497
sugar-labeled, preparation, **82**, 347
synthesized *in vitro*, isolation, **82**, 99
Proinsulin
radioiodinated, preparation by polyacrylamide gel electrophoresis, **92**, 330
Prolactin
antibody production by lymph node inoculation, **93**, 7
methionine sulfoxide reduction, **91**, 557
Proline
assay with *o*-phthalaldehyde, **91**, 17
Prolyl 3-hydroxylase, *see* Procollagen-proline,2-oxoglutarate 3-dioxygenase
Prolyl 4-hydroxylase, *see* Procollagen-proline,2-oxoglutarate 4-dioxygenase
Pronuclei
gene transfer by microinjection for production of transgenic mice, **101**, 411
Properdin
human
assays, **93**, 386
precursor purification, **93**, 385
purification, **93**, 383
Prostacyclin
dinor urinary metabolites, quantitation by gas chromatography–mass spectrometry, **86**, 585
effect on platelet aggregation as bioassay, **86**, 652
handling, **86**, 458
identification criteria, **86**, 99
physical chemistry, **86**, 453
sodium salts, synthesis from prostaglandin $F_{2\alpha}$, **86**, 459
stability
solid-state, **86**, 456
in solution, **86**, 453
Prostacyclin receptors
human platelet, binding assays, **86**, 179
Prostacyclin synthase
bovine aortic
immunoradiometric assay, **86**, 243
monoclonal antibodies
assay, **86**, 242
preparation, **86**, 241
purification, **86**, 240
porcine aortic
assays, **86**, 94
with [5,6-^{3}H]prostaglandin H_2, **86**, 99
purification, **86**, 93
Prostaglandin D_2
effect on platelet aggregation as bioassay, **86**, 652
handling, **86**, 458
physical chemistry, **86**, 452
stability, **86**, 452
urinary metabolite quantification by stable isotope dilution mass spectrometry, **86**, 559
Prostaglandin-D_2 15-dehydrogenase ($NADP^+$)
porcine brain
assay, **86**, 148
properties, **86**, 150
purification, **86**, 149
reaction sequence, **86**, 151
Prostaglandin-D_2 11-ketoreductase
rabbit liver
properties, **86**, 121
purification, **86**, 119
Prostaglandin D_2 receptors
human platelet, binding assays, **86**, 179

Prostaglandin E_1
antisera
binding properties, **86**, 260
specificity problems, **86**, 258
arylsulfatase effects, **86**, 20
effect on platelet aggregation as bioassay, **86**, 652
Prostaglandin E_1 receptors
bovine corpus luteum, distribution in intracellular organelles, **86**, 192
Prostaglandin E_2
antisera
binding properties, **86**, 263
specificity problems, **86**, 258
in biological samples, 15-keto-13,14-dihydro-prostaglandin E_2 as parameter, **86**, 319
handling, **86**, 458
physical chemistry, **86**, 436
plasma metabolite quantification by gas chromatography–mass spectrometry, **86**, 571
stability
in aqueous systems, **86**, 438
in nonaqueous systems, **86**, 441
solid-state, **86**, 448
15(*S*)-Prostaglandin E_2
^3H-labeled
in assay for NAD^+-dependent 15-hydroxyprostaglandin dehydrogenase, **86**, 133
preparation, **86**, 132
Prostaglandin endoperoxide synthetase
human platelet, assay, **86**, 653
Prostaglandin $F_{2\alpha}$
alkaline phosphatase-linked immunoassay, **86**, 272
arylsulfatase effects, **86**, 20
β-galactosidase-linked immunoassay, **86**, 269
handling, **86**, 458
major metabolite
in blood, radioimmunoassay, **86**, 339
in plasma, radioimmunoassay, **86**, 320
in urine
quantification by gas chromatography–mass spectrometry, **86**, 579
radioimmunoassay, **86**, 339
physical chemistry, **86**, 450
in plasma, 15-keto-13,14-dihydro-prostaglandin $F_{2\alpha}$ as parameter, **86**, 337
in prostacyclin sodium salt synthesis, **86**, 459
stability, **86**, 452
urinary catabolites
deuterium-labeled
gas chromatography–mass spectrometry, **86**, 557
preparation, **86**, 556
preparation, **86**, 552
Prostaglandin $F_{2\alpha}$ receptors
bovine corpus luteum
determination, **86**, 202
distribution in intracellular organelles, **86**, 192, 208
properties, **86**, 204
ovine corpus luteum
determination, **86**, 202
properties, **86**, 204
Prostaglandin G_2
effect on platelet aggregation as bioassay, **86**, 651
generation and extraction, **86**, 378
purification, **86**, 380
structural identification, **86**, 382
TLC, **86**, 379
Prostaglandin H_2
effect on platelet aggregation as bioassay, **86**, 651
generation and extraction, **86**, 378
^3H-labeled, in prostacyclin synthase assay, **86**, 99
purification, **86**, 380
structural identification, **86**, 382
TLC, **86**, 379
Prostaglandin-H_2 D-isomerase
rat brain
assay, **86**, 73
properties, **86**, 76
purification, **86**, 74
rat spleen
assay, **86**, 78
properties, **86**, 83
purification, **86**, 80
Prostaglandin-H_2 E-isomerase
ovine seminal vesicle
assay, **86**, 85

properties, **86**, 89
purification, **86**, 87
Prostaglandin H synthase
bovine seminal vesicle
assay, **86**, 55
properties, **86**, 59
purification, **86**, 58
human platelet, quantification with [acetyl-^{3}H]aspirin, **86**, 397
ovine seminal vesicle
apoenzyme
assay, **86**, 69
preparation from holoenzyme, **86**, 71
assay, **86**, 61
immunocytofluorescence, **86**, 218
immunoelectron microscopy, **86**, 220
immunoprecipitation, **86**, 216
immunoradiometric assay, **86**, 236
monoclonal antibodies
preparation, **86**, 230
purification, **86**, 235
properties, **86**, 65
purification, **86**, 63
rabbit antisera
applications, **86**, 216
IgG isolation, **86**, 215
preparation, **86**, 214
radioimmunoassay, **86**, 222
rat brain, inactivation by microwave irradiation, **86**, 635
reaction products, HPLC, **86**, 518
Prostaglandin ω-hydroxylase
rabbit liver and lung
assays, **86**, 168
product analysis by
gas chromatography, **86**, 174
gas chromatography–mass spectrometry with selected ion monitoring, **86**, 174
HPLC, **86**, 171
with radioactive substrates, **86**, 177
TLC, **86**, 171
with radioactive substrates, **86**, 177
Prostaglandin I_2, *see* Prostacyclin
Prostaglandin I_2 synthase, *see* Prostacyclin synthase
Prostaglandin 9-ketoreductase, *see* Carbonyl reductase (NADPH)
Prostaglandins
argentation-HPLC, **86**, 530
autoradiography, **86**, 498
extraction techniques, **86**, 477
with octadecylsilyl silica, **86**, 467
^{18}O derivatives, preparation for gas chromatography–mass spectrometry, **86**, 547
production in plasma and urine, 11-ketotetranor metabolites as parameters, **86**, 351
radioimmunoassay with iodinated derivatives as tracers, **86**, 300
radioiodinated derivatives
binding properties, **86**, 301
preparation, **86**, 298
related enzymes
assays, **86**, 3, 213
purification, **86**, 3
TLC, **86**, 484, 493
underivatized, HPLC, **86**, 518
Prostate
rat, spermidine synthase purification, **94**, 274
Proteases, *see also specific proteases*
in analysis of membrane insertion, **96**, 121
commercially available, properties, **96**, 137
in isolation of paramyxovirus mutants with defects in glycoprotein processing, **96**, 442
in peptide mapping of
membrane proteins in gels, **96**, 191
polypeptides and proteins in gels, **100**, 424
SDS-solubilized proteins, **96**, 222
as probes of rhodopsin structure, **81**, 236
yeast mitochondrial, matrix-localized
assays, **97**, 315, 339
properties, **97**, 315
purification, **97**, 313, 338
Protein A
Staphylococcus aureus
enzyme-labeled, in immunoassay for haptens and immunoglobulins, **92**, 411

radioiodinated
in immunoassay for haptens and immunoglobulins, **92**, 403
in indirect assay for monoclonal antibodies to cell-surface antigens, **92**, 160
preparation by polyacrylamide gel electrophoresis, **92**, 335
Protein activators
for ganglioside G_{M1} and G_{M2} hydrolysis
assay, **83**, 591
properties, **83**, 594
purification, **83**, 592
for β-glucoside hydrolysis
assay, **83**, 597
properties, **83**, 601
purification, **83**, 599
Proteinase inhibitors
peptidic, inhibitory effects on lysosomes, **96**, 756
Protein binding assays
ankyrin, **96**, 324
competitive, for methotrexate, **84**, 447
applications, **84**, 454
corrinoids, **84**, 397
clinical applications, **84**, 401
design, **84**, 379
spectrin, **96**, 324
Protein-glutamine γ-glutamyltransferase
–trimethylacetyl complex
bond nature and acyl group position, **87**, 41
formation and isolation, **87**, 37
Protein kinase inhibitors
heat-stable cAMP-dependent
effect on enzyme autophosphorylation, **99**, 179
rabbit skeletal muscle
assay, **99**, 77, 81
physiological role, **99**, 91
properties, **99**, 90
purification, **99**, 78, 84
in rabbit tissues, levels, **99**, 92
regulation, **99**, 92
Protein kinases
Ca^{2+}-activated phospholipid-dependent, rat brain
assay, **99**, 290
properties, **99**, 294
purification, **99**, 292
Ca^{2+}-dependent, purification and properties, **99**, 243
Ca^{2+}-independent, purification and properties, **99**, 308
cyclic AMP-dependent
bovine heart
activity assay, **99**, 52
catalytic subunit
bound substrates, NMR, **99**, 98
interaction with regulatory subunit, **99**, 115
magnetic resonance spectroscopy, **99**, 94
purification, **99**, 53
substrate exchange kinetics, **99**, 113
regulatory subunit
interaction with catalytic subunit, **99**, 115
NMR, **99**, 116
photoaffinity labeling, **99**, 154
bovine skeletal muscle
activation, kinetic assay, **99**, 163
cyclic nucleotide exchange, kinetic assay, **99**, 163
deactivation
equilibrium assay, **99**, 164
kinetic assay, **99**, 163
CHO cell, physiological role, genetic analysis, **99**, 197
porcine heart
ATP-binding site, affinity labeling, **99**, 141
cAMP-binding site, affinity labeling, **99**, 147
porcine kidney
assay, **90**, 201
catalytic subunit purification from
cytosol, **90**, 203
particulate fraction, **90**, 205
properties, **90**, 206
from rat tissues, activity ratio determination, **99**, 227
cyclic GMP-dependent
bovine lung
assay, **99**, 63
purification, **99**, 64

cyclic nucleotide-dependent
in amphibian oocytes, analytical microinjection techniques, **99**, 219
assays, **99**, 3
with synthetic peptide substrates, **99**, 134
binding cooperativity and selectivity, analysis with analogs, **99**, 168
bovine rod outer segment
assay, **81**, 497
properties, **81**, 501
purification, **81**, 499
grasshopper
assay, **99**, 72
properties, **99**, 76
purification, **99**, 73
stability, **99**, 75
in phosphate removal from proteins, **99**, 14
protein modulation, **99**, 206
cyclic nucleotide-independent, purification and properties, **99**, 308
microtubule protein-associated, assay, **85**, 431
polyamine-dependent, *Physarum polycephalum*
assay, **94**, 148; **99**, 368
properties, **99**, 372
purification, **94**, 149; **99**, 369
type I cyclic AMP-dependent
in intact cultured cells, regulatory subunit metabolism, analytical techniques, **99**, 233
rabbit skeletal muscle, regulatory subunit purification, **99**, 55
type II cyclic AMP-dependent
bovine brain
immunoprecipitation, **99**, 190, 191
purification, **99**, 189
radioimmunoassay, **99**, 191
regulatory subunit
antiserum preparation, **99**, 190
purification, **99**, 189
bovine heart
autophosphorylation, **99**, 177
application in analysis of enzyme structure and regulation, **99**, 185
intra- and intermolecular, distinction, **99**, 179
reversal, **99**, 183
dephosphorylation, **99**, 183
immunoprecipitation, **99**, 191
purification, **99**, 56, 176
regulatory subunit autophosphorylation, **99**, 178
subclass identification, **99**, 192
tyrosine-specific, *see* Protein-tyrosine kinase
Proteins
actin-binding
purification from macrophages, **85**, 483
sedimentation assay for crosslinking activity, **85**, 481
acyl CoA-binding, photoaffinity labeling, **91**, 639
adhesion, role in cell–collagen interactions, **82**, 503
amidination, **91**, 559
reversal, **91**, 566
amino acid analysis by single hydrolysis method, **91**, 26
amino-terminal acetylamino acids, microanalysis by HPLC, **91**, 84
amino-terminal analysis with dimethylaminoazobenzene isothiocyanate, **91**, 79
anion transport, from erythrocytes
antiserum production, **96**, 263
immunoprecipitation, **96**, 265
processing analysis *in vitro*, **96**, 257
protease fragments, preparation, **96**, 260
purification, **96**, 262
L-arabinose-binding, *Escherichia coli*
assay, **90**, 464
purification, **90**, 465
attachment to erythrocytes
facilitation by conjugate succinylation, **92**, 265
with methyl-*p*-hydroxybenzimidate, **92**, 269

with *N*-succinimidyl 3-(2-pyridyldithio)propionate, **92**, 259
with synthetic lipopolysaccharide, **92**, 274
with 1,3,5-trichlorotriazine, **92**, 271
band 3, from erythrocytes, purification of cytoplasmic domain, **96**, 321
blood coagulation, vitamin K-dependent
antibodies
interactions with protein antigens, **84**, 76
preparation, **84**, 64
radioimmunoassay, **84**, 75
bound phosphate
content determination, **99**, 7
removal by reverse reaction with protein kinases, **99**, 14
carbohydrate-binding
glycolipid ligand detection by protein binding to thin-layer chromatograms, **83**, 235
in isolation of specific sugar sequences, **83**, 241
in cell extracts
electrophoretic transfer from gels to paper, **96**, 205
immunologic detection, **96**, 212
cerebroside transfer, *see* Cerebroside transfer protein
chloroplast, cytoplasmically synthesized, assembly reconstitution *in vitro*, **97**, 507
cleavage at
aspartic acid residue with hydrochloric acid and formic acid, **91**, 324
tryptophan residue with
dimethyl sulfoxide–hydrochloric acid and cyanogen bromide, **91**, 318
o-iodosobenzoic acid, **91**, 311
coat, *see* Coat proteins
with collagen-like sequences, preparation, **82**, 319
collagenous, size estimation by
gel chromatography, **82**, 420
SDS-gel electrophoresis, **82**, 411
contractile
electron micrographs, optical diffraction and filtering, **85**, 511
electron paramagnetic resonance, **85**, 594
saturation-transfer methods, **85**, 599
spin probe–spin label methods, **85**, 621
falling-ball viscometry, **85**, 572
fluorescence spectroscopy
accessibility probes, **85**, 580
conformation probes, **85**, 575
distance probes, **85**, 590
molecular rotation detection, **85**, 586
orientation probes, **85**, 581
kinetics, analytical techniques and instrumentation, **85**, 698
non-Newtonian viscosity measurements, **85**, 572
preparation for photon correlation spectroscopic and light-scattering studies, **85**, 562
quasi-elastic light scattering, **85**, 573
sedimentation analysis, **85**, 570, 571
structure–function relationships, optical activity measurements, **85**, 677
in tissue culture cells, immunofluorescent localization, **85**, 514
turbidimetric ultracentrifugation, **85**, 572
cotranslational processing, analytical techniques, **97**, 77
with amino acid analogs, **96**, 777
crosslinked complexes
analytical techniques, **91**, 607
cleavage, **91**, 606
crosslinking reagents
amino-specific, **91**, 591
carboxyl, **91**, 601
guanidino-specific, **91**, 600
indole-specific, **91**, 601
nonspecific, **91**, 601
photoactivable, **91**, 602
selection, **91**, 580
sulfhydryl, **91**, 599
in cultured cells, image intensification techniques for detection, **98**, 266
DNA-binding, recovery in SDS gels, **91**, 273
–DNA interaction, immunoassay for specific binding sequences, **92**, 138

docking, from endoplasmic reticulum
activity assay, **96**, 693
purification, **96**, 696
–dye complexes in gels, analytical techniques, **91**, 245
electrophoretic transfer to diazobenzyloxymethyl paper, **100**, 326
elution from
polyacrylamide gels, **91**, 222
for amino acid composition determination, **91**, 36
in microgram quantities for amino acid sequencing, **91**, 227
SDS gels for amino acid sequencing, **91**, 450
fatty acid-acylated, in eukaryotic cells
detection, **96**, 797
distribution, **96**, 796
site and mechanism, **96**, 799
fucose-binding, rat liver
assay, **98**, 314
properties, **98**, 318
purification, **98**, 315
functionally related, immunological recognition of modifications, **98**, 159
functional, recovery in SDS gels, **91**, 263
D-galactose-binding, *Escherichia coli*
assay, **90**, 464
purification, **90**, 465
gene *A*, bacteriophage ϕX174
functional analysis with *in vitro* DNA replication systems, **100**, 219
properties, **100**, 217
purification, **100**, 219
glycosylation
inhibition of dolichol pathway, **83**, 432
role of intracellular membrane systems, **98**, 91
glyoxysomal, purification, **96**, 700
heme-binding, immunoassays, **84**, 148
hexokinase-binding, isolation from outer mitochondrial membrane, **97**, 469
high-mobility-group, preparative polyacrylamide gel electrophoresis, **91**, 214
HPLC, **91**, 137
imino acid analysis with *o*-phthalaldehyde, **91**, 17
insertion in
bacterial membranes, role of membrane potential, **97**, 146
subcellular membranes, analysis with proteases, **96**, 121
Int, bacteriophage λ
assay, **100**, 210
properties, **100**, 215
purification, **100**, 212
intracellular
integration into membranes, **96**, 672
signal sequences for translocation, **96**, 667
topogenic sequences, **96**, 664
ion-exchange chromatography, **91**, 170
labeling by reductive alkylation, **91**, 570
LamB, *see* Bacteriophages, lambda
–ligand or –protein binding stoichiometry, determination by continuous variation method, **87**, 509
lipid transfer, *see* Lipid transfer proteins
lysosomal membrane, iodination for composition and recycling analyses, **98**, 404
maltose-binding, *Escherichia coli*
assay, **90**, 459
properties, **90**, 462
purification, **90**, 460
M-band, structure–activity relationship, optical activity measurements, **85**, 695
membrane
immunochemical identification after gel electrophoresis, **96**, 192
insoluble
antibody preparation, **97**, 306
immunoprecipitation, **97**, 310
in polyacrylamide gels, analytical techniques, **96**, 189
recycling analysis with electron-opaque tracers, **98**, 203
structure, review, **97**, 571
two-dimensional polyacrylamide gel electrophoresis, **96**, 184
metal-binding, immunoassays, **84**, 121
metal ions, X-ray fluorescence spectrometry, **91**, 119

methionine sulfoxide reduction, **91**, 556
methylated
amino acid resolution, **91**, 8
properties, **91**, 575
microinjection into tissue culture cells, **101**, 482
microtubule-associated, *see* Microtubule-associated proteins
mitochondrial membrane, assembly in *Neurospora crassa*, analytical techniques, **97**, 275
M-line, preparation and properties, **85**, 139
myelin-specific, *see* Myelin-specific proteins
nascent secretory
biological activity, **96**, 110
gel electrophoretic analysis, **96**, 105
nonsecreted, intracellular localization in hepatocytes, **98**, 171
nuclear pore complex–lamina, purification, **96**, 597
nucleotide-binding
cellular localization, **91**, 629
photoaffinity labeling with 8-azidopurine analogs, **91**, 613
OmpA and OmpF, assembly in *Escherichia coli*, analytical techniques, **97**, 112
oncofetal, immunoassays, **84**, 3
pancreatic exocrine, radiolabeled
analysis in body fluids, **98**, 469
preparation, **98**, 468
pancreatic islet
assay, **98**, 187
biochemical characterization, **98**, 190
Ca^{2+}-stimulated phosphorylation, assay, **98**, 196
peroxisomal, analytical techniques, **96**, 727
phosphatidylcholine transfer, *see* Phosphatidylcholine transfer protein
phosphorylation
in amphibian oocytes, analytical microinjection techniques, **99**, 219
hormone-stimulated, measurement in intact cells, **99**, 20
phosphotyrosine detection and quantification, **99**, 387, 402
photoelectric current-generating, single turnover study, **88**, 35
plant, *in vitro* processing, **96**, 716
plasma membrane, recycling, **98**, 379
in polyacrylamide gels
detection methods, **91**, 211, 213, 237, 247
fixation and long-term storage, **91**, 244
posttranslational modifications, analysis with amino acid analogs, **96**, 777
purple membrane
conformational changes, monitoring during photocycle, **88**, 265
distance measurement by fluorescence energy transfer, **81**, 668
isolation by HPLC, **88**, 202
primary structure determination, **88**, 723
quantitative two-dimensional gel electrophoresis, **100**, 411
radioiodinated, preparation by polyacrylamide gel electrophoresis, **92**, 309
*rec*A, *Escherichia coli*
activity assay, **100**, 198
properties, **100**, 209
purification, **100**, 203
relatedness, assessment by amino acid composition, **91**, 60
retinal rod outer segment
antibody preparation, **81**, 240
purification by light-regulated binding to membranes, **81**, 556
retinoid-binding, bovine retina
assay, **81**, 820
properties, **81**, 823
purification, **81**, 820
retinol-binding, from serum
properties, **81**, 847
purification, **81**, 841, 845
reversed-phase chromatography, **91**, 180
preparative, **91**, 187
ribose-binding, *Escherichia coli*
assay, **90**, 467
properties, **90**, 472
purification, **90**, 469

RNA-binding, recovery in SDS gels, **91**, 273
SDS-solubilized
detergent removal by
ion-pair extraction, **91**, 254
ion-retardation chromatography, **91**, 259
elution from gels by
diffusion, **91**, 454
electrophoresis, **91**, 450
peptide mapping in one dimension, **96**, 222
Semliki Forest virus membrane, from cloned cDNA, expression analysis, **96**, 496
separation by
preparative polyacrylamide gel electrophoresis, **91**, 214
two-dimensional electrophoresis on giant gels, **91**, 190
sequence homologies
alignment scores, ALIGN program, **91**, 537
scoring matrices, **91**, 544
segment comparison scores, RELATE program, **91**, 540
selection using
computer database, **91**, 528
printed tabulations, **91**, 527
sequencing with
commercial spinning-cup sequenator, efficiency improvements, **91**, 434, 443
gas-phase sequenator, **91**, 399
silver-stained in gels, quantitative analysis, **96**, 234
size-exclusion chromatography, **91**, 147
non-ideal, **91**, 167
preparative, **91**, 161
thick filament C, H, and X
function, **85**, 138
purification, **85**, 131
thin-layer peptide mapping with sequencing at nanomole level, **91**, 466
thiol content, determination with 5,5′-dithiobis(2-nitrobenzoic acid), **91**, 58
total, quantitation methods
comparison, **91**, 116
Coomassie blue dye-binding, **91**, 108
Folin phenol, **91**, 96
o-phthalaldehyde fluorescent, **91**, 110
uv absorption, **91**, 105
transfer of *N*-asparagine-linked oligosaccharides *in vitro*, **83**, 408
translated in cell-free system
amino-terminal acetylation, prevention, **96**, 156
analytical techniques, **82**, 234; **96**, 47, 63
identification with *N*-formylmethionyl-tRNA, **96**, 150
immunoprecipitation, **96**, 111
viral, in infected MDCK cells
immunoelectron microscopy, **98**, 497
immunofluorescence, **98**, 495
D-xylose-binding, *Escherichia coli* periplasmic
assay, **90**, 473
properties, **90**, 475
purification, **90**, 474
yeast mitochondrial
pet gene-encoded, identification and characterization, **97**, 352
pulse-labeling, **97**, 326
determination and kinetics, **97**, 327
Protein solutions
desalting in centrifuge column, **91**, 278
Protein synthesis
acetylcholine receptor subunits in cell-free system, **96**, 369
by bacterial membrane-bound and free polysomes, **97**, 62, 74
bacterial membrane proteins in
semi-*in vitro* systems, **96**, 79
in vitro systems, **96**, 81
in vivo systems, **96**, 75
chloroplast proteins of cytoplasmic origin, *in vitro* reconstitution, **97**, 502
erythrocyte anion transport protein *in vitro*, **96**, 257
glycophorin A in cell-free system, **96**, 281
glyoxysomal
kinetic analysis *in vivo*, **96**, 708

in vitro methods, **96**, 710
hepatocyte endoplasmic reticulum proteins, **96**, 530
inhibitors, **98**, 153
lysosomal, in cultured cells, **96**, 729
maize endosperm membrane proteins *in vitro*, **96**, 716
membrane
characteristics, **96**, xxix
consequences of monopolies, **96**, xxxvi
sites, **96**, xxxiii
myelin-specific proteins, **96**, 378
nuclear-coded mitochondrial membrane proteins in *Neurospora crassa*, **97**, 275
OmpA and OmpF in *Escherichia coli*, **97**, 112
pancreatic zymogen granule membrane proteins, **98**, 74
peripheral membrane proteins, **96**, xlix
peroxisomal
in vitro methods, **96**, 726
in vivo methods, **96**, 721
photoreceptor membrane proteins, immunochemical analysis, **96**, 609
plant mitochondrial proteins *in vitro*, **97**, 476
in polarized epithelia, immunochemical analysis, **98**, 379
in rabbit reticulocyte lysates, **96**, 59
retinal membrane proteins, immunocytochemistry at electron microscopic level, **96**, 485
sarcoplasmic reticulum proteins in
myoblast cultures, **96**, 570
in vitro system, **96**, 578
secretory proteins in
cell-free system, **98**, 460, 465
pancreatic lobules, **98**, 20
serum protein in ascitic fluid of mice and guinea pigs, **93**, 12
Sindbis virus membrane glycoproteins *in vitro*, **96**, 513
thylakoid membrane proteins in isolated chloroplasts, **97**, 487
in wheat germ system, **96**, 38
yeast *pet* gene products *in vitro*, **97**, 352

Protein-synthesizing systems
cell-free
coupled DNA-directed, in analysis of gene expression based on di- and tripeptide formation, **101**, 690
preparation from
Bacillus subtilis, **101**, 598
Chinese hamster ovary cells, **101**, 629
Ehrlich ascites tumor cells, **101**, 616
Escherichia coli, **101**, 598
Saccharomyces cerevisiae, **101**, 644
wheat germ, **96**, 50; **101**, 635
prokaryotic, with coupled transcription–translation
applications, **101**, 678
preparation, **101**, 675
in recombinant DNA identification, **101**, 650
Protein transport
into chloroplasts, *in vitro* reconstitution, **97**, 503
cotranslational
analysis with proteases, **96**, 136
microsomal membranes for, preparation, **96**, 84
overview, **96**, 122
into glyoxysomes, *in vitro* analysis, **96**, 710
intracellular
across membrane, **96**, 665
control, **96**, 663
mechanisms, **96**, xxxvii
coupled oligosaccharide processing *in vitro*, assays, **98**, 42, 45
effects of monovalent ionophores, **98**, 51
endocytic pathway, *see* Endocytosis
from endoplasmic reticulum to Golgi complex, **98**, 8
exocytic pathway, *see* Exocytosis
from Golgi complex to lysosomes, **98**, 8
mechanisms, **96**, 669
monensin-induced arrest site, **98**, 52
in pancreatic lobules, pathway analysis, **98**, 21

phylogeny, **96**, 676
from plasmalemma to
Golgi complex, **98**, 3
lysosomes, **98**, 4
posttranslocational pathways, **96**, 678
into mitochondria
kinetic analysis in *Neurospora crassa* cells, **97**, 276
overview, **97**, 245
into mitochondrial matrix *in vivo*, **97**, 402
posttranslational
analysis with proteases, **96**, 136
overview, **96**, 129
secretory
in *Escherichia coli*
genetic analysis, **97**, 3, 11
role of membrane potential, **97**, 146
in lacrimal and salivary glands, cytoskeleton role, **98**, 175
in liver, colchicine effects *in vivo*, **98**, 169
transcellular, **98**, 7
high-voltage analytical techniques, **98**, 28
in inside-out thyroid follicles, **98**, 447
pancreatic exocrine proteins *in vivo*, **98**, 466
vesicular, **96**, xlv
control, **96**, xlvi
Protein-tyrosine kinase
Abelson murine leukemia virus-encoded
assay, **99**, 374
characterization, **99**, 373
purification, **99**, 376
epidermal growth factor-related, purification from A-431 cells, **99**, 379
Proteoglycans
analytical techniques, **82**, 794
cartilage
antiserum preparation, **83**, 221
immunoassays, **83**, 223
applications, **83**, 232
purification, **83**, 217
definition, **82**, 769
extraction, **82**, 779
properties, **82**, 779
purification and separation, **82**, 784
structures and functions, **82**, 772
Proteus vulgaris
aldehyde dehydrogenase purification, **89**, 480
Prothrombin
antibodies
interactions with protein antigens, **84**, 76
preparation, **84**, 64
radioimmunoassay, **84**, 75
Protoplast fusion
in high-efficiency transfer of bacterial DNA into eukaryotic cells, **101**, 402
Protoplasts
plant, vacuole preparation, **96**, 581
Pseudomonas
MSU-1, purification of
D-*xylo*-aldonate dehydratase, **90**, 303
galactonate dehydratase, **90**, 295
2-keto-3-deoxy-D-xylonate aldolase, **90**, 270
D-xylose dehydrogenase, **89**, 226
P. acidovorans, hydroxypyruvate reductase purification, **89**, 342
P. aeruginosa, purification of
aldehyde dehydrogenase, **89**, 485, 489
lectins PA-I and PA-II, **83**, 381
P. fluorescens
cultivation, **89**, 177, 263
purification of
D-galactose dehydrogenase, **89**, 178
gluconate 2-dehydrogenase, **89**, 189
glucose-6-phosphate dehydrogenase, **89**, 264, 267
membrane-bound D-glucose dehydrogenase, **89**, 150
6-phospho-D-gluconate dehydrogenase, **89**, 279
P. oxalaticus
cultivation, **89**, 532
formate dehydrogenase purification, **89**, 532

P. putida, purification of
4-hydroxy-4-methyl-2-ketoglutarate aldolase, **90**, 274
6-phosphogluconate dehydratase, **89**, 99
P. saccharophila, 2-keto-3-deoxy-galactonate-6-phosphate aldolase purification, **90**, 265
Puromycin
inhibitory effects on lysosomes, **96**, 763
Purple membranes
from halophilic bacteria
biogenesis, analytical techniques, **97**, 218
crystal forms
native hexagonal and orthorhombic, cooperativity of photocycle, **88**, 193
two- and three-dimensional, preparation, **88**, 112
formation reconstitution in nicotine cells, **88**, 394
kinetic resonance Raman spectroscopy with rotating sample, **88**, 643
lipid composition, analytical techniques, **88**, 98
reconstituted from brown membrane, X-ray diffraction analysis, **88**, 268
infrared absorption spectroscopy, **88**, 706
light-induced volume changes, **88**, 272
metal decoration, **88**, 235
from methanogenic bacteria, lipid composition, analytical techniques, **88**, 105
orientation by electric field, **88**, 287
phospholipid substitution methods, **88**, 26
planar monolayers, preparation for spectroscopy and electron microscopy, **88**, 232
protein primary structure, analytical techniques, **88**, 723
resonance Raman spectroscopy, **88**, 625
retinal isolation and purification for mass spectral analysis, **88**, 542
surface potential changes, analysis with spin-label probes, **88**, 690
thin layers, bacteriorhodopsin photocycles and orientation, **88**, 180
tritium–hydrogen exchange kinetics, **88**, 673
Putreanine
synthesis, **94**, 418
Putrescine aminopropyltransferase, *see* Spermidine synthase
Putrescine carbamoyltransferase
Streptococcus faecalis
assay, **94**, 340
properties, **94**, 342
purification, **94**, 340
Putrescine oxidase
Micrococcus rubens
assay, **94**, 301
purification, **94**, 302
Putrescine synthase
Lathyrus sativus seedling
assay, **94**, 336
properties, **94**, 338
purification, **94**, 337
α-Putrescinyldeoxythymidine
isolation from φW-14 DNA, **94**, 426
α-Putrescinylthymine
chemical synthesis, **94**, 425
isolation from φW-14 DNA, **94**, 423
Pyrazolone derivatives
antibody production, **84**, 597
radioimmunoassay, **84**, 599
Pyridine nucleotide-linked enzymes
stereospecificities, **87**, 101
Pyridoxal 5′-phosphate
inhibitory effects on casein kinases I and II, **99**, 314, 328
γ-(3-Pyridyl)-γ-oxo-*N*-methylbutyramide
radioimmunoassay, **84**, 636
applications, **84**, 639
Pyrophosphate–fructose-6-phosphate 1-phosphotransferase
Entamoeba histolytica
assay, **90**, 98
properties, **90**, 101
purification, **90**, 99
mung bean
assay, **90**, 91
properties, **90**, 96
purification, **90**, 94
Pyruvate decarboxylase
sweet potato root
assay, **90**, 528

properties, **90**, 531
purification, **90**, 529
Pyruvate dehydrogenase kinase
bovine kidney
assay, **90**, 196; **99**, 332
properties, **90**, 199; **99**, 335
purification, **90**, 197; **99**, 333
Pyruvate dehydrogenase (lipoamide)
Bacillus
assay, **89**, 401
properties, **89**, 405
purification, **89**, 402
bovine kidney and heart
assay, **89**, 376
properties, **89**, 382, 386
purification, **89**, 377, 383
broccoli mitochondrial
assay, **89**, 408
properties, **89**, 412
purification, **89**, 411
cauliflower mitochondrial
assay, **89**, 408
properties, **89**, 410
purification, **89**, 409
Escherichia coli
assay, **89**, 392
behavior on ethanol–Sepharose 2B, **89**, 398
purification, **89**, 393
Hansenula miso
assay, **89**, 420
properties, **89**, 423
purification, **89**, 421
Neurospora crassa
assay, **89**, 387
properties, **89**, 390
purification, **89**, 389
pigeon breast muscle
assay, **89**, 414
properties, **89**, 418
purification, **89**, 416
Pyruvate dehydrogenase phosphatase
bovine heart
assay, **90**, 402
properties, **90**, 406
purification, **90**, 404
Pyruvate kinase
bovine liver and muscle
assay, **90**, 141
hybridization, **90**, 147
properties, **90**, 149
purification, **90**, 143, 146
Escherichia coli types I and II
assay, **90**, 171
properties, **90**, 177
purification, **90**, 173
human erythrocyte and liver
assay, **90**, 131
properties, **90**, 135
purification, **90**, 132
isozymes, rat
assay, **90**, 152
properties, **90**, 162
purification, **90**, 156
separation, **90**, 155
tissue distributions, **90**, 153
rat liver, substrate-directed regulation of cAMP-dependent phosphorylation, **99**, 213
Streptococcus lactis
assay, **90**, 165
properties, **90**, 168
purification, **90**, 166
Pyruvate,orthophosphate dikinase
assay, **87**, 56
ATP/AMP site, covalent labeling, **87**, 61
enzyme-phosphoryl intermediate
isolation, **87**, 61
properties and functions, **87**, 61
enzyme-pyrophosphoryl intermediate
formation, **87**, 57
isolation, **87**, 57
properties and functions, **87**, 59
phosphoenolpyruvate/pyruvate site, covalent labeling, **87**, 65
phosphorylhistidine-containing peptide, sequence, **87**, 63
preparation from *Bacillus symbiosus*, **87**, 55
properties, **87**, 56
pyruvate enolization, monovalent anion requirement, **87**, 64

R

Radioimmunoassays
2-acetylaminofluorene–DNA adducts, **84**, 607
acetylcholine, **84**, 555
antibodies to
cholera toxin, **84**, 251

DNA–RNA hybrids of naturally occurring sequence, **84**, 210
Escherichia coli enterotoxin, **84**, 251
1-β-D-arabinofuranosylcytosine, **84**, 470
affecting factors, **84**, 479
applications, **84**, 477
1-β-D-arabinofuranosyluracil, **84**, 482
bacterial lipopolysaccharide, **84**, 270
benzodiazepines, **84**, 490
bleomycin, **84**, 463
clinical applications, **84**, 469
n-butylbiguanide, **84**, 577
calmodulin, **84**, 138
carcinoembryonic antigen, **84**, 32
in C3b and Fc receptor detection on human lymphoblastoid cell surfaces, **93**, 155
chlordiazepoxide, **84**, 497
chlorinated dibenzo-*p*-dioxins, **84**, 619
cholera toxin, **84**, 238
clonazepam, **84**, 499
competitive, in determination of antihapten antibody affinity and specificity, **92**, 589
confidence limits, measurement methods, **92**, 601
cotinine, **84**, 628
applications, **84**, 639
N-demethyldiazepam, **84**, 510
11-deoxy-15-keto-13,14-dihydro-11β, 16ξ-cyclo-prostaglandin E_2, **86**, 306
diazepam, **84**, 491
digoxin and other cardiac glycosides, **84**, 558
5α,7α-dihydroxy-11-ketotetranorprostane-1,16-dioic acid, **86**, 339
ecdysteroids, **84**, 350
egasyn, **96**, 568
eicosanoid, with iodinated derivatives as tracers, **86**, 297
Escherichia coli enterotoxin, **84**, 238
fentanyl, **84**, 526
ferritin, **84**, 166
α-fetoprotein, **84**, 3
fibrinopeptides, **84**, 102
flurazepam, **84**, 507
haloperidol, **84**, 532
haptens, with antigen-coated plastic tubes and radiolabeled or enzyme-labeled protein A, **92**, 409
in hybridoma immunoglobulin quantitation and light-chain loss variant selection, **92**, 147
in hybridoma rapid screening for detection of cell surface-specific antibodies, **92**, 194
12-hydroxyeicosatetraenoic acid, **86**, 248
immunoglobulins, with antigen-coated plastic tubes and radiolabeled protein A, **92**, 404
15-keto-13,14-dihydro-prostaglandin E_2, **86**, 306
15-keto-13,14-dihydro-prostaglandin $F_{2\alpha}$, **86**, 320
6-ketoprostaglandin $F_{1\alpha}$, **86**, 273
leucovorin, **84**, 409
metallothionein, **84**, 121
methotrexate, **84**, 409, 422
5-methyltetrahydrofolate, **84**, 409
metyrapone and reduced form, **84**, 585
morphine, **84**, 551
motilin, **84**, 359
myoglobin, **84**, 172
neocarzinostatin, **84**, 483
clinical applications, **84**, 487
nicotinamide nucleotide analogs of nicotine and cotinine, **84**, 650
nicotine, **84**, 628
applications, **84**, 639
nitrazepam, **84**, 505
N'-nitrosonornicotine and *N'*-acylnornicotine analogs, **84**, 641
normeperidine, **84**, 516
applications, **84**, 523
oxazepam, **84**, 513
phenylpiperidine drugs, **84**, 516
applications, **84**, 523
phosphatidylcholine transfer protein, **98**, 589
applications, **98**, 591
pimozide, **84**, 542
platelet factor 4, **84**, 83
prostaglandin H synthase, **86**, 222
proteoglycan aggregate components, **83**, 225
pyrazolone derivatives, **84**, 596
γ-(3-pyridyl)-γ-oxo-*N*-methylbutyramide, **84**, 628
applications, **84**, 639

separation of free and bound fractions with chromatography tubes, **92**, 366
serum bile acids, **84**, 321
staphylococcal enterotoxin C_2, **84**, 254
6-sulfido-peptide-leukotrienes, **86**, 252
β-thromboglobulin, **84**, 93
thromboxane B_2, **86**, 286
thyroxine and related compounds, **84**, 272
type-II cAMP-dependent protein kinase, **99**, 191
applications, **99**, 192
Radioiodination
antibodies to cholera toxin, **84**, 241
bleomycin, **84**, 464
calmodulin, **84**, 140
carcinoembryonic antigen, **84**, 42
cholera toxin, **84**, 240
derivatized lipopolysaccharides, **84**, 269
ecdysones, **84**, 353
egasyn, **96**, 567
endosome membranes, **98**, 548
localization, **98**, 550
in epitope distinction, **92**, 250
erythrocyte membrane proteins, **96**, 322
Escherichia coli enterotoxin, **84**, 241
ferritin, **84**, 157
α-fetoprotein, **84**, 12
fibrinopeptide analogs, **84**, 107
glycopeptides, **98**, 156
IgG_1, **86**, 235
with iodine monochloride, **92**, 277
iodothyronines, **84**, 275
low-density lipoproteins, **98**, 247
lysosomal membranes, **98**, 404
application to membrane recycling, **98**, 414
methotrexate, **84**, 429
motilin, **84**, 362
myoglobin, **84**, 173
neocarzinostatin, **84**, 485
oligosaccharides, **98**, 156
phosphatidylcholine transfer protein, **98**, 590
platelet factor 4, **84**, 86
polypeptide hormones and proteins, **92**, 309
rhodopsin, **81**, 269
serum bile acids, **84**, 339
staphylococcal enterotoxin C_2, **84**, 255
β-thromboglobulin, **84**, 96
Radioisotopic assays
galactocerebrosides and sulfatides, **83**, 191
Ragweed
antigen E, polymerization by glutaraldehyde, application in immunotherapy, **93**, 50
Receptor–ligand interactions
diffusion regulation role, **93**, 95
Resins
ion-retardation, *see* Ion-retardation resins
Resolvase
transposon $\gamma\delta$
assay, **100**, 194
gene
cloning, **100**, 192
sequence, **100**, 195
properties, **100**, 196
purification, **100**, 193
Resonance Raman spectroscopy
bacteriorhodopsin
with isotopic derivatives, **88**, 633
with microsampling rotating cells, **88**, 659
overview, **88**, 561
bacteriorhodopsin intermediates
with rotating sample, **88**, 643
with spinning sample at 77°K, **88**, 648
by time-resolved technique, **88**, 617
photolabile samples by
flow technique, **88**, 627
pump-probe technique, **88**, 626
rhodopsin
with isotopic derivatives, **88**, 633
with microsampling rotating cells, **88**, 659
overview, **88**, 561
Resorcinol
in colorimetric assay for gangliosides, **83**, 175
Restriction endonucleases
in M13mp7 phage DNA cutting and cloned single-strand sequence excision, **101**, 90
type II, guide to use as tools, **100**, 3

Reticulocyte lysates
rabbit
assay for hemoglobin synthesis, **101**, 609
preparation, **96**, 53, 72, 100; **101**, 607
properties, **96**, 66; **101**, 608
purification of
casein kinase I and II, **99**, 310, 324
dsRNA-dependent eIF-2α protein kinase, **99**, 351
translation of exogenous mRNA, **96**, 59, 100; **101**, 611
Reticulocytes
rabbit
hexokinase Ia and Ib purification, **90**, 8
^{31}P NMR profiles, **85**, 674
Retina
adenylate cyclase characterization and localization, **81**, 517
albumin embedding, **96**, 486
advantages, **96**, 492
bovine, purification of
cellular retinoid-binding proteins with bound endogenous ligands, **81**, 820
rhodopsin kinase, **99**, 364
cyclic AMP levels, measurement by radioimmunoassay, **81**, 89, 551
cyclic GMP levels, measurement by
quick-freeze techniques, **81**, 547
radioimmunoassay, **81**, 89, 551
cytosolic and membrane fractions, immunochemical analysis, **96**, 615
degeneration-associated changes in biosynthesis, **81**, 794
dissection for biochemical studies, **81**, 37
fixation, **96**, 486
ganglioside content, determination methods, **81**, 304
generation of rhodopsin and artificial visual pigments in isolated samples, **81**, 452
guanylate cyclase characterization and localization, **81**, 519
interphotoreceptor matrix
analytical techniques, **81**, 77
component identification, **81**, 83
definition, **81**, 77
glycosaminoglycan and glycoprotein composition, **81**, 85
isolation and purification, **81**, 77
isolation and mounting for optical studies, **81**, 29
lipid biosynthesis, analytical techniques, **81**, 788
membrane protein biosynthesis, immunocytochemistry at electron microscopic level, **96**, 485
microdissection techniques, **81**, 86
neural, primary culture preparation, **81**, 42
opsin biosynthesis, immunocytochemical analysis, **96**, 495
pigment epithelium
fractionation of subcellular organelles, **81**, 105
isolation methods, **81**, 90, 95
primary culture preparation, **81**, 40
retinol analog introduction into isolated samples with lipid vesicles, **81**, 447
rhodopsin biosynthesis
glycosylation analysis, **81**, 783
in isolated samples, **81**, 772
rod and cone pigments, chromatographic separation, **81**, 160
rod outer segments, *see* Rod outer segments
sectioning, **96**, 487
subcellular fractionation, **96**, 609
Retinals
analogs
cis/*trans* isomers, identification by high-performance proton NMR, **88**, 546
photoisomerization kinetics, **88**, 533
synthetic, preparation, **88**, 496
in bacteriorhodopsin, heavy atom labeling, **88**, 178
energy levels, determination by two-photon spectroscopy, **88**, 522
extraction methods, **88**, 552
applications, **88**, 558
fluorine-labeled
NMR analysis, **81**, 696
synthesis, **81**, 695

nomenclature, **88**, 507
photoisomerization
photosensitivity and quantum efficiency, **81**, 384
in presence of rod outer segments or phosphatidylethanolamine, **81**, 461
retinochrome-catalyzed, **81**, 196, 833
–protein interactions as spectroscopic and photochemical determinants, **88**, 462
purple membrane, isolation and purification for mass spectral analysis, **88**, 542
in rhodopsin regeneration, kinetics and mechanism, **81**, 445
separation and identification by HPLC, **88**, 491
spin-labeled, preparation, **88**, 175
stereochemistry, analytical techniques, **88**, 513
synthesis, **88**, 508
Retinochrome
cephalopod
in light isomerization of retinals, **81**, 196, 833
localization, **81**, 191, 827
preparation, **81**, 192, 828
properties, **81**, 193, 830
Retinol, *see* Vitamin A
Retinylopsin
^{3}H-labeled, preparation, **81**, 211
preparation from rod outer segment membranes, **81**, 211
proteolytic digestion, **81**, 212
soluble peptides
characterization, **81**, 213
purification, **81**, 212
Rhabdoms
crayfish, isolation, **81**, 34
squid, isolation and purification, **81**, 43
Rhodopseudomonas palustris
fructose-bisphosphatase purification, **90**, 379
Rhodopsin
absorption spectra, pressure effect, **88**, 719
amino-terminal residue, identification, **81,** 247
antibody preparation, **81**, 240
biosynthesis in isolated retinas, **81**, 772
bleaching intermediates
absorption spectra, pressure effect, **88**, 719
kinetics
lipid dependence, **81**, 378
picosecond time-resolved spectroscopy, **81**, 368
spectrophotometric analysis, **81**, 374
low-temperature circular dichroism, **81**, 634
low-temperature spectrophotometry, **81**, 333
carbohydrate moiety, structural analysis, **81**, 214
chemical modification with imidoesters, **81**, 275
chromophores
^{13}C NMR spectroscopy, **81**, 698
extraction methods, **88**, 557
-containing membrane systems, fast photovoltages, **81**, 431
correlation with spectral intermediates, **81**, 438
in deuterated and hydrogenated detergents, neutron small-angle scattering, **81**, 742
in disk membranes and detergent solution, uv circular dichroism, **81**, 629
distances, measurement by fluorescence energy transfer, **81**, 668
emission spectroscopy, **88**, 217
energy storage mechanism, **88**, 466
fluorine-labeled
NMR analysis, **81**, 696
preparation, **81**, 696
generation in electrophysiologically active photoreceptors, **81**, 452
glycosylation, biosynthetic studies, **81**, 783
α helices, orientation in membrane, determination by polarized infrared spectroscopy, **88**, 709
hydrogen exchange
kinetics, measurement, **88**, 673
methodology, comparison, **88**, 681
in peptide groups, infrared measurement, **88**, 676

incorporation in lipid bilayers, **81**, 149
insect, absorbance properties, measurement, **81**, 187
interactions with
chromophore as spectroscopic and photochemical determinants, **88**, 462
lipids in rod outer segment membranes
analysis with diffusible spin labels, **88**, 762
NMR analysis, **81**, 709
invertebrate, spectral changes associated with photolysis, rapid scan spectrophotometry, **81**, 392
and iodopsin, comparison, **81**, 159
isotopic analogs, resonance Raman spectroscopy, **88**, 633
kinetic properties, measurement by kinetic infrared spectroscopy, **88**, 729
light-induced electrical potentials, analysis with spin-label probes, **88**, 682
molecular structure, **88**, 453
comparison with bacteriorhodopsin, **88**, 458
peptide fragments, preparation, **81**, 251
phosphorylation, assays, **81**, 489
photocalorimetric measurements, **88**, 667
photocycle molecular aspects, **88**, 470
comparison with bacteriorhodopsin, **88**, 484
photodichroism and rotational motions, analytical techniques, **88**, 741
photoelectric response, measurement, **88**, 38
photoisomerization, photosensitivity and quantum efficiency, **81**, 384
proteolysis, application in structural analysis, **81**, 236
purification by
agarose column chromatography, **81**, 141
concanavalin–Sepharose affinity chromatography, **81**, 150, 167, 197
DEAE-cellulose column chromatography, **81**, 167
hydroxyapatite column chromatography, **81**, 144
reduction by cyanoborohydride, **81**, 288
regeneration with 11-*cis*-retinal, kinetics and mechanism, **81**, 443
resonance Raman spectroscopy, **88**, 561
retinal–opsin linkage, reduction by borane dimethylamine, **81**, 290
spin-labeled, saturation-transfer electron paramagnetic resonance, **81**, 703
sulfhydryl chemistry, analytical techniques, **81**, 223
synthetic analogs, preparation, **88**, 496, 516
thermal stability in detergents, **81**, 256
tryptophan residues, reactivity with *N*-bromosuccinimide, **81**, 285
Rhodopsin kinase
bovine rod outer segment
assay, **81**, 498; **99**, 362
properties, **81**, 506; **99**, 365
purification, **81**, 499; **99**, 364
Rhodospirillum rubrum
ribulose-bisphosphate carboxylase purification, **90**, 524
Ribonuclease inhibitors
human placental, isolation, **96**, 98
Ribonucleases
functional, recovery in SDS gels, **91**, 271
methionine-containing peptides, purification, **91**, 371
Ribonucleic acid
binding proteins, recovery in SDS gels, **91**, 273
denatured, transferred or dotted to nitrocellulose paper, hybridization method, **100**, 255
–DNA hybrids of naturally occurring sequence
antibody radioimmunoassay, **84**, 216
preparation, **84**, 211
electrophoretic transfer to diazobenzyloxymethyl paper, **100**, 326
intermolecular joining with RNA ligase, **100**, 52
intracellular delivery
with liposomes as carriers, **101**, 512
microinjection techniques, **101**, 482

maize chloroplast
isolation, **97**, 530
translation *in vitro*, **97**, 549
mammalian mitochondrial
–DNA hybridization, **97**, 455
fractionation and analysis, **97**, 439
isolation, **97**, 428, 438
labeling, **97**, 436
metabolism, **97**, 462
structural analysis, **97**, 446
messenger
acetylcholine receptor subunit
isolation from *Torpedo californica* electroplax, **96**, 368
translation in cell-free systems, **96**, 369
cell-free translation
in reticulocyte lysate system, **101**, 606
system preparation from
Chinese hamster ovary cells, **101**, 629
Saccharomyces cerevisiae, **101**, 644
in wheat germ system, **96**, 38
elastin
cell-free translation, **82**, 734
isolation, **82**, 732
erythrocyte anion transport protein, isolation from spleen cells, **96**, 266
extraction methods, **96**, 24, 97
fractionation methods, **96**, 31
glycophorin A
isolation from K562 cells, **96**, 284
translation *in vitro*, **96**, 285
histocompatibility antigen, isolation, **96**, 326
interstitial procollagen
isolation from connective tissues, **82**, 218
translation *in vitro*, **82**, 222
lysosomal enzyme
cell-free translation, **96**, 773
isolation, **96**, 770
major lens membrane polypeptide
cell-free translation, **98**, 515
isolation, **98**, 511
Neurospora crassa middle-abundant, molecular cloning, **97**, 254
recovery from gels, **96**, 35
for sarcoplasmic reticulum proteins
isolation, **96**, 576
translation *in vitro*, **96**, 578
secretory protein, cell-free translation, **98**, 460, 465
5′-terminal sequences, cloning, **100**, 285
sequence analysis methods, **100**, 431
species selection by hybridization to DNA covalently coupled to macroporous supports, **100**, 400
synthesis *in vitro* with bacterial and bacteriophage RNA polymerases, **101**, 540
transfer, *N*-formylmethionyl-, *see N*-Formylmethionyl-tRNA
Ribophorins
antibody preparation, **96**, 525
hepatocyte endoplasmic reticulum, biosynthetic mechanism, **96**, 540
isolation from
inverted rough microsomes, **96**, 523
solubilized rough microsomes, **96**, 521
microsomal content, correlation with ribosome binding capacity, **96**, 526
Ribose
binding protein, *Escherichia coli*
assay, **90**, 467
properties, **90**, 472
purification, **90**, 469
enriched with oxygen isotopes, chemical synthesis, **89**, 82
Ribose isomerase
Mycobacterium smegmatis
assay, **89**, 547
properties, **89**, 549
purification, **89**, 547
Ribose 5-phosphate
preparation from
AMP, **89**, 118
6-phosphogluconate, **89**, 117
in ribulose 1,5-bisphosphate preparation, **89**, 108

Ribose-5-phosphate isomerase
bacterial
assay, **89**, 572
properties, **89**, 576
purification, **89**, 574
regulation and inhibition, **89**, 577
Ribosomes
microsome binding capacity, correlation with ribophorin content, **96**, 526
Ribulose 1,5-bisphosphate
assays, **89**, 114
enriched with carbon isotopes, enzymatic synthesis, **89**, 87
synthesis, routes from
AMP via ribose 5-phosphate, **89**, 108
glucose 6-phosphate via 6-phosphogluconate, **89**, 108
Ribulose-bisphosphate carboxylase
activated, measurement, **89**, 54
activation, **89**, 51
assay, **89**, 47
Rhodospirillum rubrum
assay, **90**, 522
properties, **90**, 526
purification, **90**, 524
spinach leaf
purification, **90**, 515, 518
reactivation with dithiothreitol, **90**, 520
storage and preparation for assay, **90**, 519
tobacco leaf, purification and crystallization, **90**, 520
tomato leaf, purification, **90**, 521
Ribulose-1,5-bisphosphate carboxylase/oxygenase, *see* Ribulose-bisphosphate carboxylase
Ricin
A chain, antibody-coupled
preparation, **93**, 334
structural and functional analysis, **93**, 336
RNA, *see* Ribonucleic acid
RNA ligase, *see* Polyribonucleotide synthase (ATP)
RNA polymerase
bacterial and bacteriophage
application for extensive transcription *in vitro*, **101**, 562
assays, **101**, 549, 559
purification, **101**, 541, 556
transcription factors, isolation from HeLa cells, **101**, 592
Rod outer segments
birefringence and birefringence gradients, measurement, **81**, 655
calcium release upon rhodopsin photoexcitation, detection and properties, **81**, 569
cyclic nucleotide cascade, **96**, 617
disk membranes
cyclic-GMP phosphodiesterase and controller enzymes, real-time assay, **81**, 532
intact, preparation, **81**, 116
phospholipid organization, analysis with parinaric acid probes, **81**, 685
phospholipid transmembrane topology, determination methods, **81**, 309
rhodopsin–lipid interactions, NMR analyses, **81**, 709
tritium–hydrogen exchange, kinetics, **88**, 673
uv circular dichroism of rhodopsin, **81**, 629
visual pigment lateral diffusion, analytical techniques, **81**, 660
disk size and shape, determination by light scattering methods, **81**, 616
ganglioside content, determination methods, **81**, 304
guanylate cyclase localization and assay, **81**, 522
isolation, rapid freezing, and freeze-fracture methods, **81**, 585
lipid renewal, analytical techniques, **81**, 800
membrane proteins, antibody preparation, **81**, 240
Mg^{2+}-ATPase characterization and assay, **81**, 509, 514
oriented suspensions, linear dichroism studies, **81**, 605
phospholipid content, analytical techniques, **81**, 297, 806

preparation of
intact disks by
chromatography on concanavalin A columns, **81**, 61
Ficoll flotation, **81**, 57
membranes capable of regenerating visual pigment with added 11-*cis*-retinol, **81**, 506
preparation methods, **81**, 48, 124
protein purification by light-regulated binding to membranes, **81**, 566
purification of
ambiquitous GTPase, **81**, 567
cyclic-GMP phosphodiesterase, **96**, 625
cyclic-GMP phosphodiesterase inhibitor, **81**, 542
cyclic nucleotide-dependent protein kinase, **81**, 499
light-activated phosphodiesterase, **81**, 527
rhodopsin kinase, **81**, 499
transducin–GTP complex, **96**, 621
sealed, preparation and characterization, **81**, 52, 110
sonicated disk vesicles, light release of trapped ^{45}Ca, **81**, 577
X-ray and neutron diffraction studies, **81**, 593

S

Saccharides
cell surface, ultrastructural visualization, **83**, 195
lipid-linked formation, inhibitors, **98**, 138
Salmine
tryptic digest, immobilization on Sepharose, **91**, 379
Salmonella typhimurium
purification of phosphoenolpyruvate:glycose phosphotransferase system
Enzyme I, **90**, 431
Enzyme III^{Glc}, **90**, 439
HPr, **90**, 436
Salt
removal from protein solutions in centrifuge column, **91**, 278
Salt gland
avian
basolateral membranes, enzyme markers, **96**, 634
fractionation, **96**, 636
Golgi membrane
enzyme assays, **96**, 638
preparation, **96**, 635
membrane biogenesis, regulation, **96**, 658
Na^+,K^+-ATPase, localization, **96**, 640, 643, 646
phospholipid biosynthetic enzymes, cytochemical localization, **96**, 648
plasma membrane
adaptation, **96**, 628
deadaptation, **96**, 655
enzyme assays, **96**, 638
preparation, **96**, 635
Sandhoff disease
fibroblasts, β-hexosaminidase pinocytosis measurements, **98**, 292
Sarcoplasmic reticulum
protein biosynthesis in
myoblast cultures, **96**, 570
in vitro system, **96**, 578
Scanning electron microscopy
cells attached to collagen gels, **82**, 537
collagen gels, **82**, 540
SDS, *see* Sodium dodecyl sulfate
Secretory apparatus
Escherichia coli, genetic analysis, **97**, 8
Secretory component
human and rabbit, biosynthesis, processing, and functional analysis, **98**, 458
Secretory granules
pancreatic islet
biochemical characterization, **98**, 190
morphological characterization, **98**, 192
Sedimentation analysis
actin filament networks, **85**, 215
in assay for actin-binding protein crosslinking activity, **85**, 481
high-molecular-weight proteins and large macromolecular assemblies, **85**, 570, 571

microtubules during assembly *in vitro*, **85**, 436
Sedoheptulose-bisphosphatase
wheat chloroplast
assay, **90**, 392
properties, **90**, 395
purification, **90**, 394
Sedoheptulose 7-phosphate
determination with transketolase, **89**, 19
SEM, *see* Scanning electron microscopy
Seminal vesicles
bovine, prostaglandin H synthase purification, **86**, 58
ovine, purification of
prostaglandin-H_2 E-isomerase, **86**, 87
prostaglandin H synthase, **86**, 63
Semliki Forest virus
glycoproteins, intracellular
transport analysis, **96**, 466
localization with organelle-specific lectins, **96**, 461
-infected cells, thin frozen sections, immunolabeling, **96**, 479
interactions with BHK-21 cells, analytical techniques, **98**, 260
membrane proteins, expression analysis, **96**, 496
temperature-sensitive mutants
isolation, **96**, 453
membrane glycoprotein transport analysis, **96**, 461
screening, **96**, 458
Sendai virus
fusogenic reconstituted envelopes, as vehicle for DNA introduction into viable mammalian cells, **101**, 492
Sequenators
commercial spinning-cup, efficiency improvements, **91**, 434, 443
gas-phase, design and operation principles, **91**, 399
Serine proteases
covalent compounds and intermediates, preparation and properties, **87**, 66
Serotonin
arylsulfatase effects, **86**, 20
Serratia marcescens
spermidine dehydrogenase purification by affinity chromatography, **94**, 304
trp promoter-containing plasmids, application in expressing cloned genes, **101**, 155
Serum
lipoprotein-deficient, preparation, **98**, 250
Serum bile acids
antiserum
characterization, **84**, 333
production, **84**, 323
radioimmunoassay, **84**, 341
accuracy and precision, **84**, 345
radiolabeled, preparation, **84**, 339
Showdomycin
inhibitory effects on glycoprotein lipid-linked saccharide formation, **98**, 147
Signal peptidase
canine pancreatic
extraction, **96**, 786
quantitative assay, **96**, 788
Signal recognition particle
microsomal membrane
properties, **96**, 683
purification, **96**, 684
translocation assay, **96**, 690
Silver staining
DNA on polyacrylamide gels, **96**, 235
membrane proteins on polyacrylamide gels, **96**, 189
polyacrylamide gels, **96**, 230
effects on autoradiography and fluorography, **96**, 234
sensitivity, **96**, 233
intensification, **96**, 232
proteins on giant gels, **91**, 211
Simian virus 5
-infected MDCK cells
electron microscopy, **98**, 493
infectivity assays, **98**, 491
preparation, **98**, 490
viral protein localization by
immunoelectron microscopy, **98**, 497
immunofluorescence, **98**, 495
Sindbis virus
membrane glycoproteins
affinity chromatography, **96**, 519
biosynthesis *in vitro*, **96**, 512
electrophoresis, **96**, 515
posttranslational assays, **96**, 517

Size-exclusion chromatography
proteins
ideal, **91**, 147
nonideal, **91**, 167
preparative, **91**, 161
Skeletal muscle
chicken
L-glycol dehydrogenase purification, **89**, 524
light meromyosin aggregate formation and assay, **85**, 51
rabbit
light meromyosin aggregate formation and assay, **85**, 51
myosin aggregate formation and assay, **85**, 29, 48
purification of
actin, **85**, 164
cAMP-dependent protein kinase regulatory subunit, **99**, 57
fructose-bisphosphatase, **90**, 343
glyceraldehyde-3-phosphate dehydrogenase, **89**, 306
glycerol dehydrogenase, **89**, 239
glycogen synthase kinase-3, **99**, 340
glycolytic enzymes, **90**, 479
heat-stable cAMP-dependent protein kinase inhibitor, **99**, 78, 84
mRNA, **96**, 577
myosin and subfragments, **85**, 55; **98**, 371
native thin filaments, **85**, 12
phosphorylase kinase, **99**, 246
tropomyosin, **85**, 234
troponin, **85**, 244
troponin subunits, **85**, 251
rat, purification of
hexokinase type II, **90**, 12
mRNA, **96**, 577
polysomes, **96**, 575
pyruvate kinase, **90**, 158
Slow-reacting substances
anaphylactic
guinea pig
extraction from lung, **86**, 427
ileal bioassay system, characterization, **86**, 656
purification, **86**, 428
structural analysis, **86**, 431
rat
arylsulfatase effects, **86**, 20
purification, **86**, 22
structural analysis, **86**, 22
bioassays, **86**, 660, 665
comparative responses, **86**, 658
RBL-1 cell
production, **86**, 418
properties, **86**, 423
purification, **86**, 420
response specificity, verification, **86**, 661
Smooth muscle
bovine aortic
filamin purification, **85**, 326
myosin aggregate formation and assay, **85**, 35, 50
Bufo marinus stomach, preparation of single cells, **85**, 284
chicken gizzard
myosin aggregate formation and assay, **85**, 33, 50
purification of
α-actinin, **85**, 316
desmin, **85**, 492
filamin, **85**, 324
thick filaments, **85**, 280
thin filaments, **85**, 279
guinea pig
myosin aggregate formation and assay, **85**, 35
thick filament isolation, **85**, 280
porcine aortic, myosin preparation, **85**, 292
rabbit uterine, myosin aggregate formation and assay, **85**, 34
turkey gizzard, purification of
myosin, **99**, 281
myosin light-chain kinase, **85**, 298; **99**, 281
phosphatase type I and II, **85**, 308
thick filaments, **85**, 280
Sodium bisulfite
in directed mutagenesis of DNA, **100**, 457
Sodium cholate
in extraction of visual pigments, **81**, 133
Sodium dodecyl sulfate
removal from proteins
determination, **91**, 261
after elution from gels, **91**, 226

by ion-pair extraction, **91**, 254
by ion-retardation chromatography, **91**, 259
solubilized proteins, peptide mapping by limited proteolysis, **96**, 222
Sodium dodecyl sulfate gels
functional protein recovery, **91**, 263
protein elution by
diffusion, **91**, 454
electrophoresis, **91**, 222, 450
Sodium transport
primary, demonstration in *Halobacterium halobium* envelope vesicles, **88**, 426
Soluble elastin
antiserum preparation, **82**, 747
biosynthesis
in cell culture, **82**, 719
in organ culture, **82**, 717
extraction, **82**, 565, 723
identification, **82**, 725
peptide characterization, **82**, 691
preparation, **82**, 592
Somatic cells
glycoprotein mutants, isolation, **83**, 444
Somatotropin
human
deletion peptide, antiserum production, **93**, 10
stable radioiodinated, preparation by polyacrylamide gel electrophoresis, **92**, 334
two-chain, antiserum production, **93**, 9
Sorbitol dehydrogenase, *see* Iditol dehydrogenase
L-Sorbose-1-phosphate reductase
Klebsiella pneumoniae
assay, **89**, 248
properties, **89**, 251
purification, **89**, 249
Spectrin
human erythrocyte
binding assay, **96**, 324
intermediate-sized peptides
electrophoretic elution, **96**, 309
generation, **96**, 306
identification, **96**, 307
oligomer detection by gel electrophoresis, **96**, 298
purification, **85**, 475; **96**, 306, 316
radiolabeling, **96**, 322
Spectrophotometry
in actin polymerization assay, **85**, 190
bacteriorhodopsin during photocycle
at low temperatures, **88**, 163
for protonation state determination, **88**, 254
bleaching intermediate kinetics of rhodopsin, metarhodopsin I, and metarhodopsin II, **81**, 374
in glucose-6-phosphatase assay, **89**, 44
halorhodopsin in
bacterial colonies, **88**, 437
envelope vesicles, **88**, 435
Halobacterium halobium membranes, **88**, 439
media of low ionic strength, **88**, 435
planar membrane monolayers, **88**, 230
protein–dye complexes in gels, **91**, 246
rhodopsin bleaching intermediates at low temperatures, **81**, 333
rhodopsin kinetics in rod outer segment and phospholipid membrane vesicles, **81**, 380
spectral changes in invertebrate rhodopsin photolysis, **81**, 392
Sperm
sea urchin, purification of
dynein ATPase, **85**, 472
latent dynein ATPase, **85**, 453
starfish, dynein ATPase purification, **85**, 471
trout, dynein ATPase purification, **85**, 473
Spermic acid
synthesis, **94**, 419
Spermidine
2-aminoethyl and 3-aminoethyl analogs
applications, **94**, 416
preparation, **94**, 412, 414
Spermidine N^1-acetyltransferase
bovine liver
assay, **94**, 325
properties, **94**, 328
purification, **94**, 326
rat liver
assay, **94**, 321
properties, **94**, 324
purification, **94**, 322

Spermidine aminopropyltransferase, *see* Spermine synthase
Spermidine dehydrogenase
Serratia marcescens
assay, **94**, 303
purification, **94**, 304
Spermidine synthase
assays, **94**, 257, 260
bovine brain
assay, **94**, 271
properties, **94**, 275
purification, **94**, 272
Chinese cabbage
assay, **94**, 281
properties, **94**, 285
purification, **94**, 283
Escherichia coli
assay, **94**, 265
properties, **94**, 269
purification, **94**, 267
rat prostate
assay, **94**, 271
properties, **94**, 275
purification, **94**, 274
substrate synthesis, **94**, 286
Spermidine synthase inhibitors
properties, **94**, 294
synthesis, **94**, 286
Spermine
activation of casein kinase II, **99**, 328
2-aminoethyl and 3-aminopropyl analogs
applications, **94**, 416
preparation, **94**, 412, 414
Spermine synthase
assays, **94**, 257, 260
bovine brain
assay, **94**, 276
properties, **94**, 278
purification, **94**, 277
Saccharomyces cerevisiae, deficient mutants, mass screening, **94**, 107
substrate synthesis, **94**, 286
Spermine synthase inhibitors
properties, **94**, 294
synthesis, **94**, 286
Sphingomyelin
synthesis, phosphatidylcholine transfer protein-dependent, **98**, 596
Spinach
purification of
fructose-bisphosphatase, **90**, 373
glyceraldehyde-3-phosphate dehydrogenase, **89**, 317
phosphoglycerate kinase, **90**, 110
ribulose-bisphosphate carboxylase, **90**, 515
Spleen
bovine, purification of cerebroside transfer protein, **98**, 617
murine
application in analysis of erythrocyte anion transport protein synthesis and maturation *in vitro*, **96**, 258
cell size analysis for antigen-specific hybridoma high-frequency production, **92**, 34
rat, purification of
cathepsin D, **96**, 767
mRNA for lysosomal enzyme synthesis *in vitro*, **96**, 770
prostaglandin H_2 D-isomerase, **86**, 80
Sporotrichum pulverulentum
cellobiose oxidase purification, **89**, 130
Staining
immunofluorescent, contractile proteins in fixed cells, **85**, 514
membrane proteins on polyacrylamide gels, **96**, 189
negative, myosin for electron microscopy, **85**, 129
polyamines on thin-layer chromatograms and electrophoretograms, **94**, 7
proteins on gels
comparisons of dyes and procedures, **91**, 237
with Coomassie Blue, **91**, 211
silver, *see* Silver staining
Staphylococcus
S. aureus
cellular quantification by partition affinity ligand assay, **92**, 511
enterotoxin C_2, radioimmunoassay, **84**, 254
purification of
D-galactose-6-phosphate isomerase, **89**, 563

phosphoenolpyruvate:glycose phosphotransferase system components, **90**, 450, 453
D-tagatose-1,6-bisphosphate aldolase, **90**, 230
D-tagatose-6-phosphate kinase, **90**, 88
vaccine preparation with glutaraldehyde, **93**, 38
trans-Stilbene oxide
induction of endoplasmic reticulum membrane components, **96**, 544
Stomach
Bufo marinus, preparation of single smooth muscle cells, **85**, 284
Streptococcus
S. cremoris, L-lactate dehydrogenase purification, **89**, 363
S. faecalis, putrescine carbamoyltransferase purification, **94**, 340
S. lactis, purification of
phosphofructokinase, **90**, 78
pyruvate kinase, **90**, 166
S. mutans, purification of
glucokinase, **90**, 27
sucrose-6-phosphate hydrolase, **90**, 557
Streptomyces plicatus
purification of endo-β-*N*-acetylglucosaminidase L, **83**, 606
Succinate dehydrogenase
avian salt gland, assay, **96**, 638
N-Succinimidyl 3-(2-pyridyldithio)propionate
in attachment of haptens and proteins to erythrocytes, **92**, 259
Succinylation
hapten–protein conjugates, in facilitation of conjugate coupling to erythrocytes, **92**, 265
Sucrase, *see* Sucrose α-glucosidase
Sucrose
–D_2O gradients, in Golgi complex isolation from rat liver, **98**, 62
–H_2O gradients, in Golgi complex isolation from rat liver, **98**, 61
Sucrose α-glucosidase
–isomaltase complex, small-intestinal brush border membrane
assembly, **96**, 399
biosynthesis
in cell-free system, **96**, 400
mechanism, **96**, 405
isomaltase-anchoring segment, orientation of N-terminal amino acid, **96**, 392
lipid bilayer-embedded subunit, identification, **96**, 388
Sucrose laurates
in extraction of visual pigments, **81**, 133
Sucrose-6-phosphate hydrolase
Streptococcus mutans
assay, **90**, 556
properties, **90**, 559
purification, **90**, 557
Sugar chains
asparagine-linked, hydrazinolysis to produce free oligosaccharides, **83**, 263
Sugars
in biological materials, isolation with carbohydrate-binding proteins, **83**, 241
enantiomeric mixtures, resolution by gas chromatography, **89**, 3
nucleotide-linked, isolation and synthesis, **83**, 332
Sulfate
demonstration in glycophorin A, **96**, 291
Sulfatides
radioisotopic assay, **83**, 191
Sulfhydryl groups
peptide, alkylation with *N*-ethylmaleimide for peptide isolation by diagonal electrophoresis, **91**, 392
protein, specific crosslinking reagents, **91**, 599
rhodopsin, chemical modification studies, **81**, 223
Swainsonine
inhibitory effects on glycoprotein processing, **98**, 151
Sweet potato roots
purification of
L-lactate dehydrogenase isozymes, **89**, 346
pyruvate decarboxylase, **90**, 529
Synaptosomes
preparation from rat forebrain, **98**, 603

T

D-Tagatose-1,6-bisphosphate aldolase
 class I, *Staphylococcus aureus*
  assay, **90**, 228
  properties, **90**, 231
  purification, **90**, 230
 class II, *Klebsiella pneumoniae*
  assay, **90**, 232
  properties, **90**, 234
  purification, **90**, 232
D-Tagatose 6-phosphate
 synthesis, **89**, 95
D-Tagatose-6-phosphate kinase
 Staphylococcus aureus
  assay, **90**, 87
  properties, **90**, 90
  purification, **90**, 88
Tagaturonate reductase
 Escherichia coli
  assay, **89**, 210
  properties, **89**, 213
  purification, **89**, 212
Tay–Sachs disease
 fibroblast secretions
  preparation of [2-^{3}H]mannose-labeled glycopeptides, **98**, 293
  purification of β-hexosaminidase B, **98**, 291
TEM, *see* Transmission electron microscopy
Terminal deoxynucleotidyltransferase, *see* DNA nucleotidylexotransferase
Testis
 bovine, purification of
  calmodulin, **99**, 280
  myo-inositol-1-phosphate synthase, **90**, 310
 murine, purification of
  glucose-6-phosphate dehydrogenase, **89**, 256
  L-lactate dehydrogenase X, **89**, 355
  phosphoglycerate kinase, **90**, 103, 122
 ovine, calmodulin purification, **99**, 280
Test tube inversion
 in actin filament network characterization, **85**, 213
2,3,4,5-Tetranor-thromboxane B_2
 urinary
  derivative preparation for gas chromatography–mass spectrometry, **86**, 605
  preparation, **86**, 604
Theophylline
 colorimetric immunoassay with FAD as label, **92**, 422
Thermoplasma acidophilum
 cultivation, **88**, 398
 diglyceryl tetraethers
  isolation, **88**, 399
  quantitation, **88**, 405
Thermospermine
 preparation from spermidine, **94**, 416
Thermus thermophilus
 purification of
  glyceraldehyde-3-phosphate dehydrogenase, **89**, 336
  polyamines, **94**, 403
Thick filaments
 hybridization and reconstitution, **85**, 20
 native, preparation, **85**, 17
 nonmuscle, myosin phosphorylation effect on stabilization, **85**, 364
 preparation of phosphofructokinase and C-, H-, and X-proteins, **85**, 130
 smooth muscle, isolation, **85**, 280
Thin filaments
 hybridization and reconstitution, **85**, 15
 native, preparation, **85**, 12
 smooth muscle, isolation, **85**, 279
Thin-layer chromatography
 arachidonic acid metabolites, **86**, 484
 gangliosides, **83**, 167
 glycolipid ligands of carbohydrate-binding proteins, **83**, 235
 hydroxyputrescine, **94**, 433
 and ion-exchange chromatography, in peptide mapping, **91**, 384
 5-lipoxygenase products, **86**, 32
 polyamines and derivatives, **94**, 3, 16
 prostaglandin ω-hydroxylase products, **86**, 177
 prostaglandin ω-oxidation products, **86**, 171
 prostaglandins and related compounds, **86**, 493

rod outer segment phospholipids, **81**, 810
Thin-layer electrophoresis
phosphoamino acids, **99**, 397
polyamines and derivatives, **94**, 5
Thiobacillus thioparus
ribose-5-phosphate isomerase purification, **89**, 576
Thiobarbituric acid
in colorimetric assay for gangliosides, **83**, 178
Thioglycolic acid
reduction of
free methionine sulfoxide, **91**, 550
methionine sulfoxide residues in
peptides, **91**, 554
proteins, **91**, 556
Thioglycosides
incorporation into polyacrylamide gels, **83**, 299
6-Thioinosine 5′-monophosphate
in active-site trapping of myosin metal nucleotides, **85**, 93
Thiols
–mercury affinity system, in generation of overlapping labeled fragments for DNA sequencing, **100**, 116
protein, content determination with 5,5′-dithiobis(2-nitrobenzoic acid), **91**, 58
Thoracic integument
grasshopper, cyclic nucleotide-dependent protein kinase purification, **99**, 73
D-Threose
deuterated, chemical synthesis, **89**, 81
β-Thromboglobulin
radioimmunoassay, **84**, 93
Thromboxane A_2
carbocyclic, *see* Carbocyclic thromboxane A_2
effect on platelet aggregation as bioassay, **86**, 651
stable analogs
biological properties, **86**, 403
synthesis, **86**, 400
Thromboxane B_2
dinor urinary metabolite, quantitation by gas chromatography–mass spectrometry, **86**, 592
radioimmunoassay, **86**, 286
tetranor urinary metabolite
derivative preparation for gas chromatography–mass spectrometry, **86**, 605
preparation, **86**, 604
Thromboxanes
extraction with octadecylsilyl silica, **86**, 467
Thromboxane synthase
bovine platelet
assay, **86**, 107
properties, **86**, 109
purification, **86**, 108
human platelet, assay, **86**, 653
Thromboxane synthase inhibitors
assay methods, **86**, 110
Thymus
bovine, myosin preparation, **85**, 364
Thyroglobulin
transcellular transport in inside-out thyroid follicles, **98**, 447
Thyroid gland
follicle cells
inside-out, preparation, **98**, 447
isolation, **98**, 480
organizational consequences, **98**, 479
function, **98**, 478
Thyroxine
chemiluminescence immunoassay in serum, **84**, 303
enzyme immunoassay with activated thiol-Sepharose in separation method, **92**, 352
induction of mitochondria, **96**, 555
radioimmunoassay, **84**, 272
Tissues
in culture, cellular microinjection techniques, **101**, 482
freezing and sectioning, **96**, 470
immunolabeling, **96**, 477
Titin
molecular properties, **85**, 274
purification, **85**, 264
TLC, *see* Thin-layer chromatography
Tomato
lectin purification, **83**, 364
Toxins
cholera, *see* Cholera toxin

Clostridium botulinum A, B, and E, ELISA, **84**, 223
diphtheria, vaccine preparation with glutaraldehyde, **93**, 25
Escherichia coli enterotoxin, radioimmunoassay, **84**, 238
microbial and plant, antibody-linked
preparation, **93**, 281
purification and characterization, **93**, 328
staphylococcal
enterotoxin C_2, radioimmunoassay, **84**, 254
vaccine preparation with glutaraldehyde, **93**, 36
tetanus, vaccine preparation with glutaraldehyde, **93**, 27
Transcription factors
for 5 S genes, isolation from *Xenopus laevis* ovary, **101**, 586
for RNA polymerase II and III, isolation from HeLa cells, **101**, 592
Transducin
retinal rod outer segment
complex with GTP
properties, **96**, 622
purification, **96**, 621
homology with G protein, **96**, 620
properties, **96**, 619
α-subunit
ADP-ribosylated, preparation, **96**, 625
cGMP phosphodiesterase activation, **96**, 623
purification and properties, **96**, 623
Transglutaminase, *see* Protein-glutamine γ-glutamyltransferase
Transketolase
assay via ferricyanide reduction, **89**, 43
in determination of fructose 6-phosphate, sedoheptulose 7-phosphate, and xylulose 5-phosphate, **89**, 19
human erythrocyte
assay, **90**, 223
purification, **90**, 225
porcine liver
assay, **90**, 220
properties, **90**, 221
purification, **90**, 220
rat liver
assay, **90**, 217
properties, **90**, 219
purification, **90**, 217
yeast
assay, **90**, 209
properties, **90**, 215
purification, **90**, 212
Transmission electron microscopy
cells attached to collagen gels, **82**, 538
collagen gels, **82**, 540
Transposons
γδ
in analysis of cloned genes, **101**, 362
resolvase protein
activity assay, **100**, 194
gene cloning, **100**, 192
properties, **100**, 196
purification, **100**, 193
Tn10, insertion in *Escherichia coli*, **97**, 38
application, **97**, 39
1,3,5-Trichlorotriazine
in attachment of haptens and proteins to erythrocytes, **92**, 271
Tridecyltrimethylammonium bromide
in extraction of visual pigments, **81**, 133
3,3′,5′-Triiodothyronine
radioimmunoassay, **84**, 272
3,5,3′-Triiodothyronine
enzyme immunoassay with activated thiol–Sepharose in separation method, **92**, 352
induction of mitochondria, **96**, 555
radioimmunoassay, **84**, 272
with direct partition affinity ligand assay technique, **92**, 519
Trimethylacetate
–transglutaminase complex
bond nature and acyl group position, **87**, 41
formation and isolation, **87**, 37
Trinitrotoluene
bioluminescent immunoassay, **92**, 426
Triose-phosphate isomerase
chicken muscle
assay, **89**, 579; **90**, 490
properties, **89**, 583
purification, **89**, 580; **90**, 492

rabbit muscle
assay, **89**, 579
properties, **89**, 583
purification, **89**, 582
Tritium
–hydrogen exchange in
bacteriorhodopsin
analytical techniques, **88**, 676
exchange-in, **88**, 674
exchange-out, **88**, 675
membrane-bound rhodopsin, comparison with deuterium exchange, **88**, 681
rhodopsin
analytical techniques, **88**, 676
exchange-in, **88**, 674
exchange-out, **88**, 675
kinetic effects
in deduction of enzyme reaction mechanisms, **87**, 711
on initial rates of enzyme reactions, **87**, 607
Triton X-100
in extraction of visual pigments, **81**, 133
removal from proteins eluted from gels, **91**, 226
Tropoelastin
antibody preparation, **82**, 762
identification in cell-free translation assay, **82**, 739
incorporation into insoluble elastin cell culture, **82**, 630
isolation from
copper-deficient aorta, **82**, 652, 661
lathyritic aorta, **82**, 668
production in copper-deficient swine, **82**, 657
purity evaluation, **82**, 664
quantitation in cell-free translation system, **82**, 742
Tropomyosin
cytoplasmic
properties, **85**, 374
purification, **85**, 374
α- and β-forms
chromatography on hydroxyapatite, **85**, 237
preparation, **85**, 234, 239
separation, **85**, 239
spin labeling, **85**, 609
structure–activity relationship, optical activity measurements, **85**, 694
–troponin complex, preparation, **85**, 15
Troponin
assay, **85**, 262
purification from
cardiac muscle, **85**, 249
skeletal muscle, **85**, 244
reconstitution from subunits, **85**, 260
structure–function relationship, optical activity measurements, **85**, 687
subunits
Ca^{2+}-binding, spin labeling, **85**, 609
associated conformational changes, **85**, 611
purification from
cardiac muscle, **85**, 253
skeletal muscle, **85**, 251
–tropomyosin complex, preparation, **85**, 15
Trypan blue
in tumor cell cytotoxicity assay, **93**, 233
Trypanosoma b. brucei
infection in mice, treatment with ornithine decarboxylase inhibitors, **94**, 210
Trypsin
in analysis of
membrane insertion, **96**, 136
protein interaction with organelle membranes, **96**, 143
properties, **96**, 137
Tryptophan promoters
bacterial, introduction into plasmids and application in expressing cloned genes, **101**, 155
Tryptophan residues
protein
analysis by hydrolysis, **91**, 26
cleavage with
dimethyl sulfoxide–hydrochloric acid and cyanogen bromide, **91**, 318
o-iodosobenzoic acid, **91**, 311
specific crosslinking reagents, **91**, 601
rhodopsin, reactivity with *N*-bromosuccinimide, assay, **81**, 285

Tsushimycin
inhibitory effects on glycoprotein lipid-linked saccharide formation, **98**, 144
Tubulin
brain
physical properties, **85**, 397
preparation methods, **85**, 376, 381, 394
–guanine nucleotide complexes
preparation, **85**, 417
properties, **85**, 423
nonneural, isolation, **85**, 385
Tubulin:tyrosine ligase
microtubule-associated, assay, **85**, 432
Tumor cells
A-431, purification of epidermal growth factor-related tyrosyl kinase, **99**, 379
AH-130, purification of pyruvate kinase type M_2, **90**, 159
associated antigens, polyclonal antiserum preparation with agarose-entrapped immune complexes as immunogens, **93**, 78
cytotoxicity of
humoral factors, assessment with
^{51}Cr, **93**, 238
[^{125}I]iododeoxyuridine, **93**, 235
methyl[^{3}H]thymidine, **93**, 239
trypan blue, **93**, 234
immunocompetent cells, assessment with
^{51}Cr, **93**, 240
methyl[^{3}H]thymidine, **93**, 242
Ehrlich ascites
preparation of cell-free translation system, **101**, 616
purification of phosphofructokinase, **90**, 36
EL.4 and C14, biosynthetic labeling in tissue culture, **91**, 414
HeLa
preparation of whole-cell extracts for DNA transcription, **101**, 570
purification of DNA topoisomerase I, **100**, 134
HT29, biosynthesis of secretory component *in vitro*, **98**, 463
K562, biogenesis and processing of glycophorin A *in vitro*, **96**, 281
mastocytoma, purification of leukotrienes, **86**, 409
metabolic responses to cytotoxic T-cell attack, analysis by cell immobilization to poly-L-lysine-coated surfaces, **93**, 260
RBL-1
production of slow-reacting substance, **86**, 418
purification of
leukotriene-forming enzymes, **86**, 32
slow-reacting substance, **86**, 420
thymoma SL2, biosynthesis and assembly of H-2 antigens *in vitro*, **96**, 326
treatment with glutaraldehyde, **93**, 56
Tunicamycin
inhibitory effects on
glycophorin A *N*-glycosylation, **96**, 289
glycoprotein lipid-linked saccharide formation, **98**, 138
Turbidimetry
microtubules during assembly *in vitro*, **85**, 433
Two-photon spectroscopy
visual chromophore energy levels, **88**, 522

U

UDP-*N*-acetyl-D-glucosamine
in assay for transport-coupled oligosaccharide processing *in vitro*, **98**, 45
UDP-*N*-acetylglucosamine 2-epimerase
Escherichia coli
assay, **83**, 515
properties, **83**, 518
purification, **83**, 516
UDP-*N*-acetyl-D-glucosamine:α-D-mannoside β1-2-*N*-acetyl-D-glucosamine transferases, *see* *N*-Acetyl-D-glucosamine transferases

UDP-N-acetyl-D-mannosamine dehydrogenase
Escherichia coli
assay, **83**, 519
properties, **83**, 521
purification, **83**, 520
UDPgalactose:N-acetyl-D-glucosamine β1-3-galactosyltransferase
porcine trachea and rat intestine
assay, **98**, 125
properties, **98**, 126
UDPgalactose:N-acetyl-D-glucosamine β1-4-galactosyltransferase
bovine milk
assay, **98**, 122
properties, **98**, 124
purification, **98**, 123
UDPgalactose galactosyltransferase, *see* Lactose synthase
UDPgalactose:ovomucoid galactosyltransferase, *see* Glycoprotein 4-β-galactosyltransferase
UDPglucose 4-epimerase
Saccharomyces cerevisiae
assay, **89**, 589
properties, **89**, 592
purification, **89**, 590
UDPglucose-hexose-1-phosphate uridylyltransferase
Escherichia coli
characterization, **87**, 31
exchange reactions, **87**, 27
purification, **87**, 30
steady-state kinetics, **87**, 22
stereochemistry, **87**, 36
uridylylation stoichiometry, **87**, 31
Ultracentrifugation
in actin polymerization assay, **85**, 206
turbidimetric, high-molecular-weight proteins and large macromolecular assemblies, **85**, 572
Ultraviolet absorption
in protein quantitative determination, **91**, 105, 239
Urine
isolation of
arachidonic acid and metabolites with octadecylsilyl silica, **86**, 467
prostaglandin D_2 metabolite, **86**, 561
prostaglandin $F_{2\alpha}$ catabolites, **86**, 552
quantitative determination of
prostacyclin dinor metabolites by gas chromatography–mass spectrometry, **86**, 585
prostaglandin D_2 metabolite by stable isotope dilution mass spectrometry, **86**, 565
prostaglandin $F_{2\alpha}$ catabolites by gas chromatography–mass spectrometry, **86**, 557
prostaglandin $F_{2\alpha}$ major metabolite by
gas chromatography–mass spectrometry, **86**, 579
radioimmunoassay, **86**, 339
thromboxane B_2 dinor metabolite by gas chromatography–mass spectrometry, **86**, 592
UTP–hexose-1-phosphate uridylyltransferase
Entamoeba histolytica
assay, **90**, 552
properties, **90**, 555
purification, **90**, 553
Saccharomyces cerevisiae
assay, **89**, 584
properties, **89**, 588
purification, **89**, 586

V

Vaccines
antibody responses to, analytical techniques, **93**, 60
preparation by glutaraldehyde action on
allergens, **93**, 50
bacterial cells, **93**, 38
toxins, **93**, 25
tumor cells, **93**, 56
venoms, **93**, 44
viruses, **93**, 41
Vacuoles
plant
intact, properties, **96**, 584
preparation
from lysed protoplasts, **96**, 581
by tissue mechanical shearing, **96**, 583
Vanadate
inhibitory effects on lysosomes, **96**, 760

in myosin active-site trapping, **85**, 116
applications, **85**, 122
Veillonella alcalescens
lactate-oxaloacetate transhydrogenase purification, **89**, 370
Venoms
vaccine preparation with glutaraldehyde, **93**, 44
Vesicles
coated
general features, **98**, 337
from human placenta
assay, **98**, 321
components, **98**, 325
isolation, **98**, 321
in intracellular transport, **98**, 9
isolation methods, comparative studies, **98**, 337
from rat liver
characterization, **98**, 330
isolation, **98**, 327
Halobacterium halobium envelope
bacteriorhodopsin and halorhodopsin
assay, **88**, 13
reconstitution, **88**, 14
light-induced pH gradients, measurement with spin-labeled amine and carboxylic acid probes, **88**, 334
preparation, **88**, 11, 427
properties, **88**, 430
proton movement and membrane potential changes, simultaneous measurement, **88**, 356
phospholipid, *see* Phospholipid vesicles
seminal, *see* Seminal vesicles
sonicated disk
preparation, **81**, 578
trapped ^{45}Ca, light release, **81**, 579
Vesicular stomatitis virus
G protein, transport-coupled processing *in vitro*, reconstitution assays, **98**, 37
interaction with MDCK cells, **98**, 264
polarized assembly from cultured epithelial cells, analytical techniques, **98**, 486
Vimentin
antiserum
production, **85**, 495
specificity analysis, **85**, 497
identification and properties in different cell types, **85**, 490
immunofluorescence, **85**, 503
polyacrylamide gel electrophoresis, **85**, 494
purification from tissue culture cells, **85**, 494
Vinblastine
effects on secretory processes in lacrimal and salivary glands, **98**, 178
inhibitory effects on lysosomes, **96**, 762
Viruses, *see also specific viruses*
microinjection into tissue culture cells, **101**, 482
Viscoelastometry
actin filament networks, **85**, 228, 231
Viscosimetry
capillary, in actin polymerization assay, **85**, 185
falling-ball, high-molecular-weight proteins and large macromolecular assemblies, **85**, 572
microtubules during assembly *in vitro*, **85**, 435
rotational, actin filament networks, **85**, 223
Viscosity
effect on bacteriorhodopsin photocycle, **88**, 297
Visual pigments, *see also specific visual pigments*
artificial, generation in electrophysiologically active photoreceptors, **81**, 452
chromophore extraction methods, **88**, 552
extraction detergents, evaluation, **81**, 133
gecko, nonrhodopsin-like properties, **81**, 171
insect
absorbance properties, measurement, **81**, 187
extraction, **81**, 183
microspectrophotometric analysis, **81**, 729
kinetic resonance Raman spectroscopy with microsampling rotating cells, **88**, 659

lateral diffusion in rod disk membrane, measurement, **81**, 660
orientation and concentration, polarized microspectrophotometry, **81**, 642
picosecond spectroscopy, **81**, 354
pressure effects, analytical techniques, **88**, 714
regeneration
associated vitamin A cycle, measurement, **81**, 463
with 11-*cis*-retinol, rod outer segment membrane preparation for, **81**, 506
in rod outer segment bleached by light, **81**, 459
resonance Raman spectroscopy
flow technique, **88**, 627
pump-probe technique, **88**, 626
rod and cone, chromatographic separation, **81**, 160
Vitamin A
analogs, technique for introduction into isolated retina, **81**, 447
binding protein, serum
properties, **81**, 847
purification, **81**, 841
cycle associated with visual pigment regeneration, measurement, **81**, 463
energy levels, determination by two-photon spectroscopy, **88**, 522
storing cells, detection and localization, **81**, 834
Vitamin A_2
synthesis and photochemistry, **88**, 515
Vitamin B_{12}
radioassay, **84**, 369
Voltammetry
differential pulse anodic stripping, in heterogeneous immunoassay, **92**, 437
Von Willebrand factor
immunoradiometric assay, **84**, 51

W

Water
and CO_2, ^{18}O exchange
carbonate dehydratase-catalyzed, **87**, 746
kinetic expressions, **87**, 739
uncatalyzed, **87**, 743
kinetic expressions, **87**, 735
–sucrose gradients, in Golgi complex isolation from rat liver, **98**, 60
Wheat germ
preparation of cell-free protein-synthesizing system, **96**, 39; **101**, 635

X

X-ray diffraction
collagen fibrils, **82**, 138
Halobacterium halobium purple membranes, **88**, 124, 268
light interactions with bacteriorhodopsin, **88**, 282
microtubules during assembly *in vitro*, **85**, 437
nanosecond, application to bacteriorhodopsin, **88**, 276
polysaccharides, **83**, 87
retinal rod outer segments, **81**, 593
X-ray fluorescence spectrometry
metal ions in proteins, **91**, 119
Xylan 1,4-β-xylosidase
Bacillus pumilus
assay, **83**, 632
properties, **83**, 636
purification, **83**, 633
Penicillium wortmanni
assay, **83**, 639
properties, **83**, 642
purification, **83**, 641
D-Xylose
binding protein, *Escherichia coli*
assay, **90**, 473
properties, **90**, 475
purification, **90**, 474
D-Xylose dehydrogenase
pseudomonad MSU-1
assay, **89**, 226
properties, **89**, 227
purification, **89**, 226
β-D-Xylosidase, *see* Xylan 1,4-β-xylosidase
Xylulose 5-phosphate
determination with transketolase, **89**, 19

Y

Yeast
ADCI promoter, in gene expression regulation, **101**, 192
cellular quantification by partition affinity ligand assay, **92**, 515
centromere DNAs, selection procedure for isolation, **101**, 300
chromosomal DNA
construction of specific rearrangements, **101**, 269
contiguous sequences, cloning by integrative transformation, **101**, 290
library construction, **97**, 348
CYC1 promoter fusions, design for analysis of cloned gene expression, **101**, 181
gene disruption techniques, **101**, 202
high-copy vectors, construction using 2-μm circle sequences, **101**, 307
lacZ gene fusions
for cloning regulated genes, **101**, 253
construction and application, **101**, 167
design for analysis of cloned gene expression, **101**, 181
in gene expression analysis, **100**, 293
linear plasmids, construction, **101**, 245
mutant genes
cloning by integrative transformation, **101**, 290
eviction and transplacement, **101**, 211
mutants in polyamine biosynthetic pathway, mass screening, **94**, 104
pet genes
mutant transformation, **97**, 349, 355
confirmation, **97**, 360
product translation *in vitro*, **97**, 352
preparation of cell-free translation system, **101**, 644
purification of
S-adenosylmethionine decarboxylase, **94**, 232
aldehyde dehydrogenase, **89**, 469
galactokinase, **90**, 31
galactose-1-phosphate uridylyltransferase, **89**, 586
glyceraldehyde-3-phosphate dehydrogenase, **89**, 330
ornithine decarboxylase, **94**, 136
phosphofructokinase, **90**, 50
phosphoglucomutase, **89**, 600
phosphoglycerate kinase, **90**, 111, 118
transketolase, **90**, 212
UDPglucose 4-epimerase, **89**, 590
secretory mutants
classification, **96**, 806
cytologic and biochemical characterization, **96**, 808
electrophoretic analysis, **96**, 812
isolation, **96**, 804
STE genes, cloning in 2-μm vectors, **101**, 325
transformation with linear and gapped plasmids, genetic applications, **101**, 228
vectors with negative selection, construction, **101**, 278
Yolk sac
fetal rabbit
anatomic relationships, **93**, 191
preparation of membrane Fc receptor, **93**, 192
Yonetani–Theorell graphical method
for analysis of enzyme active center overlapping subsites, **87**, 500
application to hepatic alcohol dehydrogenase, **87**, 504

Z

Zein
maize, *in vitro* processing, **96**, 716
Zinc
in fructose-bisphosphatase, analysis and removal, **90**, 328
Zinostatin
radioimmunoassay, **84**, 483
clinical utilization, **84**, 487
Zymogen granules
pancreatic
isolation, **98**, 68
membrane characterization, **98**, 72
subfractionation, **98**, 69

Contributor Index

Boldface numerals indicate volume number.

A

Abdulaev, N. G., **88**, 723
Abrahamson, E. W., **81**, 509, 698
Achenjang, Fidelis M., **91**, 633
Ackermann, Karen E., **89**, 593
Adachi, Osao, **89**, 154, 159, 198, 203, 291, 450, 491
Adams, Alois J., **81**, 61
Adams, Andrea C., **93**, 233
Adams, Christopher M., **91**, 633
Addis, John S., **96**, 627
Adelstein, Robert S., **85**, 298, 308
Aders, L., **81**, 90
Adesnik, Milton, **96**, 530
Adiga, P. R., **94**, 335, 429
Aguire, G., **81**, 90
Ahl, Patrick L., **88**, 741
Ahlem, Clarence, **90**, 473
Akhtar, M., **84**, 254
Akino, Toyoaki, **81**, 315
Albersheim, Peter, **83**, 3
Albro, Phillip W., **84**, 619
Alhonen-Hongisto, L., **94**, 247
Allen, S. H. George, **89**, 367
Ally, Ariff, **86**, 511
Alonzo, Nicholas, **91**, 8
Alvarez, Richard A., **81**, 463
Åman, Per, **83**, 3
Ameyama, Minoru, **89**, 20, 141, 149, 154, 159, 187, 194, 198, 203, 291, 450, 491
Ammerer, Gustav, **101**, 192
Anchors, James Michael, **90**, 396
Anderson, Carl W., **101**, 635
Anderson, David J., **96**, 111, 367
Anderson, Gary L., **94**, 286
Anderson, N. Leigh, **92**, 196
Anderson, Richard L., **89**, 93, 248, 275, 562; **90**, 87, 91, 228, 232, 299
Anderson, Robert E., **81**, 297, 800
Andersson, Lars, **89**, 435
Andersson, Leif C., **96**, 281
Angelos, Karen L., **99**, 250
Annamalai, A. E., **90**, 340
Anwar, Rashid A., **82**, 606
Appella, Ettore, **91**, 559
Applebury, M. L., **81**, 354
Argade, Pramod V., **88**, 643
Argan, Carole, **97**, 396
Ariel, Naomi, **84**, 32
Armstrong, Frank B., **89**, 296
Aromatorio, Debra K., **91**, 384
Aronsson, Anne-Charlotte, **90**, 535
Arpin, Monique, **96**, 121; **97**, 408
Arvan, Peter, **98**, 75
Asato, Alfred E., **88**, 506
Ash, John F., **85**, 514
Ashraf, Jaweed, **83**, 311
Astaldi, Giulia C. B., **92**, 39
Atmar, Valerie J., **94**, 147; **99**, 366
Attardi, Giuseppe, **97**, 435
Aumick, Georgina Evans, **83**, 603
Austen, K. Frank, **86**, 252
Aveldaño, Marta I., **86**, 518
Avrameas, Stratis, **92**, 489
Awerbuch, Tamara E., **83**, 444
Ayers, Arthur R., **89**, 129
Azarnia, Roobik, **98**, 537

B

Bacchi, Cyrus J., **94**, 209
Baenziger, Jacques U., **98**, 154
Bai, N. Jayanthi, **90**, 241
Bailey, Allen J., **82**, 360
Baker, Barbara N., **81**, 374
Baker, John R., **83**, 216

Baker, P. F., **98**, 28
Baker, Raymond M., **83**, 444
Bakker, Evert P., **88**, 26
Balch, William E., **98**, 37
Bale, William F., **92**, 277
Balogh-Nair, Valeria, **88**, 496
Baltimore, David, **99**, 373
Bankert, Richard B., **92**, 182, 257
Bárány, Michael, **85**, 624
Barber, George A., **83**, 522
Barden, Roland E., **91**, 633
Barel, Monique, **93**, 155
Barenholz, Y., **81**, 678
Barker, R., **83**, 58; **89**, 64, 73, 79, 83
Barker, Winona C., **91**, 524
Barnes, Wayne M., **101**, 98
Barnett, Leslie, **101**, 3
Barnett, Philip, **82**, 325
Barrnett, Russell J., **96**, 627
Barton, Kenneth A., **101**, 527
Basinger, Scott F., **81**, 772
Basu, Sandip K., **98**, 241
Baumann, Christian, **81**, 29
Baxter, Robert C., **92**, 601
Bayer, Edward A., **83**, 195
Bayley, Hagan, **88**, 74
Baylor, D. A., **81**, 403
Bazán, Nicolás G., **81**, 788
Beasley, Alfreda, **99**, 55
Becher, Brian, **88**, 265
Becker, A., **100**, 183
Beckwith, Jonathan, **97**, 3, 11
Been, Michael D., **100**, 137; **101**, 90
Befani, O., **94**, 314
Beham, Richard A., **99**, 51
Bélanger, Luc, **84**, 19
Bell, Christopher W., **85**, 450
Beltz, Gerald A., **100**, 266
Bender, Rudolf, **90**, 283
Benedict, Charles D., **83**, 530, 535
Ben-Efraim, S., **93**, 24
Benkovic, Stephen J., **90**, 327
Bennett, J. Claude, **91**, 434, 450
Bennett, John, **97**, 487
Bennett, Marylyn Hoy, **82**, 535
Bennett, Pauline M., **85**, 511
Bennett, R. A., **94**, 69
Bennett, Vann, **96**, 313
Benz, Roland, **97**, 286
Berenstein, Elsa H., **92**, 17
Berent, Susan L., **83**, 596
Berg, Richard A., **82**, 372
Bergmeyer, Jurgen, **88**, 89
Berman, Elaine R., **81**, 77, 95
Bernard, Gary D., **81**, 752
Bernlohr, Robert W., **90**, 70, 384
Bernström, Kerstin, **86**, 38
Bevan, Michael, **101**, 98
Bewley, Glenn C., **89**, 296
Bey, Philippe, **94**, 199
Beyer, Thomas A., **83**, 458
Bhakdi, Sucharit, **93**, 409
Bhown, Ajit S., **91**, 434, 450
Bird, Colin R., **94**, 344
Birge, Robert R., **88**, 522
Birnboim, H. C., **100**, 243
Bishayee, Subal, **98**, 555
Bissett, Donald L., **89**, 93, 562; **90**, 87, 228
Bitensky, Mark W., **81**, 526
Bizzini, Bistra, **86**, 258
Black, W. J., **90**, 340
Blair, A. Huntley, **93**, 280
Blake, Diane A., **83**, 127
Blakeley, Robert L., **91**, 49
Blakesley, Robert W., **100**, 3, 368
Blanchard, Rita A., **84**, 60
Blatchly, Richard A., **88**, 491
Blaurock, A. E., **88**, 124
Blobel, Günter, **96**, 38, 84, 111, 367, 589, 663, 682; **98**, 458
Bloj, Bernabé, **98**, 574
Boccù, Enrico, **89**, 282
Boeker, Elizabeth A., **94**, 125, 180
Bogomolni, Roberto A., **88**, 379, 434
Bogorad, Lawrence, **97**, 524
Böhlen, Peter, **91**, 17
Böhni, Peter C., **97**, 311
Boime, Irving, **96**, 777
Bojanovski, Marina, **90**, 115
Bok, Dean, **81**, 290, 763
Bologna, Marcia L., **96**, 305
Bonatti, Stefano, **96**, 512
Bonavida, Benjamin, **93**, 270
Bond, Martha W., **91**, 318
Bonner, William M., **96**, 215
Boos, Winfried, **89**, 59
Bornstein, Paul, **82**, 96
Borys, Tom, **81**, 509
Botstein, David, **100**, 457; **101**, 167

Bownds, Deric, **81**, 211
Boyle, Stephen M., **94**, 117
Bradley, Thomas P., **93**, 270
Braell, William A., **96**, 257
Braiman, Mark, **88**, 648
Bramson, H. Neil, **99**, 93
Brandan, Enrique, **98**, 326
Brandt, Richard B., **89**, 35
Brash, Alan R., **86**, 579, 585
Braun, Dietmar G., **91**, 41
Brennan, Catherine A., **100**, 38
Brenner, Sydney, **101**, 3
Breton, J., **81**, 605
Brewer, Bonita J., **100**, 137
Bridges, C. D. B., **81**, 65, 133, 463
Briggs, J., **92**, 458
Broach, James R., **101**, 307
Brockhaus, Manfred, **83**, 235
Brocklehurst, Keith, **87**, 427
Brodsky, Barbara, **82**, 127
Brody, Richard S., **87**, 213
Brot, Nathan, **101**, 690
Broughton, Alan, **84**, 463
Brown, David R., **100**, 217
Brown, John C., **84**, 359
Brown, Joseph P., **92**, 160
Brown, Michael F., **81**, 709
Brown, Michael S., **98**, 241
Brown, Patricia A., **101**, 278
Brown, William E., **91**, 36, 384
Brunner, Josef, **96**, 386
Brusilow, William S. A., **97**, 188
Buc, Henri, **90**, 60
Buchanan, John M., **87**, 76
Buchwald, Stephen L., **87**, 279
Buckler, Robert T., **92**, 413
Buczyłko, Janina, **90**, 259
Bünemann, Hans, **100**, 400
Burgos, Justino, **89**, 516, 523
Bürk, Robert R., **91**, 247
Burns, Robert A., **96**, 815
Burr, B., **96**, 239, 716
Burr, F. A., **96**, 239, 716
Buss, Janice E., **99**, 7
Butkowski, Ralph J., **82**, 410
Butler, Vincent P., Jr., **84**, 102, 558
Butler, William T., **82**, 339
Buttin, Gérard, **92**, 175
Butz, F., **89**, 278
Buytenhek, M., **86**, 60, 84
Byers, Floyd M., **89**, 29
Byers, Larry D., **89**, 326

C

Cacciapuoti, Giovanna, **94**, 57
Cadman, E., **89**, 83
Cafiso, David S., **88**, 682
Cais, Michael, **92**, 336, 445
Callender, Robert, **88**, 625
Cameron, Richard S., **98**, 75
Candau, Pedro, **88**, 751
Canellakis, Evangelos S., **94**, 193
Canfield, Robert E., **84**, 102
Cantoni, Giulio L., **94**, 365
Caperelli, Carol A., **90**, 327
Caplan, S. Roy, **88**, 26
Carbon, John, **101**, 300
Cardenas, Janet M., **90**, 140
Carlsen, William F., **81**, 668
Carlson, Francis D., **85**, 562
Caroline, Dina F., **89**, 386
Cartenì-Farina, Maria, **94**, 57, 355
Casadaban, Malcolm J., **100**, 293
Casara, Patrick, **94**, 199
Castle, J. David, **98**, 75
Castonguay, Andre, **84**, 641
Caterson, Bruce, **83**, 216
Caulfield, Michael P., **97**, 62
Cenatiempo, Yves, **101**, 690
Ceska, Miroslav, **84**, 238
Chabre, Marc, **81**, 593, 605, 742
Chae, Kun, **84**, 619
Chafouleas, James G., **84**, 138
Chambard, Marianne, **98**, 477
Chambaut-Guérin, A. M., **98**, 175
Chamberlin, M., **101**, 540
Chambliss, Glenn H., **101**, 598
Champoux, James J., **100**, 137; **101**, 90
Chan, Julia Y., **89**, 219
Chan, K.-F. Jesse, **99**, 259
Chandrasekaran, E. V., **89**, 121, 228
Chang, Chin C., **84**, 121
Chang, David Guey-Bin, **86**, 142
Chang, Jui-Yoa, **91**, 41, 79, 455
Chantrapromma, Kan, **94**, 416
Charriaut, Christiane, **93**, 155
Chassy, Bruce M., **90**, 25, 556
Chatterjee, Tapati, **90**, 352

Chen, Hui-Zhu, **101**, 674
Cherbas, Peter T., **100**, 266
Cherry, Richard J., **88**, 248
Chiang, Peter K., **94**, 365
Chilton, Mary-Dell, **101**, 527
Chopra, Inder J., **84**, 272
Chou, Joany, **100**, 293
Chrétien, M., **92**, 292
Christen, Philipp, **91**, 392
Christ-Hazelhof, E., **86**, 77
Christner, James E., **83**, 216
Christopherson, Richard I., **91**, 278
Chua, Nam-Hai, **97**, 502
Chumley, Forrest, **101**, 211
Claeyssens, Marc, **83**, 631, 639
Clark, Charles C., **82**, 346
Clark, E. L., **89**, 79
Clark, George, **84**, 619
Clark, Noel A., **88**, 326, 696
Clark-Curtiss, Josephine E., **101**, 347
Clarke, Louise, **101**, 300
Claus, T. H., **89**, 101; **99**, 212
Clay, Keith L., **86**, 547
Clayton, David A., **97**, 426
Cleland, W. Wallace, **87**, 159, 353, 366, 390, 625, 641
Cleveland, Don W., **96**, 222
Cohen, Isaac, **85**, 373
Cohen, Lawrence, **101**, 650
Cohen, Philip, **99**, 243, 337
Cohen, Seymour S., **94**, 279
Cohen, Stanley, **99**, 379
Cohn, Murray S., **94**, 104, 231
Cohn, Zanvil A., **98**, 404
Colbert, Robert A., **91**, 190
Colca, Jerry R., **98**, 182
Coleman, Jack, **97**, 124
Coligan, John E., **91**, 413
Coll, R. J., **87**, 66
Collier, R. John, **100**, 342
Collins, Jimmy H., **85**, 357
Colman, David R., **96**, 378
Colowick, Sidney P., **100**, xxv
Compton, P. D., **87**, 66
Cone, Richard A., **88**, 741
Connolly, Bernard A., **87**, 197
Connor, Robert J., **84**, 607
Contreras, M. Angeles, **92**, 277
Converse, Carolyn A., **96**, 244
Cooke, Peter, **85**, 277
Cooke, Roger, **85**, 574
Cooper, Alan, **81**, 285; **88**, 667
Cooper, Herbert L., **94**, 458
Cooper, John A., **85**, 182, 211, 316
Cooper, Jonathan A., **99**, 387
Cooperman, Barry S., **87**, 526
Copeland, B. R., **90**, 223
Corbin, Jackie D., **99**, 55, 168, 227
Corder, Clinton N., **86**, 135
Corey, E. J., **86**, 252
Corless, Joseph M., **81**, 585
Cornish-Bowden, Athel, **91**, 60
Costello, M. Joseph, **81**, 585
Coward, James K., **94**, 286
Cozzarelli, Nicholas R., **100**, 171
Craig, Susan W., **85**, 316
Creeger, Elva S., **83**, 326
Creek, Kim E., **98**, 290
Crescitelli, Frederick, **81**, 171
Crespi, Henry L., **88**, 3
Cromlish, J. A., **89**, 237, 501
Crouch, Rosalie K., **88**, 175
Crouse, Gray F., **101**, 78
Crow, V. L., **90**, 165
Crowell, Richard L., **96**, 443
Crowlesmith, Ian, **97**, 112
Crowley, Jane F., **83**, 368
Cumming, D. A., **89**, 101
Cunningham, Leon W., **82**, 509
Curtis, Donna R., **81**, 251
Curtiss, Roy, III, **101**, 347
Cusanovich, Michael A., **81**, 443

D

Dabrowska, Renata, **99**, 279
Dabrowski, Janusz, **83**, 69
Dachà, Marina, **90**, 3
da Costa, Maria, **84**, 458
Daemen, Frans J. M., **81**, 110, 223, 320
Dagher, Shawky M., **89**, 310
Dagley, Stanley, **90**, 272, 277
Dahl, Gerhard, **98**, 537
Dahlberg, Albert E., **100**, 326
Dahms, A. Stephen, **89**, 226; **90**, 269, 294, 302, 473
Daiss, John L., **98**, 337
Dallner, Gustav, **96**, 542
Dalzoppo, Daniele, **91**, 311

Daniels, Charles J., **97**, 146
Daniels, Gary R., **94**, 147
Darvill, Alan G., **83**, 3
Das, Manjusri, **98**, 555
Date, Takayasu, **97**, 46, 57
Datema, Roelf, **83**, 432
Daum, Günther, **97**, 311
Daves, G. Doyle, Jr., **94**, 48
Davidson, W. S., **89**, 501
Davies, Peter J. A., **85**, 322
Davila, Mark, **89**, 228
Davis, Bernard D., **97**, 62
Davis, Rowland H., **94**, 36, 112
Day, Jeffrey S., **86**, 229
Dayhoff, Margaret O., **91**, 524
Deal, William C., Jr., **89**, 310
Dearborn, Dorr G., **91**, 570
De Bazán, Haydée E. Pascual, **81**, 788
De Blas, Angel L., **92**, 36
De Bruyne, Clement K., **83**, 631, 639
Dedman, John R., **84**, 138
Deese, Alan J., **81**, 709
Degen, Jay, **91**, 367
De Grip, Willem J., **81**, 197, 223, 256
Delaage, M. A., **84**, 350
de Leon, Stella, **96**, 570
Deleyn, François, **83**, 639
DeLisi, Charles, **93**, 95
Della Ragione, Fulvio, **94**, 57, 321
DeLuca, Marlene, **92**, 426
DeMaine, Margaret M., **90**, 327
Demetriou, Achilles A., **94**, 396
Dencher, Norbert A., **88**, 5, 31, 167, 420
Deng, Guo-ren, **100**, 96
Dennis, David T., **89**, 319
DePierre, Joseph W., **96**, 542
de Reggi, M. L., **84**, 350
De Rosa, Mario, **94**, 355
Devaux, Philippe F., **81**, 703
deVera, A., **101**, 540
De Vos, Guido, **97**, 524
De Vries, G. W., **81**, 85, 517
DeWitt, David L., **86**, 229, 240
Diamond, Alan, **100**, 431
Dickie, N., **84**, 254
Dieckmann, Carol L., **97**, 355, 361
Dignam, John D., **101**, 582
Dimond, Randall L., **96**, 815; **98**, 159
Dirden, Bernadette, **99**, 250
Distler, Jack J., **83**, 392; **98**, 304
Dixon, Ross, **84**, 490
Dobberstein, Bernhard, **96**, 325, 692
Donald, Alan, **90**, 269, 294, 302
Donehower, Ross C., **84**, 447
Donelson, John E., **100**, 116
Doran, K., **100**, 368
Douglas, Michael G., **97**, 337, 344
Douillard, J. Y., **92**, 168
Downer, Nancy W., **88**, 673, 681
Doyle, Matthew J., **92**, 432
Dratz, Edward A., **81**, 116, 124, 275, 685, 709, 806
Dray, Fernand, **86**, 258, 297
Drenthe, Erik H. S., **81**, 320
Dreyer, William J., **91**, 399
Dreyfus, H., **81**, 304
Dryburgh, Jill R., **84**, 359
Drzymala, Robert E., **81**, 660
DuBois, Garrett C., **91**, 559
Dudock, Bernard, **100**, 431; **101**, 635
Duggleby, Ronald G., **89**, 319
Dundas, Ian, **88**, 360
Dunn, William A., **98**, 225
Dunphy, William H., **98**, 37

E

Eakin, Richard M., **81**, 17
Easterby, John S., **90**, 11
Ebrey, Thomas G., **88**, 462, 516
Eckfeldt, John H., **89**, 474
Eckstein, Fritz, **87**, 197
Edström, Jan-Erik, **96**, 496
Edwards, N. S., **86**, 552, 604
Edwards, Ross B., **81**, 39
Egge, Heinz, **83**, 69
Eibl, Hansjorg, **98**, 623
Eichner, R. D., **89**, 359
Eickbush, Thomas H., **100**, 266
Eikenberry, Eric F., **82**, 127
Eisenberg, Evan, **85**, 709
Eisenstein, Laura, **88**, 297
Elbein, Alan D., **89**, 547; **98**, 135
Elfvin, Myra J., **85**, 149
Eling, Thomas, **86**, 511
Elings, V. B., **92**, 458
Elliott, Bruce E., **93**, 113
Ellis, Keith J., **87**, 405
El-Maghrabi, M. R., **89**, 101; **99**, 212

Eloranta, Terho, **94**, 257, 270, 276
El-Sayed, M. A., **88**, 617
Elting, James J., **83**, 408, 429
Elzinga, Marshall, **91**, 8
Endo, Yasuo, **94**, 42
Engelman, D. M., **88**, 81
Englander, Joan J., **88**, 673
Engvall, Eva, **82**, 803; **84**, 3
Enoch, Harry G., **89**, 537
Eppenberger, Hans M., **85**, 139
Erickson, Ann H., **96**, 38
Eriksson, Karl-Erik, **89**, 129
Erlichman, Charles, **84**, 447
Erlichman, Jack, **99**, 176, 187
Eschenbruch, Margret, **91**, 247
Eshdat, Yuval, **83**, 386
Esmon, Brent, **96**, 802
Esnouf, M. P., **89**, 579
Eubank, William B., **90**, 97
Evanczuk, Ann Tavormina, **81**, 532

F

Fager, Lei Yen, **81**, 160
Fager, Roger S., **81**, 160, 288, 577
Fairwell, Thomas, **91**, 502
Falardeau, Pierre, **86**, 585
Falkenhein, Sandra F., **86**, 416, 655
Fambrough, Douglas M., **96**, 331
Farber, Debora B., **81**, 496, 551
Faris, Barbara, **82**, 615
Farnsworth, Christopher C., **81**, 124
Farquhar, Marilyn Gist, **98**, 1, 203
Farrell, Kevin W., **85**, 385
Fay, F. S., **85**, 284
Feeney-Burns, Lynette, **81**, 95
Feingers, J., **92**, 366
Feinman, Jeffrey, **98**, 555
Feldman, Ricardo A., **96**, 3
Felgner, Philip L., **97**, 469
Feracci, Hélène, **96**, 406
Feramisco, James R., **85**, 514
Ferenci, Thomas, **90**, 459
Ferrendelli, J. A., **81**, 85, 517
Ferri, G., **89**, 316
Ferro, Adolph J., **94**, 361
Ferro-Novick, Susan, **96**, 802
Fiebach, Helmar, **92**, 227
Field, Charles, **96**, 802
Fink, A. L., **87**, 66
Fink, Gerald R., **101**, 211
Fink, Mary Lynn, **94**, 347
Finne, Jukka, **83**, 269
Fire, Andrew, **101**, 568
Fischer, Edmond H., **94**, 180
Fischer, H. David, **98**, 290
Fischer, Itzhak, **101**, 629
Fischer, Stuart G., **100**, 424
Fishbein, James C., **91**, 466
Fisher, Knute A., **88**, 230
Fisher, Paul A., **96**, 589
Fitzpatrick, F. A., **86**, 286, 306
Fleer, Eduard A. M., **98**, 623
Fleischer, Becca, **98**, 60, 326
Fleischer, Norman, **99**, 187
Fleischer, Sidney, **98**, 326, 599, 623
Fleischman, Darrell, **81**, 522
Flockhart, David A., **99**, 14
Floss, Heinz G., **87**, 126
Flower, Roderick J., **86**, 91, 477
Flynn, T. G., **89**, 237, 501
Foellmer, Harald, **90**, 498
Folk, J. E., **87**, 36; **94**, 347, 451, 458
Fong, Shao-Ling, **81**, 65, 133, 251
Fontana, Angelo, **89**, 282; **91** ,311
Forde, B. G., **97**, 476
Fordyce, Alison M., **90**, 77
Fornaini, Giorgio, **90**, 3
Forsyth, James M., **88**, 276
Foster, Judith Ann, **82**, 559, 665, 731, 762
Fox, Philip C., **92**, 17
Fozard, John R., **94**, 213
Frade, Raymond, **93**, 155
Franke, Werner W., **96**, 597
Frankel, Robert D., **88**, 276
Franzblau, Carl, **82**, 588, 615
Franzén, Lars-Erik, **83**, 3
Fraser, Clarence L., **85**, 450
Frederiksen, Dixie W., **85**, 292, 371
Freeman, David S., **84**, 516
Frei, Erich, **100**, 309
Freimüller, Barbara, **89**, 301
Freitag, Helmut, **97**, 286
Frey, Perry A., **87**, 20, 213
Fries, Erik, **98**, 37
Frischauf, Annemarie, **101**, 78
Fuchs, Joachim, **90**, 498
Fuchs, Roy, **100**, 3
Fujita, Kazunobu, **94**, 185

Fujita, Shinobu C., **89**, 335
Fukasawa, T., **89**, 584
Fukuda, Michiko N., **81**, 214
Fukumoto, Joseph M., **88**, 306
Fuller, G. M., **92**, 577
Fung, Bernard Kwok-Keung, **81**, 269; **96**, 617
Furie, Barbara C., **84**, 60
Furie, Bruce, **84**, 60
Furlong, Clement E., **90**, 223, 467
Furthmayr, Heinz, **96**, 268
Furukawa, Ken, **83**, 373
Furuta, Shuichi, **89**, 414

G

Gabriel, Othmar, **83**, 332
Gabrion, Jacqueline, **98**, 477
Gaetani, Sancia, **96**, 3, 121
Gahmberg, Carl G., **96**, 281
Galletti, Patrizia, **94**, 57, 73
Galli, Claudio, **86**, 635
Gallop, Paul M., **82**, 571
Galloway, Cynthia J., **98**, 545
Gambacorta, Agata, **94**, 355
Gamon, Konrad, **97**, 112
Ganem, Bruce, **94**, 416
Garoff, Henrik, **96**, 496
Garrels, James I., **100**, 411
Garrison, James C., **99**, 20
Gärtner, Wolfgang, **88**, 546
Garvey, Justine S., **84**, 121
Gasior, Eugeniusz, **101**, 644
Gaskin, Felicia, **85**, 433
Gasser, Susan M., **97**, 245, 329
Gates, Frederick T., III, **91**, 413
Gauger, John A., **86**, 229
Gay, Nicholas J., **97**, 195
Gay, Steffen, **82**, 3
Gee, Adrian P., **92**, 403; **93**, 339, 375
Gee, Robert, **90**, 515
Gehring, Heinz, **91**, 392
Georgatsos, John G., **94**, 162
Gerber, Gerhard E., **88**, 56
Gergely, John, **85**, 1
Gerrard, Jon M., **86**, 642
Gheţie, V., **92**, 523
Ghose, Tarun I., **93**, 280
Gibbons, I. R., **85**, 450
Gibson, Frank, **97**, 176
Gierow, Peter, **89**, 44
Gilboa-Garber, Nechama, **83**, 378
Gill, Gordon N., **99**, 206
Gilman, M., **101**, 540
Gilula, Norton B., **98**, 510
Ginsburg, Victor, **83**, 235
Gjika, Hilda B., **84**, 516
Glass, David B., **99**, 119
Gleeson, Paul, **98**, 98
Glonek, Thomas, **85**, 624
Glorioso, Joseph, **101**, 402
Glössl, Josef, **83**, 559
Godchaux, Walter, III, **81**, 52, 564
Goebel, Werner, **88**, 374
Gold, M., **100**, 183
Goldberg, Erwin, **89**, 351
Goldin, Alan L., **101**, 402
Goldman, A., **88**, 81
Goldman, David, **96**, 230
Goldsmith, Timothy H., **81**, 34
Goldstein, Irwin J., **83**, 127, 368
Goldstein, Joseph L., **98**, 241
Goodman, Joel M., **97**, 130
Goodno, Charles C., **85**, 116
Gordon, Jon W., **101**, 411
Goss, Neil H., **87**, 51
Gottesman, Michael, **99**, 197
Gottschalk, Gerhard, **90**, 82, 283
Gowland, Peter, **100**, 309
Gracy, Robert W., **89**, 550
Graessmann, A., **101**, 482, 492
Graessmann, M., **101**, 482
Graff, Gustav, **86**, 376, 386
Grandi, Claudio, **91**, 311
Granger, B. L., **85**, 488
Granot, Joseph, **99**, 93
Granström, Elisabeth, **86**, 306, 320, 339, 493
Gratzer, W. B., **85**, 475
Graves, Donald J., **99**, 259, 268
Gray, Gary R., **83**, 248
Green, Pamela, **96**, 74
Greene, Lois E., **85**, 709
Greenfield, Lawrence, **100**, 342
Greengard, Paul, **99**, 154
Grez, Manuel, **100**, 285
Griffiths, G., **96**, 466
Grimm, Elizabeth A., **93**, 270
Grinnell, Frederick, **82**, 499, 535
Gruner, S., **88**, 282

Guarente, Leonard, **101**, 181
Gubbins, Earl J., **97**, 524
Guerrillot, L., **89**, 484
Guesdon, Jean-Luc, **92**, 489
Gumport, Richard I., **100**, 38
Gunsalus, Robert P., **97**, 188
Guo, Li-He, **100**, 60
Guranowski, Andrzej B., **94**, 365
Gurdon, J. B., **101**, 370
Gutowski, Jan A., **84**, 264
Guyer, Mark S., **101**, 362

H

Hack, E., **97**, 476
Hafner, Edmund H., **94**, 83
Hafner, Edmund W., **94**, 91, 117, 219
Hagenaars, A. M., **84**, 223
Hagins, Frances M., **81**, 43
Haid, Albert, **96**, 192
Haigh, Wallace B., Jr., **96**, 298
Haigler, Harry T., **98**, 283
Haimovich, Joseph, **93**, 147
Haley, Boyd E., **91**, 613
Hall, Michael O., **81**, 290
Hallenbeck, Patrick, **99**, 250
Halliday, June W., **84**, 148
Halsall, H. Brian, **92**, 432
Hama, Takao, **89**, 469
Hamanaka, Toshiaki, **88**, 268
Hammarström, Sven, **86**, 38, 202
Han, Peter F., **90**, 334
Hanahan, Douglas, **100**, 333
Hanfland, Peter, **83**, 69
Hanna, Ragy, **89**, 121
Hansen, Bruno, **92**, 309
Hansen, David E., **87**, 279
Hansen, Harald S., **86**, 156
Hanson, Barbara A., **83**, 137
Hanson, William L., **94**, 209
Hara, Reiko, **81**, 190, 827
Hara, Tomiyuki, **81**, 190, 827
Harada, T., **89**, 420
Haralson, Michael A., **82**, 225
Harding, Roy W., **89**, 386
Hardy, Simon J. S., **97**, 70
Hargrave, Paul A., **81**, 211, 214, 251; **90**, 259
Häring, P., **92**, 242
Harkins, Richard N., **91**, 367
Hárosi, Ferenc I., **81**, 642
Harper, Timothy, **86**, 30
Harris, Ben G., **90**, 21, 44, 251
Harris, Edward D., Jr., **82**, 423
Harris, R. P., **89**, 579
Harth, S., **81**, 304
Hartley, James L., **100**, 116
Hartman, Fred C., **90**, 522
Hartshorne, David J., **99**, 279
Hartwig, John H., **85**, 480
Hascall, Vincent C., **82**, 769
Hashimoto, Takashi, **89**, 414
Hassett, Annemarie, **87**, 279
Hathaway, Gary M., **99**, 308, 317
Hatton, Mark W. C., **89**, 172
Hauschka, Peter V., **82**, 65
Hauser, Hansjörg, **100**, 285
Hauswald, Charles L., **90**, 299
Hay, Rick, **97**, 245
Hayaishi, Osamu, **86**, 73, 147
Hayashi, Shin-Ichi, **94**, 154, 185
Hayashi, Yoko, **86**, 269
Hayes, M. L., **89**, 64, 83
Haylett, Thomas, **83**, 357
Hayman, Edward G., **82**, 803
Hays, John B., **90**, 447
Hays, Sally E., **84**, 532
Heaney-Kieras, J., **94**, 441
Hebda, Patricia A., **83**, 522
Heineman, William R., **92**, 432
Heinz, Fritz, **89**, 301, 497
Helenius, Ari, **98**, 260
Heller, John S., **94**, 193
Hellerqvist, Carl G., **82**, 530
Hellström, Ingegerd, **92**, 160
Hellström, Karl Erik, **92**, 160
Hemmings, Brian A., **99**, 337
Henderson, Lou, **91**, 254
Hendriks, R., **84**, 542
Henkin, Tina M., **101**, 598
Hennig, Bernd, **97**, 261
Henshaw, Edgar C., **101**, 616
Hermodson, Mark, **91**, 352
Herrmann, Steven H., **92**, 86
Hertzberg, Elliot L., **98**, 501
Herzog, Volker, **98**, 203, 447
Hess, Benno, **88**, 180, 193, 254
Hewick, Rodney M., **91**, 399
Heykants, J., **84**, 542
Heyn, Maarten P., **88**, 5, 31

Higgins, Paul J., **93**, 78
Hildebrand, Eilo, **88**, 420
Hill, Robert L., **83**, 458; **98**, 309
Hillier, Alan J., **89**, 362
Hinkins, Susan, **99**, 279
Hipps, Paul P., **89**, 593
Hirabayashi, T., **89**, 420
Hiraki, Kenji, **88**, 268
Hirn, M. H., **84**, 350
Hirs, C. H. W., **91**, 3
Hirschberg, Carlos B., **83**, 444
Ho, Hsu-Tso, **87**, 213
Ho, Yen-sen, **101**, 123
Hofer, H. Werner, **90**, 44
Hoffman, Jerald L., **94**, 223
Hoffman, Rosemary T., **81**, 772
Hoffman, T., **92**, 168
Hoffmann, R., **85**, 284
Hofmann, Eberhard, **90**, 49
Hogg, Robert W., **90**, 463
Höjeberg, Bo, **88**, 74
Holt, G., **97**, 301
Hölttä, Erkki, **94**, 306
Hong, Keelung, **81**, 144
Honig, Barry, **88**, 462
Hood, Leroy E., **91**, 227, 318, 399, 486
Hopewell, William D., **88**, 306
Höpner, Th., **89**, 531
Horecker, B. L., **90**, 340, 345
Horrocks, Lloyd A., **86**, 518
Hortin, Glen, **96**, 777
Hosey, M. Marlene, **90**, 352
Hosomi, Saburo, **89**, 232
Hough, C. J., **100**, 368
Houghten, Richard A., **91**, 549
Howard, Gary C., **91**, 36
Howles, Phillip N., **88**, 426
Howley, Peter M., **101**, 387
Hoyer, Leon W., **84**, 51
Hsiao, Chu-Lai, **101**, 300
Hsieh, Tao-shih, **100**, 161
Huang, Charles Y., **87**, 509
Huang, Henry V., **91**, 318
Huang, Jungi, **101**, 433
Huang, Kuo-Sen, **88**, 74
Hubbard, Ann L., **98**, 225
Hubbard, S. Catherine, **83**, 399
Hubbard, Walter C., **86**, 366, 571, 607
Hubbell, Wayne L., **81**, 144, 269; **88**, 682
Huber, Mary M., **86**, 416, 655
Hübscher, Ulrich, **91**, 263
Hudson, Billy G., **82**, 410
Hugo, Hasso V., **90**, 82
Huisman, William, **90**, 294, 473
Hultin, Tore, **91**, 359
Hunkapiller, Michael W., **91**, 227, 318, 399, 486
Hunt, Lois T., **91**, 524
Hunt, Tim, **96**, 50
Hunter, Tony, **99**, 387
Hurley, James B., **81**, 542; **96**, 617
Hurwitz, Jerard, **100**, 217
Hutner, Seymour H., **94**, 209

I

Ikawa, Miyoshi, **89**, 145
Imahori, Kazutomo, **89**, 335; **90**, 126, 179; **94**, 301, 303
Imamura, Kiichi, **90**, 150
Inbar, D., **92**, 366
Inglis, A. S., **91**, 26, 324, 443
Inman, John K., **91**, 559
Inoue, Makoto, **86**, 370
Inouye, Masayori, **96**, 74; **97**, 124
Inouye, Sumiko, **97**, 124
Isenberg, Irvin, **91**, 214
Ishii, Shin-Ichi, **91**, 378
Ishimoto, Nobutoshi, **83**, 515, 519
Israel, Mervyn, **94**, 411
Ito, Eiji, **83**, 515, 519
Ito, Koreaki, **97**, 57
Ito, Makoto, **83**, 619
Ito, Shosuke, **94**, 463
Ito, Yoichiro, **91**, 335
Ivatt, Raymond J., **83**, 399
Iwasa, Tatsuo, **88**, 163
Izumori, Ken, **89**, 547

J

Jackson, Richard J., **96**, 50
Jackson, Robert C., **96**, 784
Jacobs, Diane M., **84**, 264
Jacobs, Kenneth A., **100**, 266
Jago, G. Richard, **89**, 362
Jakschik, Barbara A., **86**, 30
Jänne, J., **94**, 247

Jans, David A., **97**, 176
Jarabak, Joseph, **86**, 126, 152, 163
Järnefelt, Johan, **83**, 311
Jentoft, Neil, **91**, 570
Jergil, Bengt, **89**, 44
Jeyaseelan, Kandiah, **89**, 399
Ji, Tae H., **91**, 580
Jimeno-Abendano, J., **90**, 288
Johnson, Joe, Jr., **90**, 334
Johnson, Roy A., **86**, 459
Jokinen, Mikko, **96**, 281
Jongeneel, Cornelis Victor, **100**, 144
Jordan, Kevin B., **96**, 815
Josefsson, Lars-Göran, **97**, 77
Joshi, J. G., **89**, 599
Jou, Yi-Her, **92**, 257
Jourdian, George W., **83**, 392; **98**, 304
Juhl, Henning, **99**, 37
Junge, Wolfgang, **81**, 569
Just, Wilhelm W., **91**, 281, 299
Juszczak, Elizabeth, **81**, 251
Juy, Dominique, **92**, 175

K

Kääriäinen, Leevi, **96**, 453
Kafatos, Fotis C., **100**, 266
Kagan, Herbert M., **82**, 588, 637
Kahn, Axel, **90**, 131
Kaiser, E. T., **99**, 93
Kameji, Takaaki, **94**, 154
Kamel, R. S., **84**, 422
Kamo, Naoki, **88**, 356
Kamp, H. H., **98**, 581
Kämpe, Olle, **96**, 281
Kanda, Yoshikazu, **81**, 840
Kanfer, Julian N., **83**, 191
Kaplan, Donald A., **100**, 342
Kaplan, Henry S., **92**, 3
Kaplan, Karen L., **84**, 83, 93
Kaplan, Michael W., **81**, 655
Kaplan, Nathan O., **100**, xxv
Kapp, Oscar H., **91**, 259
Käpyaho, K., **94**, 247
Karhi, Kimmo K., **96**, 281
Karn, Jonathan, **101**, 3
Karnovsky, Manfred L., **90**, 396
Karr, Steven B., **82**, 731
Karr, Timothy L., **85**, 416, 439
Karsten, Uwe, **92**, 227
Kates, M., **88**, 98
Kato, Kanefusa, **92**, 345
Kato, Keitaro, **96**, 764
Kaupp, U. Benjamin, **81**, 569
Kawamura, Takashi, **83**, 515, 519
Kawashima, Seiichi, **94**, 301, 303
Kay, Cyril M., **85**, 677
Keen, James H., **98**, 359
Keith, David A., **82**, 571
Keller, Gilbert A., **98**, 368
Kellermann, Odile K., **90**, 459
Kelly, Grahame J., **90**, 371
Kendrick-Jones, John, **85**, 364
Kennedy, Eugene P., **98**, 596
Keränen, Sirkka, **96**, 453
Kerbel, Robert S., **93**, 113
Kerlavage, Anthony R., **99**, 140
Kersters-Hilderson, Hilda, **83**, 631
Kester, Harry, **89**, 399
Keszthelyi, Lajos, **88**, 287
Khorana, H. Gobind, **88**, 56, 74
Kikkawa, Ushio, **99**, 288
Kilbride, Paul, **81**, 547
Killion, Robert B., **90**, 490
Kimball, Edward S., **91**, 413
Kimura, James H., **82**, 769
Kimura, Kinuko, **89**, 258; **90**, 235
Kindahl, Hans, **86**, 306, 320, 339
Kindl, H., **96**, 700
King, Glen I., **88**, 241
Kingston, R., **101**, 540
Kinsey, Robert A., **88**, 310
Kintanar, Augustin, **88**, 310
Kitamikado, Manabu, **83**, 610, 619
Kito, Yuji, **81**, 43, 167
Kivirikko, Kari I., **82**, 245
Klee, Claude B., **85**, 298
Klein, Udo, **83**, 559
Kleinman, Hynda K., **82**, 503
Klenk, Hans-Dieter, **96**, 434
Kliger, David S., **88**, 533
Klinman, Judith P., **87**, 711
Klotz, Joan L., **84**, 194
Knecht, David A., **96**, 815; **98**, 159
Knecht, René, **91**, 41
Knight, D. E., **98**, 28
Knight, Peter J., **85**, 9, 12
Knowles, Jeremy R., **87**, 279; **89**, 108
Knowles, William J., **96**, 305

Knudsen, Peter J., **81**, 144
Knutson, Victoria P., **96**, 423
Kobata, Akira, **83**, 105, 263, 625
Kobayashi, T., **81**, 368
Kochetov, G. A., **89**, 19, 43; **90**, 209
Kochibe, Naohisa, **83**, 373
Kochman, Marian, **90**, 254, 259
Kohn, Leonard D., **89**, 341
Kondor-Koch, Claudia, **96**, 496
Konigsberg, William H., **91**, 254
Konish, M., **98**, 301
Konishi, Tetsuya, **88**, 202
Kono, Tetsuro, **98**, 431
Kontoyiannidou, Voula, **83**, 132
Kopperschläger, Gerhard, **90**, 49
Koren, Hillel S., **93**, 245
Korenbrot, Juan I., **88**, 45
Korenstein, Rafi, **88**, 180, 193
Koretz, Jane F., **85**, 20
Korff, Jeffrey M., **86**, 152
Korn, Edward D., **85**, 357
Kornberg, Arthur, **101**, xxv
Kornfeld, Stuart, **83**, 416
Kosman, Daniel J., **89**, 163
Kotagal, Nirmala, **98**, 182
Kotlarz, Denise, **90**, 60
Kozaki, S., **84**, 223
Krah, David L., **96**, 443
Krantz, Mark J., **84**, 32
Krause, Sonja, **85**, 160
Krebbers, Enno, **97**, 524
Krebs, Edwin G., **99**, 162
Kreibich, Gert, **96**, 378, 520, 530, 764
Kresse, Hans, **83**, 559
Kreutz, W., **88**, 729
Kreuzer, Kenneth N., **100**, 144
Krietsch, Wolfgang K. G., **90**, 103, 110
Kristofferson, David, **85**, 439
Krohne, Georg, **96**, 597
Krolick, K. A., **93**, 333
Kropf, Allen, **81**, 384
Kropinski, A. M. B., **94**, 422
Krueger, Judy H., **100**, 501
Krupp, Michael N., **96**, 423
Kruse, C., **96**, 700
Krusius, Tom, **83**, 269
Krutzsch, Henry C., **91**, 511
Kuehn, Glenn D., **94**, 147; **99**, 366
Kühn, Hermann, **81**, 489, 556
Kukuruzinska, Maria A., **90**, 417, 431
Kula, Maria-Regina, **89**, 424, 527; **90**, 319
Kulbe, Klaus D., **90**, 115, 498
Kulczycki, Anthony, Jr., **93**, 178
Kulkarni, Padmaja N., **93**, 280
Kumagai, Akira, **86**, 17
Kumar, Nalin M., **98**, 510
Kumazaki, Takashi, **91**, 378
Kuntz, Günter W. K., **90**, 103, 110
Kurylo-Borowska, Z., **94**, 441
Kurz, G., **89**, 176, 261, 278
Kuschmitz, Dietrich, **88**, 254
Kushwaha, S. C., **88**, 98
Kvist, Sune, **96**, 325
Kwiatkowska, Danuta, **90**, 254
Kyriakidis, Dimitrios A., **94**, 162, 193
Kyte, Jack, **91**, 367

L

Lacy, Paul E., **98**, 182
Laferté, Suzanne, **84**, 32
Laine, Roger A., **83**, 311
Lancashire, Christine L., **85**, 316
Lancaster, Jack R., Jr., **88**, 412
Land, Hartmut, **100**, 285
Landau, Burton J., **96**, 443
Landick, Robert C., **97**, 146
Landon, J., **84**, 422
Lands, William E. M., **86**, 69
Lane, M. Daniel, **96**, 423
Lange, Gary, **82**, 744
Langer, Helmut, **81**, 182, 729
Langone, John J., **84**, 409, 628, 657; **92**, 403
Langworthy, Thomas A., **88**, 396
Lanyi, Janos K., **88**, 439
Larrinua, Ignacio M., **97**, 524
Larrivee, Denis, **81**, 34
Larsson, Per-Olof, **89**, 457
Latzko, Erwin, **90**, 371
Law, Ming-Fan, **101**, 387
Lazarides, E., **85**, 488
Lazarow, Paul B., **96**, 721
Leaver, C. J., **97**, 476
Leavitt, Alan L., **89**, 3, 9
Lebherz, Herbert G., **90**, 490
Leclair, S., **85**, 284
Ledeen, Robert W., **83**, 139
Lee, Chi-Yu, **89**, 252, 296, 351, 445, 559; **90**, 16, 121

Lee, James C., **85**, 376
Lee, Rehwa H., **81**, 496
Lee, Reiko T., **83**, 289, 299
Lee, Yuan Chuan, **83**, 278, 289, 294, 299, 306
Legrain, Pierre, **92**, 175
Lehrach, Hans, **96**, 496; **101**, 78
Lehrman, Mark A., **98**, 309
Leibach, Fredrich H., **90**, 200, 407,
Leicht, Wolfgang, **89**, 497
Leighton, Terrance, **88**, 369
Leissing, Nancy, **89**, 135
Lemke, Horst-Dieter, **88**, 89
Lennarz, William J., **83**, 408, 429; **98**, 91
Leonardi, C. L., **96**, 184
Lernmark, Åke, **92**, 309
Lessmann, D., **89**, 261
Lester, Robert L., **83**, 132, 137; **89**, 537
Leuthard, Peter, **91**, 247
Leventhal, Judith M., **101**, 598
Levin, Daniel H., **99**, 346
Levine, Lawrence, **86**, 246, 252
Levine, Myron, **101**, 402
Levine, Rhea J. C., **85**, 149
Levine, Rodney L., **99**, 402
Levy, Abraham, **100**, 309
Lewis, Aaron, **88**, 217, 561, 659
Lewis, Robert A., **86**, 252
Li, Choh Hao, **91**, 549
Li, Su-Chen, **83**, 588, 610
Li, Yu-Teh, **83**, 588, 610, 619
Liao, Mei-June, **88**, 74
Libby, Paul R., **94**, 325, 329
Liebman, Paul A., **81**, 532, 660
Light, Nicholas D., **82**, 360
Lin, Mow, **88**, 542
Lincoln, Thomas M., **99**, 62
Lind, Christina, **88**, 74
Linde, Susanne, **92**, 309
Lindenmaier, Werner, **100**, 285
Lindley, Edward V., **88**, 426
Linnane, Anthony W., **97**, 294
Liou, Gregory I., **81**, 133
Litman, Burton J., **81**, 57, 150, 309, 577, 629, 678
Liu, Guiling, **101**, 433
Liu, Leroy F., **100**, 133
Liu, Robert S. H., **81**, 694; **88**, 506
Livi, George P., **96**, 815
Lizardi, Paul M., **96**, 24
Lobelle-Rich, Patricia, **90**, 97, 552
Lodish, Harvey F., **96**, 257
Loew, Ellis R., **81**, 647
Lolley, Richard N., **81**, 37, 496, 551
London, Erwin, **88**, 74
London, Irving M., **99**, 346
Long, Mary V., **90**, 522
Louvard, Daniel, **98**, 379
Lovell, Stephen J., **85**, 12
Lowey, Susan, **85**, 55
Lowry, O. H., **81**, 85
Loyter, A., **101**, 492
Lozier, Richard H., **88**, 133, 282
Lujan, Eva, **91**, 227
Lusis, Aldons J., **96**, 557
Luster, Michael I., **84**, 619
Lusty, Carol J., **97**, 396

M

Maas, Richard L., **86**, 592
MacDonald, Russell E., **88**, 426
MacElroy, Robert D., **88**, 772; **89**, 571
MacGregor, John S., **90**, 340
MacKay, Vivian L., **101**, 325
MacLennan, David H., **96**, 570
Maclouf, Jacques, **86**, 273, 612
MacNeal, Robert K., **85**, 416
Magnani, John L., **83**, 235
Magnani, Mauro, **90**, 3
Magolda, Ronald L., **86**, 400
Mahler, Henry R., **97**, 374
Mahoney, Walter C., **91**, 352
Maier, E., **89**, 176
Majerus, Philip W., **86**, 11
Malcovati, Massimo, **90**, 170
Maley, Frank, **83**, 603
Maller, James L., **99**, 219
Maloy, W. Lee, **91**, 413
Maltman, K. L., **94**, 422
Mamas, Suzanne, **86**, 258
Mandel, P., **81**, 304
Manley, James L., **101**, 568
Mannervik, Bengt, **87**, 370; **90**, 535
Månsson, Mats-Olle, **89**, 457
Mäntele, Werner, **88**, 729
Manthey, Alice E., **100**, 38
Mapoles, John, **96**, 443
Marcantonio, Eugene E., **96**, 520

Marchesi, Vincent T., **96**, 268
Marchitto, Kevin S., **94**, 361
Marciani, Dante J., **83**, 320
Marcus, Frank, **90**, 352
Margossian, Sarkis S., **85**, 55
Marie, Joëlle, **90**, 131
Markham, George D., **94**, 117, 219, 228
Markwell, John P., **89**, 275; **90**, 232
Maroux, Suzanne, **96**, 406
Marschke, Charles K., **90**, 70
Marsh, Derek, **88**, 762
Marsh, Mark, **98**, 260
Martensen, Todd M., **99**, 402
Martin, George R., **82**, 831
Martin, Paul L., **101**, 582
Martin, Susan R., **100**, 137
Martinez-Arias, Alfonso, **100**, 293
Maruta, Hiroshi, **85**, 357
Marzuki, Sangkot, **97**, 294
Marzullo, Marie-Therese, **82**, 762
Masseyeff, René, **84**, 19
Mateescu, Gheorghe D., **81**, 698
Mathews, W. Rodney, **86**, 409
Mathies, Richard, **88**, 633, 648
Matlin, Karl, **98**, 260
Matsumoto, Hiroyuki, **81**, 154, 694
Matsushita, Kazunobu, **89**, 149, 187
Matsuura, Shiro, **97**, 408
Mattiasson, Bo, **92**, 498
Mauchamp, Jean, **98**, 477
Mauck, Linda A., **90**, 309
Maurer, P., **89**, 261
Mayers, George L., **92**, 257
Mays, Carol, **82**, 325
Maytin, Edward V., **91**, 190
Mazurkiewicz, Joseph E., **96**, 627
Mazzaferro, Paul K., **92**, 257
McAda, Phyllis C., **97**, 337
McCann, Peter P., **94**, 209
McCloskey, Michael A., **83**, 540
McConaughy, Betty L., **100**, 137
McCoubrey, William K., Jr., **100**, 137
McCubbin, William D., **85**, 677
McCurry, Stephen D., **89**, 47, 108; **90**, 515
McDaniel, Michael L., **98**, 182
McDowell, J. Hugh, **81**, 211, 251
McGregor, W. C., **90**, 345
McGuinness, E. T., **89**, 135
McIntyre, J. Oliver, **98**, 623
McKay, Ronald, **92**, 138
McKinney, James D., **84**, 619
McNeil, Michael, **83**, 3
McVittie, J. D., **89**, 579
Meadow, Norman D., **90**, 417, 423, 439
Means, Anthony R., **84**, 138
Mecham, Robert P., **82**, 744
Mehlhorn, Rolf J., **88**, 334, 751, 772
Meikle, A. Wayne, **84**, 585
Meldolesi, Jacopo, **98**, 67
Mellman, Ira, **98**, 545
Melloni, E., **90**, 345
Meloche, H. Paul, **89**, 98; **90**, 263
Mendicino, Joseph, **89**, 121, 228; **90**, 200, 407
Merriam, P., **85**, 284
Merrick, William C., **101**, 606
Merril, Carl R., **96**, 230
Mescher, Matthew F., **92**, 86
Meselson, Matthew, **100**, 333
Messer, Janice L., **97**, 469
Messing, Joachim, **101**, 20
Metcalf, Brian, **94**, 199
Metz, Raymond J., **98**, 613
Meyer, David I., **96**, 692
Meza-Basso, Luis, **101**, 690
Michael, Herman, **98**, 555
Michalopoulos, George, **82**, 544
Micheel, Burkhard, **92**, 227
Michel, Hartmut, **88**, 111
Michel-Villaz, M., **81**, 605
Michiels, M., **84**, 542
Middaugh, C. R., **89**, 571
Mierendorf, Robert C., Jr., **98**, 159
Miggiano, V., **92**, 26, 242
Miki, Naomasa, **81**, 526
Mildvan, Albert S., **99**, 93
Miljanich, George P., **81**, 275, 806
Miller, Edward J., **82**, 3, 33
Miller, Jacqueline S., **101**, 650
Miller, Susan M., **87**, 711
Minakuchi, Ryoji, **99**, 288
Minta, Joe O., **93**, 375
Mitchell, Diane C., **98**, 304
Mitchell, John L. A., **94**, 140
Mitra, S. B., **86**, 192
Mitsui, Toshio, **88**, 268
Mizunuma, Hideo, **90**, 357
Mizuochi, Tsuguo, **83**, 105, 263
Moczar, Madeleine, **82**, 839
Model, Peter, **97**, 138

Modest, Edward J., **94**, 411
Moen, John E. T., **93**, 253
Moffatt, John G., **84**, 470
Moldave, Kivie, **101**, 629, 644
Mondovì, B., **94**, 314
Monson, Janet M., **82**, 218
Montague, Charles, **85**, 562
Montoya, Julio, **97**, 435
Moonen, P., **86**, 84
Moore, C. H., **90**, 77
Moraru, I., **92**, 523
Morgan, Richard A., **86**, 246
Morimoto, Takashi, **96**, 3, 121; **97**, 408
Morris, David L., **92**, 413
Morris, David R., **94**, 125, 373
Morris, H. R., **86**, 426
Morrison, John F., **87**, 405
Morrow, Jon S., **96**, 298
Mosbach, Klaus, **89**, 435, 457
Mosimann, James E., **92**, 36
Mostov, Keith E., **98**, 458
Motto, Michael G., **88**, 178
Muhly, Marion, **93**, 409
Mukohata, Yasuo, **88**, 407
Muller, P., **98**, 175
Müller, Rolf, **92**, 589
Müller, U., **89**, 531
Muller, William A., **98**, 404
Mullet, John E., **97**, 502
Mulligan, Bernard J., **97**, 524
Mulligan, R. Michael, **89**, 47
Muniyappa, K., **90**, 200, 407
Munson, Nancy Letko, **85**, 160
Munson, Peter J., **92**, 543
Murphy, Robert C., **86**, 30, 409, 547
Murray, Andrew W., **101**, 253
Murray, John M., **85**, 15
Murray, Louann W., **82**, 65
Murthy, P. Suryanarayana, **90**, 241
Muskavitch, Karen M. T., **97**, 524
Muto, Yasutoshi, **81**, 840
Myerowitz, Rachel, **96**, 729
Myers, Charles E., **84**, 447
Myllylä, Raili, **82**, 245

N

Na, George C., **85**, 393
Nabedryk-Viala, Eliane, **88**, 676
Nachbar, Martin S., **83**, 363
Nagakura, S., **81**, 368
Naito, Takayuki, **81**, 167
Nakagawa, Hiroki, **83**, 610
Nakajima, Hiroshi, **90**, 179
Nakamura, Izumi, **91**, 378
Nakanishi, Koji, **88**, 178, 491, 496
Nakazato, Hiroshi, **84**, 210
Nakazawa, Atsushi, **90**, 436
Narasimhan, Saroja, **98**, 98
Narita, Kozo, **81**, 247; **91**, 84
Narumiya, Shuh, **86**, 45
Nash, Howard A., **100**, 210
Nashima, Kazuko, **81**, 167
Nathan, Henry C., **94**, 209
Neilands, J. B., **94**, 437
Nelson, David R., **97**, 571
Nelson, Nathan, **97**, 510
Nemes, Peter P., **81**, 116, 275
Neslund, Gerald, **90**, 473
Nettleman, Mary, **86**, 607
Neufeld, Elizabeth F., **83**, 573; **96**, 729
Neugebauer, Dorothea-Ch., **88**, 235; **97**, 218
Neupert, Walter, **97**, 261, 275, 286
Nichols, Brian P., **101**, 155
Nicholson, Bruce J., **98**, 519
Nickerson, J. M., **92**, 577
Nicolaou, K. C., **86**, 400
Nicolas, J. C., **92**, 359
Nicoli, D. F., **92**, 458
Niebieski, Regina, **85**, 81
Nielsen, Jennifer B. K., **97**, 153
Niesel, David W., **89**, 296
Nika, Heinz, **91**, 359
Nikaido, Hiroshi, **97**, 85
Nilsson, Bo, **83**, 46
Nimmo, H. G., **90**, 330
Nishizuka, Yasutomi, **99**, 288
Noelken, Milton E., **82**, 410
Nogi, Y., **89**, 584
Nöll, G. N., **81**, 447
Noll, Markus, **100**, 309
Nordlie, Robert C., **87**, 319
North, Paula, **98**, 599
Northrop, Dexter B., **87**, 607
Norton, I. Lucile, **90**, 522
Norton, Scott J., **90**, 542, 547
Nossel, Hymie L., **84**, 102
Notermans, S., **84**, 223
Novick, Peter, **96**, 802

Nugteren, D. H., **86**, 49, 77, 84
Nunez, H. A., **83**, 58; **89**, 64
Nunn, B., **81**, 403
Nwokoro, Ngozi A., **89**, 219

O

Ôba, Kazuko, **89**, 345; **90**, 528
O'Brian, Catherine A., **99**, 93
O'Brien, David F., **81**, 378
O'Brien, Paul J., **81**, 141, 783
O'Brien, William A., **83**, 444
Öbrink, Björn, **82**, 513
O'Connell, Edward L., **89**, 98; **90**, 263
O'Day, Kathleen, **92**, 237
Oertel, Wolfgang, **97**, 226
Oesterhelt, Dieter, **88**, 10, 89, 111, 349, 417; **97**, 218, 226
Offer, Gerald, **85**, 130
Ogletree, Martin L., **86**, 607
Ohad, I., **97**, 554
Oka, Takami, **94**, 389
Okabayashi, Tadashi, **84**, 470
Okada, Masato, **94**, 301, 303
Okita, Thomas W., **83**, 549
Okuyama, Harumi, **86**, 370
Oldfield, Eric, **88**, 310
Oliva, Adriana, **94**, 73
Olsnes, Sjur, **83**, 357
Olsson, Lennart, **92**, 3
O'Malley, Karen, **97**, 344
Opheim, Dennis J., **90**, 384
Oppenheim, J. D., **83**, 363
Oppenheimer, Catherine L., **83**, 458
Oray, Bedii, **90**, 542, 547
Örning, Lars, **86**, 38
Orr, Elizabeth A., **97**, 524
Orr-Weaver, Terry L., **101**, 228
Osber, Lynn, **100**, 197
Osborne, H. Beverley, **81**, 742; **88**, 676, 681
Oshima, Tairo, **89**, 335; **94**, 401
Ostrander, Frank, **91**, 227
Ott, Ursula, **82**, 831
Otter, Richard, **100**, 171
Ötting, F., **84**, 577
Ottolenghi, Michael, **88**, 470
Ovchinnikov, Yu. A., **88**, 723
Overath, P., **97**, 158
Owen, John, **84**, 83, 93
Owens, J. C., **97**, 554
Oxender, Dale L., **97**, 146

P

Pace-Asciak, C. R., **86**, 552, 604
Packer, Lester, **88**, 356, 751, 772
Pai, M. Ramachandra, **90**, 241
Pajula, Raija-Leena, **94**, 257, 276
Pak, William L., **81**, 423
Palade, George E., **96**, xxix
Palmiter, Richard D., **96**, 150
Panagiotidis, Christos A., **94**, 162
Panniers, Richard, **101**, 616
Papahadjopoulos, Demetrios, **101**, 512
Papermaster, David S., **81**, 48, 214, 240; **96**, 244, 485, 609
Pardee, Joel D., **85**, 164
Parham, Peter, **92**, 110
Park, Myung Hee, **94**, 458
Parker, Charles W., **86**, 416, 655
Parker, Judy, **91**, 384
Parris, W., **100**, 183
Parson, William W., **88**, 272
Part, Marie-Louise, **94**, 213
Paschke, Eduard, **83**, 559
Pastan, Ira H., **85**, 322; **98**, 266
Paterson, Bruce M., **101**, 650
Pato, Mary, **85**, 308
Paulson, James C., **83**, 458
Paulus, Thomas J., **94**, 36, 112
Payne, M. Elizabeth, **99**, 299
Paz, Mercedes A., **82**, 571
Pearse, B. M. F., **98**, 320
Pearson, Terry W., **92**, 196
Pegg, Anthony E., **94**, 69, 158, 169, 206, 234, 239, 260, 294, 321
Pepperberg, David R., **81**, 452
Pepys, Elizabeth O., **93**, 164
Pepys, M. B., **93**, 164
Peralta, Jose M., **92**, 377, 391
Perham, Richard N., **91**, 559
Perlman, Philip S., **97**, 374
Perreault, G. J., **88**, 217
Perry, John W., **94**, 389
Persson, Lo, **94**, 166
Perz, Angela, **97**, 254
Petell, James K., **90**, 490
Peterkofsky, Beverly, **82**, 453

Peterson, Gary L., **91**, 95
Peterson, Per A., **96**, 281
Petryshyn, Ray, **99**, 346
Pettit, Flora H., **89**, 376; **90**, 195, 402; **99**, 331
Pfeifer, Felicitas, **88**, 374
Phares, E. F., **90**, 522
Phillips, Mark A., **86**, 366
Pick, A. J., **92**, 366
Pierce, J., **89**, 73, 83
Pierce, John W., **89**, 47
Pierschbacher, Michael, **82**, 803
Pietruszko, Regina, **89**, 428
Pilkis, J., **89**, 101
Pilkis, S. J., **89**, 101; **99**, 212
Piper, Priscilla J., **86**, 426
Pittenger, Mark, **98**, 555
Pizzolato, M. A., **92**, 220
Place, Allen R., **91**, 466
Plapp, Bryce V., **87**, 469
Pober, Jordan S., **81**, 236
Pohlmann, Regina, **83**, 559
Poirier, Miriam C., **84**, 607
Poland, Russell E., **84**, 532
Pollak, Alfred, **89**, 108
Pollard, Thomas D., **85**, 123, 182, 211, 331
Pontremoli, S., **90**, 345
Poorthuis, B. J. H. M., **98**, 586, 592
Porcelli, Marina, **94**, 73
Portalier, Raymond, **89**, 210
Porter, E. Victoria, **90**, 25, 556
Pösö, Hannu, **94**, 206, 234
Potter, James D., **85**, 241
Potter, Robert L., **91**, 613
Pouchelet, Marcel, **97**, 396
Powell, William S., **86**, 168, 467, 530
Powers, Dennis A., **91**, 466
Pratt, J. J., **84**, 369
Preiss, Jack, **83**, 549
Prescott, Stephen M., **86**, 11
Price, Paul A., **91**, 13
Prieels, Jean-Paul, **83**, 458
Pritchard, G. G., **90**, 77, 165
Probst, I., **88**, 334
Prockop, Darwin J., **82**, 305
Proia, Richard L., **96**, 729
Przybyla, Alan, **82**, 731
Puchalski, Christina M., **96**, 729
Purich, Daniel L., **85**, 416, 439; **87**, 3
Puszkin, S., **98**, 350

Q

Qadri, S. Saleheen, **90**, 11
Qian, Siying, **101**, 433
Quayle, J. Rodney, **89**, 566; **90**, 314
Quintanilha, Alexandre, **88**, 682

R

Racagni, Giorgio, **86**, 635
Racanelli, Toni, **88**, 356
Rachubinski, Richard A., **97**, 396
Radding, Charles M., **100**, 197
Radin, Norman S., **83**, 596; **98**, 613
Rafferty, Charles N., **88**, 167
Raina, Aarne, **94**, 257, 270, 276
Randall, Douglas D., **89**, 408
Randall, Linda L., **97**, 70, 77
Rangel-Aldao, Raphael, **99**, 176
Rannels, Stephen R., **99**, 55, 168
Ransil, Bernard J., **93**, 60
Rao, Ch. V., **86**, 192
Raso, Vic, **84**, 483
Ratnaparkhi, Makarand V., **92**, 36
Ray, Paul H., **83**, 525, 530, 535
Raymond, Yves, **97**, 396
Razin, A., **101**, 492
Reading, H. William, **81**, 794
Rearick, James I., **83**, 458
Redman, Colvin M., **98**, 169
Reed, Lester J., **89**, 376; **90**, 195, 402; **99**, 331
Reed, Randall R., **100**, 191
Rees, Dianne D., **85**, 292
Reeves, Richard E., **90**, 97, 552
Reggio, Hubert, **98**, 379
Regnier, Fred E., **91**, 137
Regoeczi, Erwin, **89**, 172
Reid, Graeme A., **97**, 305, 324
Reid, K. B. M., **82**, 319
Reimann, Erwin M., **99**, 51, 77
Reinberg, Danny, **100**, 217
Reiser, Jakob, **96**, 205
Reisler, Emil, **85**, 84
Relyveld, E. H., **93**, 24
Rentzepis, P. M., **81**, 354
Revel, Jean-Paul, **98**, 519
Rhodes, R. Kent, **82**, 33
Ricciardi, Robert P., **101**, 650

Rich, Celeste B., **82**, 665, 731, 762
Richard, John P., **87**, 213
Riddles, Peter W., **91**, 49
Riedel, Heimo, **96**, 496
Rigaud, Michel, **86**, 612
Risteli, Leila, **82**, 831
Ritchie, D. G., **92**, 577
Rittenhouse, Judith, **90**, 352
Rittenhouse, Susan Erika, **86**, 3
Robakis, Nikolaos, **101**, 690
Robbins, April R., **96**, 729
Robbins, Phillips W., **83**, 399
Robert, Ladislas, **82**, 839
Robert-Baudouy, J., **90**, 288
Roberts, Bryan E., **101**, 650
Roberts, L. Jackson, II, **86**, 559, 592
Roberts, Robert, **90**, 185
Robey, Pamela Gehron, **82**, 831
Robinson, Neal C., **97**, 571
Robison, David J., **84**, 60
Rodermel, Steven R., **97**, 524
Rodriguez, Raymond L., **83**, 549
Rodriguez-Boulan, Enrique, **98**, 486
Roeder, Robert G., **101**, 582
Rogers, Thomas J., **92**, 237
Rohde, Heilwig, **82**, 831
Rohr, Gerhard, **98**, 466
Rohr, Thomas E. , **83**, 540
Rokosz, Michael J., **91**, 119
Rollins, Thomas E., **86**, 213
Romaniuk, Paul J., **87**, 197; **100**, 52
Rome, Leonard H., **83**, 578, 582
Romeo, Giovanna, **94**, 355
Ronnett, Gabriele V., **96**, 423
Rose, Irwin A., **87**, 84
Rose, Mark, **101**, 167
Rose, Zelda B., **87**, 42
Roseman, Saul, **83**, 294, 306; **90**, 417
Rosen, Ora M., **99**, 176
Rosenbaum, Joel L., **85**, 409
Rosenberg, Martin, **101**, 123
Rosenberry, Terrone L., **82**, 325
Rosenbloom, Joel, **82**, 716
Rosenfeld, Melvin G., **96**, 764
Rosengren, Elsa, **94**, 166
Rosevear, Paul R., **83**, 58; **99**, 93
Roskoski, Robert, Jr., **99**, 3
Rosner, Marsha R., **83**, 399
Ross, P. E. , **84**, 321
Rossignol, B., **98**, 175
Roth, Gerald J., **86**, 222, 392
Roth, Margit, **93**, 409
Roth, Thomas F., **98**, 337
Rothenberg, Sheldon P., **84**, 458
Rothfield, Lawrence I., **83**, 326
Rothman, James E., **98**, 37
Rothschild, Kenneth J., **88**, 326, 643, 696
Rothstein, Rodney J., **101**, 202, 228
Rotundo, Richard L., **96**, 353
Rozen, Rima, **97**, 396
Rubin, Charles S., **99**, 187
Rubin, R. W., **96**, 184
Rubin, Robert T., **84**, 532
Ruby, Stephanie W., **101**, 253
Rucker, Robert B., **82**, 650
Ruddle, Frank H., **101**, 411
Ruoslahti, Erkki, **82**, 803; **84**, 3
Ruschig, U., **89**, 531
Rush, Jeffrey S., **83**, 311
Russel, Marjorie, **97**, 138
Russo, John, **89**, 226
Ryan, Clarence A., **96**, 580

S

Saari, John C., **81**, 819
Sabatini, David D., **96**, 378, 520, 530, 764
Sabatini, S., **94**, 314
Sabularse, Dario C., **90**, 91
Sack, Robert A., **81**, 506
Sadler, J. Evan, **83**, 458
Sage, Helene, **82**, 96
Sahagian, G. Gary, **83**, 392; **96**, 729; **98**, 304
Sahm, Hermann, **89**, 271, 424, 527; **90**, 319
Saibil, H., **81**, 605
Sale, Winfield S., **85**, 450
Salisbury, Jeffrey L., **98**, 368
Salmon, John A., **86**, 45, 91, 477
Saman, Eric, **83**, 631
Samejima, Keijiro, **94**, 270
Sammons, R. Douglas, **87**, 213
Samuels, Mark, **101**, 568
Sanches, R., **88**, 696
Sandberg, L. B., **82**, 657
Sandri-Goldin, Rozanne M., **101**, 402
Sandvig, Kirsten, **83**, 357
Sankarappa, Shankar K., **86**, 357
Santoro, Samuel A., **82**, 509
Saraste, Jaakko, **96**, 453

Sarkar, Dwijen, **99**, 187
Sarma, Syam, **88**, 369
Sarmiento, Roberto Martín, **89**, 516, 523
Sarver, Nava, **101**, 387
Sasaki, Shoji, **89**, 480
Sasaki, Takeshi, **84**, 201
Sawyna, Vitaly, **85**, 149
Schachter, Harry, **89**, 219; **98**, 98
Schantz, Rudi, **97**, 524
Scheek, R. M., **89**, 305
Scheele, George, **96**, 94; **98**, 17, 466
Scheffler, I. E., **94**, 108
Schekman, Randy, **96**, 802
Schell, Mark A., **90**, 30
Schellenberg, G. D., **90**, 223
Schlager, Seymour I., **93**, 233, 260
Schlamowitz, Max, **93**, 190
Schlecht, Peter, **81**, 729
Schleimer, Robert P., **84**, 526
Schlender, Keith K., **99**, 77
Schlesinger, David H., **91**, 494
Schlesinger, Kenneth, **86**, 607
Schlesinger, Milton J., **96**, 795
Schloss, John V., **90**, 522
Schmidt-Glenewinkel, Thomas, **100**, 217
Schnaar, Ronald L., **83**, 294, 306
Schneider, Barbara G., **96**, 485
Schnell, R. J., **81**, 90
Schnetkamp, Paul P. M., **81**, 110
Schoenborn, Benno P., **88**, 241
Scholey, J. M., **85**, 364
Schook, W., **98**, 350
Schowen, K. Barbara, **87**, 551
Schowen, Richard L., **87**, 551
Schreiber, Alain B., **93**, 147
Schroeder, Hartmut R., **84**, 303
Schütte, Horst, **89**, 271, 424, 527; **90**, 319
Schütz, Günther, **100**, 285
Schwartz, Maxime, **97**, 100
Schwarz, Ralph T., **83**, 432
Schwemer, Joachim, **81**, 182, 729
Scopes, Robert K., **90**, 479
Sebald, Walter, **97**, 254
Seely, James E., **94**, 158, 169, 206
Sefton, Bartholomew M., **99**, 387
Segawa, T., **89**, 584
Seglen, Per O., **96**, 737
Segrest, Jere P., **82**, 398
Seidah, N. G., **92**, 292
Seidel, John C., **85**, 594
Seiler, Nikolaus, **94**, 3, 10, 25
Seki, Takaharu, **81**, 43
Seltzer, Stanley, **88**, 542
Semenza, Giorgio, **96**, 386
Seppänen, P., **94**, 247
Sepulveda, Carmen A., **93**, 260
Serianni, A. S., **83**, 58; **89**, 64, 73, 79, 83
Seyfried, Christine E., **94**, 373
Shalongo, William H., **87**, 647
Shapira, Stuart K., **100**, 293
Sharon, Nathan, **83**, 386
Sharp, Phillip A., **101**, 568
Shastry, Barkur S., **101**, 582
Shatzman, Allan, **101**, 123
Shaw, Anita R., **93**, 190
Shen, Wei-Chiang, **84**, 650
Shepherd, V., **98**, 301
Sherman, William R., **89**, 3, 9, 593; **90**, 309
Sheu, Kwan-Fu, **87**, 20, 213
Shevach, Ethan M., **92**, 66
Shi, Jian-ping, **90**, 349
Shibata, Takehiko, **100**, 197
Shichi, Hitoshi, **81**, 61, 247, 459; **99**, 362
Shichida, Yoshinori, **81**, 333, 634
Shidoji, Yoshihiro, **81**, 840
Shimizu, Takao, **86**, 73, 147
Shinagawa, Emiko, **89**, 141, 187, 194
Shizuta, Yutaka, **85**, 322
Shore, Gordon C., **97**, 396
Shortle, David, **100**, 457
Shriver, John W., **81**, 698
Siakotos, A. N., **81**, 90
Siber, George R., **93**, 60
Sibley, David, **90**, 294
Siebert, Fritz, **88**, 729
Siegl, Adelaide M., **86**, 179
Sigel, Morton B., **93**, 3
Sigrist, Hans, **88**, 207
Sih, Charles J., **86**, 99
Silhavy, Thomas J., **97**, 3, 11
Silver, Pamela, **97**, 46, 130
Silverman, David N., **87**, 732
Simion, F. Anthony, **98**, 326
Simkins, Ronald A., **89**, 248
Simoni, Robert D., **90**, 447; **97**, 188
Simons, A. Ray, **92**, 377
Simons, Kai, **96**, 466; **98**, 260
Sindhu, Ram K., **94**, 279
Singh, Harmeet, **83**, 191
Sinha, Y. N., **93**, 3
Siraganian, Reuben P., **92**, 17
Sjoerdsma, Albert, **94**, 209

Sklar, Larry A., **81**, 685
Skulachev, V. P., **88**, 35
Skutelsky, Ehud, **83**, 195
Slater, E. C., **89**, 305
Sloboda, Roger D., **85**, 409
Sly, William S., **98**, 290
Smillie, L. B., **85**, 234
Smith, David F., **83**, 235, 241
Smith, H. Gilbert, Jr., **81**, 57, 577
Smith, Julia A., **96**, 3
Smith, Michael, **100**, 468
Smith, Ronald G., **94**, 48
Smith, Stephen B., **99**, 162
Smith, Terence A., **94**, 176, 311, 344
Smith, William L., **86**, 213, 229, 240
Soderling, Thomas R., **99**, 37, 299
Somers, Robert L., **81**, 459; **99**, 362
Son, Paula H., **101**, 98
Spanos, Ad, **91**, 263
Spar, Irving L., **92**, 277
Spector, Sydney, **84**, 551, 555
Speranza, M. L., **89**, 316
Sperling, Walter, **88**, 167
Spiker, Steven, **91**, 214
Sprecher, Howard, **86**, 357, 518
Springer, Timothy A., **92**, 147
Springgate, Clark F., **90**, 378
Sprott, G. D., **88**, 98
Spudich, Elena Negri, **88**, 213
Spudich, James A., **85**, 164
Spudich, John Lee, **88**, 213
Srivenugopal, K. S., **94**, 335, 429
Stachow, Chester S., **90**, 378
Staehelin, Th., **92**, 26, 242
Stafford, Walter F., III, **85**, 570
Stahl, P., **98**, 301
Stähli, C., **92**, 26, 242
Stallcup, Kathryn C., **92**, 86
Stalon, Victor, **94**, 339
Stamatoff, J., **88**, 282
Stanley, Pamela, **96**, 157
Starcher, Barry C., **82**, 759
Stark, George R., **96**, 205
Starling, Jane A., **90**, 44
Starr, Roger, **85**, 130
Steck, Germaine, **91**, 247
Steglich, C., **94**, 108
Stehle, Randall G., **86**, 436
Steinbach, Rolf A., **89**, 271
Steinberg, Robert A., **99**, 233
Steinman, Charles R., **84**, 181, 187
Steinman, Ralph M., **98**, 404
Steinmetz, Andre A., **97**, 524
Steitz, T. A., **88**, 81
Stellwag, Edmund J., **100**, 326
Stiles, John I., **101**, 290
Stocchi, Vilberto, **90**, 3
Stocker, J., **92**, 242
Stoeber, François, **89**, 210; **90**, 288
Stone, Marvin J., **84**, 172
Stone, P. J., **82**, 588
Storey, Kenneth B., **90**, 39, 366
Stossel, Thomas P., **85**, 480
Stoter, Ann, **90**, 479
Stournaras, C., **89**, 278
Stowell, Christopher P., **83**, 278
Strating, Marijke, **89**, 391
Straubinger, Robert M., **101**, 512
Straus, J. William, **101**, 635
Strehler, Emanuel, **85**, 139
Stringer, Claude D., **90**, 522
Strom, Stephen C., **82**, 544
Stryer, Lubert, **81**, 668; **96**, 617
Stull, James T., **99**, 7
Sue, Julia M., **89**, 108
Sugawara, Neal, **101**, 269
Sugawara, Yasutake, **89**, 480
Sugiyama, Yasuo, **88**, 407
Suissa, Mordechai, **96**, 192; **97**, 305
Sul, Hei Sook, **99**, 250
Sullivan, Cathleen P., **92**, 86
Sullivan, Kathleen A., **82**, 637
Sumi, Takahiko, **90**, 35
Sumper, Manfred, **88**, 391
Sundler, Frank, **94**, 166
Supowit, Scott C., **90**, 21
Suzuki, Koichi, **90**, 126, 179
Suzuki, Tsuneo, **93**, 219
Szent-Györgyi, Andrew G., **85**, 81
Szostak, Jack W., **101**, 228, 245, 253, 269, 278

T

Tabas, Ira, **83**, 416
Taber, Douglass F., **86**, 366, 592
Tabor, Celia White, **94**, 29, 83, 91, 104, 117, 135, 219, 228, 231, 265, 396, 418, 420, 434
Tabor, Herbert, **94**, 29, 83, 91, 104, 117, 135, 219, 228, 231, 265, 396, 418, 420, 434

Tai, Hsin-Hsiung, **86**, 99, 110, 113, 131, 142
Tai, Mindy M., **84**, 60
Tai, Phang C., **97**, 62
Tainer, Beth, **86**, 511
Takács, B., **92**, 242
Takai, Yoshimi, **99**, 288
Takasaki, Seiichi, **83**, 263
Takatori, Takehiko, **84**, 596
Tamaki, Nanaya, **89**, 469
Tamir, Y., **92**, 366
Tanaka, Minoru, **81**, 61
Tanaka, Takehiko, **90**, 150
Tang, Kuo-Chang, **94**, 286
Tang, Wen-Jing Y., **85**, 450
Tanzer, Marvin L., **82**, 65
Tapper, Douglas P., **97**, 426
Tarentino, Anthony L., **83**, 603
Tartakoff, Alan Michael, **98**, 47
Tashima, Yohtalou, **90**, 357
Taylor, G. W., **86**, 426
Taylor, K. A., **85**, 364
Taylor, Susan S., **99**, 140
Tchen, T. T., **94**, 431
Teague, W. Martin, **90**, 402
Teather, R. M., **97**, 158
Teerlink, T., **98**, 586
Terada, O., **89**, 243
Terouanne, B., **92**, 359
Terry, Brian J., **85**, 416
Thacher, Scott M., **81**, 514
Thilo, Lutz, **98**, 415
Thomas, David D., **81**, 668
Thomas, Patricia S., **100**, 255
Thompson, J. A., **100**, 368
Tibbelin, Gudrun, **90**, 535
Timasheff, Serge N., **85**, 393
Timm, Beate, **96**, 496
Timpl, Rupert, **82**, 472, 831
Tippins, J. R., **86**, 426
Tipton, K. F., **90**, 330
Titus, Bradley G., **86**, 69
Tobari, Jiro, **94**, 431
Tokumoto, Hidekado, **86**, 147
Tokunaga, Fumio, **88**, 163
Tokuyasu, K. T., **96**, 466
Tolbert, N. E., **89**, 47; **90**, 515
Tomioka, Hisao, **86**, 17
Topp, Randy, **89**, 399
Torchia, D. A., **82**, 174
Torsvik, Terje, **88**, 360
Tower, Barbara B., **84**, 532
Towner, Paul, **88**, 546
Trabold, Norma C., **84**, 51
Traniello, S., **90**, 345
Traugh, Jolinda A., **99**, 308, 317
Tressel, Paul S., **89**, 163
Trimble, Robert B., **83**, 603
Trinick, John A., **85**, 9, 17
Trissl, H.-W., **81**, 431
Troy, Frederic A., **83**, 540
Tsai, Ming-Daw, **87**, 235
Tsang, Victor C. W., **92**, 377, 391
Tse-Eng, Doris, **84**, 558
Tsin, Andrew T.C., **81**, 133
Tsuda, Motoyuki, **81**, 392; **88**, 552, 714
Tsunasawa, Susumu, **81**, 247; **91**, 84
Tuazon, Polygena T., **99**, 308
Tuderman, Leena, **82**, 305
Tung, Amar S., **92**, 47; **93**, 12
Turco, Salvatore J., **83**, 132, 137
Turini, P., **94**, 314
Turkewitz, Aaron P., **92**, 86
Tyagi, Anil K., **94**, 135
Tyma, Jennifer L., **99**, 77
Tzagoloff, Alexander, **97**, 355, 361

U

Uddman, Rolf, **94**, 166
Uehara, Kihachiro, **89**, 232
Uhl, Rainer, **81**, 509
Uhlenbeck, Olke C., **100**, 52
Uhr, J. W., **93**, 333
Ui, Michio, **90**, 35
Ujita, Susumu, **89**, 258; **90**, 235
Uotila, Marjatta, **84**, 3
Urbani, Lenore J., **98**, 37
Uritani, Ikuzo, **89**, 345; **90**, 528
Urry, Dan W., **82**, 673
Usukura, Jiro, **88**, 118
Utting, John M., **89**, 341
Uwajima, T., **89**, 243

V

Vainstein, A., **101**, 492
Valentini, Giovanna, **90**, 170
Valkenburg, Cornelius A., **94**, 48

van Dam, Karel, **88**, 17
Vandecasteele, J. P., **89**, 484
Vander Jagt, David L., **89**, 513
VanderLaan, W. P., **93**, 3
Vander Mallie, Ronald J., **84**, 121
van der Ouderaa, F. J. G., **86**, 60
van Dijck, Piet W. M., **88**, 17
Van Doorslaer, Elisa, **83**, 631
Van Etten, Richard A., **97**, 426
Van Keuren, Margaret L., **96**, 230
van Nieuwmegen, Ria, **93**, 83
Van Rollins, Mike, **86**, 518
van Rooijen, Nico, **93**, 83
van Tol, A., **90**, 340
Van Vunakis, Helen, **84**, 628, 641, 657
Vardanis, Alexander, **99**, 71
Vater, Carol A., **82**, 423
Veis, Arthur, **82**, 186
Vella, George, **98**, 98
Venkitasubramanian, T. A., **90**, 241
Veronese, Francesco M., **89**, 282
Verrier, Bernard, **98**, 477
Vevert, Jean-Paul, **94**, 199
Viebrock, Adelheid, **97**, 254
Viitala, Juha, **83**, 311
Vijay, Inder K., **83**, 540
Villafranca, Joseph J., **87**, 180
Vinogradov, Serge N., **91**, 119, 259
Viola, Ronald E., **87**, 353
Virmaux-Colin, N., **81**, 304
Virtanen, Ismo, **96**, 453
Visser, Jaap, **89**, 391, 399
Vitetta, E. S., **93**, 333
Voelker, Dennis R., **98**, 596
Vogel, Günter, **98**, 421
Vogelsang, Heike, **97**, 226
von Figura, Kurt , **83**, 559
von Wartburg, Jean-Pierre, **89**, 181, 506
Voris, Bruce P., **91**, 190

W

Wacker, Hans, **96**, 386
Wagner, D. B., **92**, 366
Wagner, Gottfried, **88**, 344
Wagner, Paul D., **85**, 72
Wagner, Renate M., **98**, 304
Wagner, Robert P., **89**, 386
Waite, J. Herbert, **82**, 65
Walker, Graham C., **100**, 501
Walker, John E., **97**, 195
Walker-Simmons, Mary, **96**, 580
Wall, Doris A., **98**, 225
Wallace, B. A., **88**, 447
Walsh, Donal A., **99**, 80, 250
Walsh, Michael P., **99**, 279
Walter, Peter, **96**, 84, 682
Walter, Ulrich, **99**, 154
Walton, Gordon M., **99**, 206
Wang, Janet K., **81**, 211
Wang, Jean Yin Jen, **99**, 373
Wang, Kuan, **85**, 264, 514
Wang, Ying-lai, **90**, 349
Wannlund, Jon, **92**, 426
Warnaar, Sven O., **93**, 253
Warnock, Robert, **86**, 511
Warren, G., **96**, 466
Warren, R. A. J., **94**, 422
Wasteson, Åke, **83**, 573
Watanabe, Kikuko, **86**, 147
Waterman, Michael R., **84**, 172
Watson, Douglas, **83**, 444
Watts, Anthony, **88**, 762
Watts, Colin, **97**, 130
Waygood, E. Bruce, **90**, 423, 431
Webb, Martin R., **87**, 301
Weber, Hans Jurgen, **88**, 369, 379, 434
Webster, Paul, **98**, 379
Wedler, Frederick C., **87**, 647
Wehmeyer, Kenneth R., **92**, 432
Weidinger, Gottfried, **88**, 374
Weigand, Rex D., **81**, 297
Weigel, Nancy, **90**, 431, 436
Weigel, Paul H., **83**, 294, 306
Weiner, Henry L., **81**, 660
Weir, Jerry P., **85**, 371
Weiss, O., **92**, 366
Weissbach, Herbert, **101**, 690
Wells, Gerald B., **83**, 132, 137
Wells, James A., **85**, 93
Wells, Robert D., **100**, 368
Weng, Jian, **101**, 433
Wenger, William C., **89**, 93
Wermuth, Bendicht, **89**, 181, 506
Werner, Rudolf, **98**, 537
Westerhausen, Andrea, **89**, 497
Westerman, Jan, **98**, 581
Westhoff, Peter, **100**, 400
Wettern, M., **97**, 554

White, Hillary D., **99**, 162
White, Howard D., **85**, 698
Whitehouse, Susan, **99**, 80
Whitesides, George M., **89**, 108
Wickens, M. P., **101**, 370
Wickner, William, **97**, 40, 46, 57, 130
Widnell, Christopher C., **98**, 395
Wiersinga, Wilmar M., **84**, 272
Wiggs, J., **101**, 540
Wilchek, Meir, **83**, 195
Wilden, Ursula, **81**, 489
Willerson, James T., **84**, 172
Williams, Edward H., **98**, 510
Williams, Richard S., **97**, 487
Williams, Robley C., Jr., **85**, 376
Williams, Theodore P., **81**, 374
Willingham, Mark C., **98**, 266
Willnow, P., **89**, 531
Wilson, Britt C., **92**, 391
Wilson, Curtis M., **91**, 236
Wilson, David B., **90**, 30
Wilson, John E., **97**, 469
Wilson, N. M., **90**, 223
Winek, David, **98**, 326
Winston, Fred, **101**, 211
Winter, William T., **83**, 87
Wirtz, K. W. A., **98**, 581, 586, 592
Wofsy, Leon, **92**, 472
Woldring, M. G., **84**, 369
Wolfe, P. B., **97**, 40
Wolt, T. B., **82**, 657
Wong, Chi-Huey, **89**, 108
Wong, Lee-Jun, **87**, 20
Wong, Patrick Y-K., **86**, 117
Wong, Yun-Hua H., **90**, 309
Wood, Harland G., **87**, 51
Woodrow, Ian E., **90**, 392
Worcester, D. L., **81**, 593
Wright, J. K., **97**, 158
Wu, Ray, **100**, 60, 96
Wu, Su Chen, **82**, 762

X

Xu, Gen-jun, **90**, 349

Y

Yamada, Eichi, **81**, 3, 834; **88**, 118
Yamamoto, Katsuhiko, **99**, 362
Yamamoto, Shozo, **86**, 55, 73, 106, 269
Yamanoha, Banri, **94**, 270
Yamashita, Katsuko, **83**, 105
Yamazaki, Akio, **81**, 526
Yang, Sue-Lein, **87**, 20
Yanofsky, Charles, **101**, 155
Ye, Yukun K., **97**, 524
Yeaman, Stephen J., **90**, 195; **99**, 331
Yeltman, Don R., **90**, 251
Yoakum, George H., **101**, 138
Yokosawa, Hideyoshi, **91**, 378
Yonetani, Takashi, **87**, 500; **89**, 474
Yoshikami, S., **81**, 447
Yoshimoto, Tanihiro, **86**, 106
Yoshizawa, Tôru, **81**, 154, 333, 634
You, Kwan-sa, **87**, 101
Young, Donald A., **91**, 190
Young, Lisa S., **100**, 137
Yount, Ralph G., **85**, 93
Yu, Hyuk, **81**, 616
Yu, Peter H., **94**, 66
Yu, Robert K., **83**, 139
Yuan, Barbara, **86**, 113
Yuan, James H., **89**, 351

Z

Zagon, Ian S., **94**, 169
Zahler, Peter, **88**, 207
Zambonin, Marcello, **91**, 311
Zappia, Vincenzo, **94**, 57, 73, 355
Zeng, Yishen, **101**, 433
Zerner, Burt, **91**, 49
Zhou, Guang-yu, **101**, 433
Zilversmit, Donald B., **98**, 565, 574
Zimmerman, William F., **81**, 52, 564
Zimmermann, Gerhard, **90**, 371
Zimmermann, Richard, **97**, 57, 275
Zingsheim, Horst-Peter, **97**, 218
Zipursky, Stephen L., **100**, 217
Zoccoli, Michael A., **90**, 396
Zoller, Mark J., **99**, 140; **100**, 468
Zopf, David, **83**, 46
Zubay, Geoffrey, **101**, 674
Zweig, Stephen E., **92**, 66
Zwerling, Hayward, **88**, 772
Zwizinski, C., **97**, 40